With much affection and joy, this book is dedicated to my mother, father, and grandmother. I am blessed by their love and kindness.

Introductory Statistics for Business and Economics

Introductory Statistics for Business and Economics

RICHARD K. MILLER
Washington and Jefferson College

ST. MARTIN'S PRESS
New York

Library of Congress Catalog Card Number: 80–51065

Manufactured in the United States of America.
43210
fedcba
For information, write St. Martin's Press, Inc.,
175 Fifth Avenue, New York, N.Y. 10010

cover design: Notovitz & Perrault

typography: Leon Bolognese

ISBN: 0–312–43451–0

ACKNOWLEDGMENTS

"The Standard Normal Distribution" reprinted with permission of Macmillan Publishing Co., Inc. from "Areas under the Normal Distribution" in *Elements of Econometrics* by Jan Kmenta. Copyright © 1971 by Macmillan Publishing Co., Inc.

"The *t* Distribution" taken with permission from Table III of Ronald A. Fisher and Frank Yates: *Statistical Tables for Biological, Agricultural and Medical Research*, 6th edition, 1974, published by Longman Group Ltd., London (previously published by Oliver and Boyd, Edinburgh), and reprinted by permission of the authors and publishers.

"The χ^2 Distribution" from Table 8, "Percentage Points of the χ^2 Distribution" in E. S. Pearson and H. O. Hartley, editors, *Biometrika Tables for Statisticians*, volume I, 3rd edition (London: Cambridge University Press for the Biometrika Trustees, 1966). Reprinted with permission.

"The *F* Distribution" ("Critical Values for the Upper .05 Area" and "Critical Values for the Upper .01 Area") from Table 18, "Upper 5% Points and Upper 1% Points of the *F* Distribution" in E. S. Pearson and H. O. Hartley, editors, *Biometrika Tables for Statisticians*, volume I, 3rd edition (London: Cambridge University Press for the Biometrika Trustees, 1966). Reprinted with permission.

"Table of Random Numbers" from The RAND Corporation, *A Million Random Digits with 100,000 Normal Deviates* (Glencoe, Ill.: The Free Press, 1955), p. 337. Reprinted with permission.

Preface

Introductory Statistics for Business and Economics is the by-product of a Business Statistics course that I have taught at Washington and Jefferson College for the past nine years. During that time, I have become aware of the large number of economics and business administration majors who view exposure to mathematics of *any* kind with a pronounced degree of trepidation and defeatism. In spite of a proliferation of statistics texts, such fears persist.

The Mathematical Association of America recently reported that introductory statistics classes often "are too technique-oriented, overemphasizing computation and underemphasizing the fundamental ideas underlying statistical reasoning."* This indictment can also be extended to many of the books associated with these courses. Although a host of texts carry the claim that "only high school algebra is required," some students are still overwhelmed by a frustrating array of symbols and equations. Other students are doubtful whether statistical procedures have a practical use.

I have written this text with two objectives in mind: (1) to illustrate how statistical analysis can be applied to "real-world" situations, and (2) to discuss the basic theory of statistics without relying on complex symbols and advanced mathematics. To accomplish the first objective, I have included an extensive collection of worked-out business and economics examples in the body of the text, as well as a variety of end-of-chapter exercises. The second objective is attained through the presentation of numerous tables and diagrams and by a clear step-by-step approach to proving key principles.

I feel that most introductory statistics courses, regardless of the type of enrollment, share a common core of subject matter. This basic material is presented in Chapters One through Eleven. Following this core, the direction of individual courses is determined by both the preferences of the instructor and the characteristics of the students. Chapters Twelve through Twenty are especially relevant for majors in business and economics. I believe that instructors of such students will find that this text offers enough "additional" topics to satisfy divergent concepts of what the introductory business statistics course should entail.

* *A Compendium of CUPM Recommendations,* vol. II, 1975, p. 475.

The text can be used in either a one- or two-semester program. The accompanying diagram indicates the organization.

	Descriptive Statistics	*Probability*	*Statistical Inference*
Core Chapters	2, 3	4, 5, 6	7, 8, 9, 10, 11
Additional Chapters	16, 19		12, 13, 14, 15, 17, 18, 20

I have been fortunate in being able to develop *Introductory Statistics for Business and Economics* with the help of St. Martin's Press. Not only have I enjoyed my friendships with the staff, but I have been impressed with their dedication and professionalism. I particularly want to express my appreciation to Bertrand W. Lummus, senior editor; Carolyn Eggleston, manuscript editor; Edward B. Cone, managing editor; and Thomas V. Broadbent, director of the college department.

The text was also enhanced by the comments of many reviewers. Dr. Ronald S. Koot of Pennsylvania State University was exceptionally helpful and articulate.

Finally, I am grateful to the literary executor of the late Sir Ronald A. Fisher, F.R.S., to Dr. Frank Yates, F.R.S., and to Longman Group Ltd., London, for permission to reprint Table III from their book *Statistical Tables for Biological, Agricultural and Medical Research* (6th edition, 1974).

RICHARD K. MILLER

Contents

Introductory Statistics for Business and Economics

1 Introduction

RELEVANCE OF STATISTICS

Because we are so persistently exposed to numerical information in our daily lives, almost everybody is a player in the "numbers game" called *statistics*. To the man on the street, the words "statistics" and "numbers" imply the same thing. There are, however, distinctions, as we shall see. Techniques for obtaining, classifying, and evaluating quantitative information fall under the heading of "statistics." Thus statistical methods are used in the construction of Census Bureau questionnaires, the selection of recipients of these questionnaires, and the tabular display of responses. We will shortly see, moreover, that "averages" are called "statistics" under certain circumstances. Throughout this text, we will illustrate the many uses of statistics in business and economics. Additionally, we will apply statistical methods to problems from many other fields, for example, politics.

Preenrollment anxiety over a class in introductory statistics may be reinforced early in the semester, perhaps when

$$\sigma^2 = \frac{\Sigma(X - \mu)^2}{N}$$

is first written on the blackboard. There is no escaping the fact that formulas play a large role in statistics. We will try, though, to develop the rationale or logic behind such relationships, which hopefully will make the formulas appear less forbidding. In addition, the mathematical equipment that is needed is intentionally restricted to algebra and/or arithmetic.

Newspapers and weekly periodicals have been giving increasing attention to statistical studies. The Gallup, Harris, and Roper polls of American opinion are printed regularly. Debates in the news media as to whether cigarette smokers are more likely than non-smokers to get particular diseases or whether various working conditions adversely affect health rely heavily on statistical evidence.

Students who complete an introductory statistics course should have acquired the ability to communicate quantitative information effectively and efficiently. They should be able to identify types of real-life situations in which statistics and statistical techniques are useful. Furthermore, they ought to understand that statistical analysis, in spite of its theoretical foundation and sophisticated form, is not without its limitations.

Some students complain, "Why must I take Statistics?" We hope to answer this by integrating the methods of statistics with interests common to business people and economists. To emphasize the text's concern with "applications," Chapter One will outline a number of problems that can be dealt with statistically. When you look over these examples, imagine how you might approach them now. Quite possibly, you would rely on intuition. Before too long, you'll begin to realize that the procedures explained in the following chapters have advantages over "hit-and-miss," "seat-of-the-pants," or "rule-of-thumb" attempts at problem solving.

DIMENSIONS OF STATISTICS

The major topics discussed in this text are descriptive statistics (Chapters Two and Three), probability (Chapters Four through Six), statistical inference (Chapters Seven through Fifteen, and Twenty), linear regression (Chapters Sixteen through Eighteen), and time series (Chapter Nineteen). Before we illustrate their principal uses, it is necessary to introduce some important terminology.

The information collected for any statistical study is associated with a *variable of interest.* Such a variable could be the salaries of corporation presidents, the advertising expenses of tobacco companies, or the debt/equity ratios of firms in the steel industry. A single piece of information (e.g., the debt/equity ratio of U.S. Steel Corp.) is called either an *item* or an *observation.* These items are the researcher's *data.* Table 1–1 illustrates these concepts.

If the six courses listed in Table 1–1 are the only business administration courses scheduled for the fall semester, then every possible item relevant to the "enrollments" variable is given in the table. Under these circumstances, we say that the column headed "Students Enrolled" (40, 20, 30, 35, 15, 40) constitutes a *statistical population.*

TABLE 1–1
Enrollments in business administration courses taught at College T during the Fall 1980 semester — VARIABLE OF INTEREST

Course	Students Enrolled
Item #1: Business Statistics	40
Item #2: Cost Accounting	20
Item #3: Management	30
Item #4: Marketing	35
Item #5: Labor Relations	15
Item #6: Corporate Finance	40

DATA

Researchers often find that they can't obtain observations for all of a particular population. Among other reasons, this task can be (1) quite expensive, (2) very time consuming, or (3) physically impossible. Practical considerations, therefore, may force researchers to examine only part of a population. We refer to parts or subsets of populations as *samples*. Table 1–2 breaks the enrollment data in Table 1–1 into three samples.

TABLE 1–2

Sample One (items 1, 2, and 4)	*Sample Two (items 2, 4, 5, and 6)*	*Sample Three (items 4 and 6)*
40	20	35
20	35	40
35	15	
	40	

How we define the variable of interest determines whether the data represent a population or a sample. Even though enrollments in the six courses in Table 1–1 form the population of "enrollments in business administration," they are just a sample of "enrollments in business administration, economics, and mathematics."

Numbers computed to *describe* a population or a sample are called *summary measures*. When summary measures derive from population data, they are said to be *parameters*. A *statistic* is a summary measure obtained from sample observations. The average enrollment per course in Table 1–1 is 30 students. In a sense, 30 "summarizes" the information supplied by the six items. Since the items form a population, 30 is a parameter. If the variable of interest is redefined as "enrollments in business administration, economics, and mathematics," the six items then constitute a sample and 30 becomes a statistic.

Parameters offer insights into populations by showing what is "average" or "typical." However, it is easier to collect data on samples than on populations. This raises a critical question: "How can researchers use a statistic to estimate an unknown parameter?" We might ask, for example, "If 100 of Company J's 6,000 employees are asked to state their feelings about working 10 hours a day for four days versus 8 hours a day for five days, will the responses be a good indication of the entire plant's preferences?"

We will first direct our attention to calculating summary measures. At the same time, we will suggest ways of displaying data. Determining parameters and statistics, as well as presenting data in an informative fashion, are objectives of *descriptive statistics.* About 40 or 50 years ago, introductory statistics courses were dominated by this topic. A descriptive technique, putting data in a table, is applied in the following example.

In the early 1970s, Senator Philip Hart of Michigan introduced an industrial reorganization bill that would have given Congress the power to break up industries composed of a few giant firms. Imagine that as part of its debate over this bill, a congressional subcommittee asked an economist to investigate whether there was a predictable relationship between corporate size and profitability. The economist believed that members of the subcommittee would be curious about the size of the firms used in her study. Consequently, she passed out Table 1–3 before testifying.

TABLE 1–3

Company Size (assets in $ billions)	*Number of Companies*
0–0.9	12
1.0–1.9	15
2.0–2.9	6
3.0–3.9	4
4.0 and over	3
	40

The central focus of statistics today is analysis of sample data for the purpose of gaining some understanding of the population from which the sample was drawn. This involves using methods of *statistical inference.* These methods enable researchers to *estimate* parameter values from sample statistics. The widely reported national unemployment rate is an estimate of a parameter.

The Bureau of Labor Statistics contacts roughly 47,000 households (approximately 100,000 people) every month. Any individual 16 years of age or older is said to be "employed" if he or she is currently working part-time or full-time. People who are "unemployed" are looking for a job, whereas someone "not in the labor force" is neither working nor seeking work. The labor force (i.e., the employed plus the unemployed) numbers over 100 million people. The statistic "unemployment rate for the sample" becomes an estimate of the parameter "unemployment rate for the country."

During presidential campaigns, the prevailing unemployment rate is frequently an issue. If the rate is high, the incumbent will argue that his administration was plagued by a series of unfortunate economic conditions and that steps are being taken to bring the rate down. The challenger will say that the high rate is just another result of the incumbent's ineffectiveness.

Although the unemployment rate is based on a sample of less than 1 percent of the labor force, presidential candidates (and others) generally will not attack the procedures used to derive it. There is a reason for this reluctance. Before a sample is selected, researchers can objectively compute the probability that their estimate will be near the "true" parameter value. When the Secretary of Labor maintains, "I am quite confident that the June unemployment rate was close to 5.8 percent," his "confidence" arises from the estimation technique itself, not from a boastful personality.

Besides estimation, statistical inference also includes *hypothesis testing*. In testing hypotheses, researchers tentatively assign a value to a parameter and then select a sample. By virtue of the characteristics of the sample, the hypothesized parameter value will either seem (1) reasonable or (2) doubtful. The researchers establish the criterion or criteria that will allow them to choose between 1 and 2. In the following example, the hypothesis "95 percent of all cars have been inspected" is subjected to a test.

Drivers in a particular state must have their cars inspected by a certified mechanic twice a year. The deadline for the current inspection period is March 31. Because the weather has been severe for many weeks, some people could not get to an inspection station. Should the deadline be extended? The Secretary of Transportation feels that if 95 percent of all cars have been inspected, there is no need to grant an extension. To determine whether this is the case, he has 1,000 of the state's registered car owners contacted. Only 90 percent of the cars in this sample were inspected. The sample "evidence" makes the secretary skeptical of the 95 percent parameter value. As a consequence, he declares an extension.

Notice that sample summary measures are involved in the inference process. *Probability theory,* or the "nature of chance," is the bridge which connects descriptive statistics and statistical inference. How we evaluate "chance" often influences our behavior, as the next example illustrates.

A handful of executives at Corporation Q play poker every Thursday night. Ted Burns, a new employee, is invited to join the poker group. Burns wins his first hand by drawing four aces. After a few executives joke about "beginner's luck," the second hand is dealt. When Burns again draws four aces, the laughing stops and one player calls him a cheater.

The group changed its attitude because the "chance" of getting two consecutive hands of four aces in a fair game is very slight; that is, two consecutive hands of four aces is *possible but improbable.* The players concluded that this outcome was "more probable" when a game was rigged.

The decision to extend the inspection deadline in the preceding example followed the same kind of thinking. Even if the parameter were 95 percent, the secretary knew that a sample having a 90 percent inspection rate could be selected. He felt, though, that a 90 percent sample outcome was "more probable" when the parameter was less than 95 percent.

Knowledge of probability theory is also essential for solving a problem such as the following.

Engineers at M. W. Electronics have put all the equipment in perfect order. Experience, nonetheless, suggests that 50 of the next 1,000 transistor components that are produced will be defective. Ms. Smith is going to receive a shipment of five transistor components. Management wants to compute the probability of this shipment's containing three defective parts. The company calculates that only 1 out of every 1,000 shipments of 5 transistor components will have 3 defectives. Therefore, management doesn't expect Smith to get three defectives.

Table 1–4 clarifies the distinction between estimation and hypothesis testing. It points out, moreover, the role that probability plays in making inferences.

In the next example, statistical inference is explicitly linked to managerial decision making. The scenario reveals that sampling can have disadvantages as well as advantages and shows the importance of probability theory.

A manufacturer produces 1,000 jar lids an hour. At the end of each hour, he must determine whether the machinery is cutting the lids to specifications. If specifications are not met, further production is curtailed until the equipment is fixed.

The manufacturer needs a strategy for detecting a malfunction. The equipment can be checked directly through inspection or indirectly by looking at all or some of the lids. Inspecting the equipment requires a substantial amount of time, during which the machines are shut down and no jar lids produced—a costly procedure. On the other hand, if the lids are defective, resources and manpower will have been wasted.

Rather than inspecting the equipment every hour, the manu-

TABLE 1–4

Estimation:	No initial statement regarding the parameter value is made.	A sample is selected.	A statistic is computed.	A parameter is estimated.	Identification of the "confidence" with which a researcher can say that the estimate is near the true parameter value.
Hypothesis Testing:	An initial statement regarding the parameter value is made.	A sample is selected.	A statistic is computed.	Assuming that the initial statement is correct, a researcher determines the probability of getting a sample like the one selected.	
				Probability low Conclude that the hypothesized parameter value is unreasonable Reject the initial statement.	Probability high Conclude that the hypothesized parameter value is reasonable Accept the initial statement.

facturer prefers to evaluate the state of the machines indirectly. When the "typical" lid produced in the last hour is close to specifications (e.g., 2.5 inches in diameter), he will feel that everything is working correctly, and 1,000 additional lids will be run off. If the "typical" lid differs greatly from specifications, repairmen will examine the equipment.

Studying 1,000 jar lids is itself time consuming. As an alternative, the manufacturer could measure, say, 50 lids and decide whether to stop or continue production on the basis of the "typical" sample lid.

The 50 lids, unfortunately, will give a false impression about the state of the equipment if the "typical" sample lid is not near the "typical" population lid. The manufacturer may shut down when he doesn't really need to, or he may fail to call for an inspection when there actually is a malfunction.

The "typical" sample and population lids may be different for many reasons. Even when the equipment is operating satisfactorily, dust particles, noise, a sudden change in temperature, or imperfections in the lid metal will cause occasional defects. On the other hand, when the machines are faulty, some proper lids will still be punched out from time to time.

After the manufacturer measures the 50 lids and computes the average diameter, his conclusion about the machines may or may not derive from "rare" lids—that is, (1) inferior lids selected by chance from a batch of mostly good lids or (2) good lids selected by chance from a batch of mostly inferior lids.

The problem facing the entrepreneur is twofold. He must decide whether a "typical" sample lid meeting specifications is due to situation 2 or to the fact that nothing is the matter. He must also decide whether a "typical" sample lid not meeting specifications can be attributed to situation 1 or to a malfunction.

For the manufacturer, developing criteria for assessing production accuracy is part of the inference process. He must ultimately determine the probability of obtaining the observed sample when the hypothesized condition of the machines is "no malfunction."

Table 1–5 relates the concepts of variable, population, sample, parameter, and statistic to this example. Most importantly, the table details some of the risks that are inevitable if actions are taken in an atmosphere of uncertainty.

In later chapters, we will seek to "explain" the values a particular variable has. In using *linear regression,* we will pair values of one variable (Y) with the values of another variable (X). Once we have determined the way in which two variables are associated, we can use the value of X to predict the value of Y. The following example illustrates the relevance of "explaining" Y.

To stimulate corporate spending, the Federal Reserve Board intends to increase the money supply by buying government securities. The Fed believes that such an increase will reduce interest rates and, consequently, encourage businesses to borrow funds for the purpose of financing plant and equipment expenditures.

The size of the spending stimulus is a policy decision. Once the size is established, however, the Fed must compute the specific change in interest rates which will bring about the desired business expansion. It therefore estimates the relationship between business investment (Y) and the rate of interest (X). For instance, it may find that a 1 percentage point drop in the current rate of interest will increase business spending by $2 billion. Once the relationship between X and Y has been computed, the Fed knows how much to reduce interest rates (X) to produce the desired increase in investment (Y).

The discussion of *time series analysis* will also examine the basis

TABLE 1–5

MANAGEMENT TASK: Devise a strategy for determining whether the production process is working correctly (i.e., whether specifications are being met)

VARIABLE OF INTEREST: Measurements of jar lids produced in an hour

Options	*Benefits*	*Costs*	*Action Deriving from Conclusion*
1. Examine every finished item (i.e., look at the *population* of lids)	Management will definitely know whether or not a malfunction has occurred by comparing the "typical" measurement (a *parameter*) to the lid specifications	Time consuming, results in excessive "down time"*	1. Continue production if the "typical" item is near specifications 2. Investigate equipment if the "typical" item is not near specifications
2. Examine some of the finished items (i.e., investigate a *sample* of lids).	Not much time is needed	Management must use the "typical" sample measurement (a *statistic*) to speculate about the "typical" population measurement (a *parameter*). The manufacturer, therefore, is exposed to the risk of drawing an incorrect conclusion. A. No unnecessary cost is incurred if	
		1. Machines are working properly and the manufacturer concludes this.	1. Continue production
		2. Machines are not working properly and the manufacturer concludes this.	2. Check the equipment

(continued)

Table 1–5 continued

Options	*Benefits*	*Costs*	*Action Deriving from Conclusion*
		B. An unnecessary cost is incurred if	
		1. Machines are working properly and the manufacturer concludes otherwise.	1. Check the equipment
		2. Machines are not working properly and the manufacturer concludes otherwise.	2. Continue production

* I am envisioning a situation where all 1,000 lids must be examined before the machines will be turned on again. Thus, the machines will be idle far less time when 50, instead of 1,000, lids are measured. In other words, whether 50 or 1,000 lids are measured, the machines won't be running for a certain amount of time.

for different Y values. The emphasis will be on predicting future Y values in the light of Y's past behavior. We will look for repetitive movements in Y over time. Our final example concentrates on production per month (Y) in an effort to discover if high-volume and low-volume periods have a regular pattern.

The president of W. M. & Son finds that labor requirements vary during the year. Workers are laid off in some months, and in others they get substantial overtime. He knows that layoffs create ill will and overtime is costly. To minimize both, the president plans to schedule employee vacations solely during slack periods. Since the schedule must be set up a year in advance, his strategy will be successful only if he can predict production peaks and troughs for the coming year. He does this by studying past monthly production (Y) figures. Data for the past five years (i.e., 60 observations of Y) show that February and October have consistently been "slow" months. The president therefore schedules vacations for those two months.

Chapter One has indicated some of the dimensions of statistics which we will explore in the remaining 19 chapters. Inasmuch as each chapter generally is the foundation for the topics that appear in the chapter after it, you should try to understand how everything "fits together" as you go along.

SUMMARY

We have stressed that the study of statistics is quite practical because statistical methods can be applied to a host of "real-world" problems. Even though we distinguished five major topics, the development of some of these (e.g., statistical inference) is dependent upon techniques originally discussed elsewhere (e.g., in connection with descriptive statistics and probability). We will continually examine variables, populations, samples, parameters, and statistics. Such concepts should be mastered quickly so that material presented in later chapters can be seen in proper perspective.

EXERCISES

1. Choose the topic—(1) descriptive statistics, (2) probability, (3) estimation, (4) hypothesis testing, (5) linear regression, or (6) time series—that is associated with a study of each of the following situations:

1–1. To reduce both traffic congestion and pollution, a large city wants to increase the number of people who travel on buses. The transit authority has analyzed the relation between Y = level of ridership and X = bus fare. By how much should fares be lowered to attract 5,000 more passengers per day?

1–2. An official of a state's welfare department believes that 20 percent of the welfare checks issued by his department are incorrect. Errors arise from over- and underpayments and from payments to ineligible people. A sample of 500 welfare checks is selected. Of these checks, 14 percent are in error. Does 20 percent appear to be a sensible parameter value?

1–3. Economic policymakers are anxious to lower the unemployment rate. They notice that the rate for June is very high compared to the March rate. Is the current employment strategy ineffective or are June unemployment rates typically high?

1–4. The chance that a Company R employee will be absent from work is small. An important order must be finished by Wednesday. If six employees are absent that day, the order won't be completed and the buyer will cancel it. What is the likelihood that six employees will be absent?

1–5. The bargaining committee of a union has just negotiated a contract. Workers will receive an hourly wage increase spread over three years, and people with 15 years seniority will receive an extra week of paid vacation. The contract must be approved by the rank and file. A sample of 100 workers is chosen. Thirty percent of the members of the sample like the contract. What is the population percentage in favor of it?

1–6. The Bureau of the Census has collected millions of pieces of information on the condition of housing in the United States. How can the bureau display its findings in a compact manner?

1–7. During the last week of a tough mayoral campaign, Candidate Webb appears tired and irritable. If more than 50 percent of the electorate is for him, he will end his speaking engagements and make a few token addresses. Otherwise, he will "fight to the wire." To test the claim "more than 50 percent of the electorate supports Webb," he commissions a poll of 200 registered voters. Based on the fact that 130 of these people support Webb, can he relax prior to election day?

1–8. State G feels that it can only finance an extensive highway construction program by raising the sales tax from 3 percent to 4 percent. If two-thirds of the 1,000 residents in a sample think that the tax hike is unjustified, what percentage of all of the state's residents object to the proposal?

1–9. According to the United States Tennis Association, a regulation tennis ball cannot weigh less than 2 ounces. A tournament director weighs 50 of the 2,000 balls that have been shipped to her. Ten of the balls are less than 2 ounces. Should she return the shipment?

1–10. The Food and Drug Administration is halting the sale of a particular medication. Ten percent of all Americans suffer from disease W, and 70 percent of these victims take this medication. If an American is chosen at random, what is the chance that he both has disease W and uses the banned medication?

1–11. To determine whether overtime must be scheduled, a production manager pairs information on Y = output per day and X = number of man-hours. How do changes in the number of man-hours affect the level of output?

1–12. A researcher is exploring the extent to which high school students are familiar with certain business practices. She chose 40 students from J. T. High School. Only five of the students knew that "3/10, net 30" indicated that a seller offered a 3 percent discount if bills were paid within ten days, but demanded full payment within 30 days. What proportion of the J. T. student body is familiar with "3/10, net 30"?

1–13. Many landlords have found that they can improve their financial position by converting apartment buildings into condominiums. Ten of the 20 "hi-rises" in City D are condominiums. If a "hi-rise" is picked at random, what is the chance that it will be a condominium?

1–14. A psychologist is investigating the effects of prolonged unemployment on family stability. She compiled the following information on 13 families:

Length of Time Head of Household Has Been Unemployed (in months)	*Number of Families*
1– 2.9	2
3– 4.9	3
5– 6.9	5
7– 8.9	2
9–10.9	1
	13

What is the average length of time a member of the sample was unemployed?

2. To confirm the relevance of statistics, look through newspapers for several days and report five instances where statistical techniques were involved.

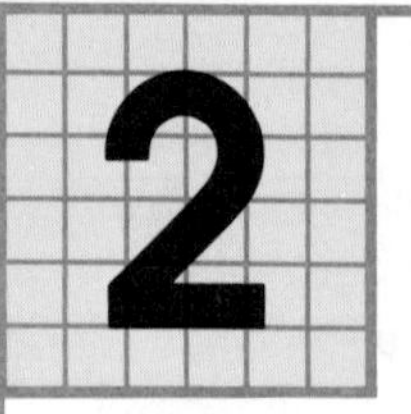

Descriptive Statistics: *Ungrouped Data*

THOUGHTS TO BEAR IN MIND

Chapters Two and Three discuss ways of organizing and presenting quantitative information. We may wish to communicate such information by means of a list, a graph, or the summary measures that we have called parameters and statistics. The manner of presentation is influenced by both (1) the goal of our research and (2) the intended audience.

Several points should be stressed before we continue. First, unless the variable of interest is explicitly defined, we can't distinguish population data from sample data. This is because populations and samples may be large or small. We have already noted that the monthly labor-force sample has as many as 100,000 observations. By contrast, consider the following.

Due to an ethics code established by Congress in 1977, senators must reveal their assets once a year. Since every state has two senators, the variable "financial holdings of United States senators" will involve a population of only 100 observations.

Second, sample summary measures, in terms of the rationale for deriving them and the way in which they are interpreted, generally parallel the corresponding parameters. Suppose that the average rate of profit for five firms is 12 percent. Because of the parallelism just noted, 12 percent is regarded as a "typical" value irrespective of whether the five firms are a population or a sample.

Third, judgments based on summary measures may or may not be accurate. In the next example, using a summary number as the sole basis for evaluating information results in an incorrect conclusion.

To relax from the pressures of a surprise audit, the president of Bank J spent the weekend out of town. The temperature reached

110 degrees on Saturday, but only 40 degrees on Sunday. When he returned to the bank, his secretary asked about the average high temperature. Told that it was 75 degrees, she failed to understand why her boss hated the weather. Her confusion came from assuming that each high temperature was close to the average.

We will soon look at three types of averages and study techniques for analyzing the extent to which the averaged items vary. This will help us overcome such problems as that faced by the secretary.

MEASUREMENT SCALES

Observations are measured on nominal, ordinal, interval, and ratio scales. We will distinguish among these because some statistical procedures aren't applicable to all scales.

Nominal Scale

A *nominal scale* classifies observations by words.

When the Federal Reserve System was created in 1913, the Federal Reserve Board began to establish United States monetary policy. Its actions are not controlled by the President or Congress. Table 2–1 indicates the way people view the Fed's "independence."

TABLE 2–1

Person	*Opinion*
A	For
B	Against
C	For
D	For
E	Against

The observations in the right column are *categorical* or *qualitative* data. Researchers often replace words with arbitrarily chosen numbers. If we let 1 = For and 2 = Against, Table 2–1 would appear as Table 2–2.

A different numbering *code* (e.g., 60 = For and 10 = Against) would be equally legitimate. Therefore, we can't compare 1 and 2 or 60 and 10 arithmetically; that is, (1) For is not 50 more than Against, (2) For is not 6 times more than Against, and (3) For is not greater than Against.

TABLE 2–2

Person	*Opinion*
A	1
B	2
C	1
D	1
E	2

Ordinal Scale

Observations in an *ordinal scale* have some sort of relation to one another and can therefore be ranked.

The large increase in merger activity during the past twenty years has caused doubt over the effectiveness of current antitrust legislation (e.g., the Clayton Act). Senator Edward Kennedy of Massachusetts recently sponsored a bill that would automatically prohibit mergers of "big" companies, even if the acquiring and acquired companies functioned in different lines of business (i.e., the merger would not lessen competition in a particular industry). Five economists were questioned about the Kennedy bill; Table 2–3 gives their responses.

TABLE 2–3

Economist	*Opinion*
A	Not for
B	Mildly for
C	Not for
D	Strongly for
E	Not for

Although the observations are categorical, we can, nonetheless, *rank* them in terms of the criterion "support for the bill." "Strongly for" represents more support than "Mildly for." By giving higher numbers to the higher ranks, we might substitute Table 2–4 for Table 2–3. In discussing Table 2–4, we are justified in saying that

TABLE 2–4

Economist	*Opinion*
A	5
B	10
C	5
D	40
E	5

"40 is more than 10." We cannot, however, attach any importance to the size of the difference.

Interval Scale

In an *interval scale,* the numbers can be ranked *and* their relationship to one another is known. The Fahrenheit and Celsius, or centigrade, temperature scales are prime examples. We can (1) rank the temperatures and (2) state that "90 is 10 degrees warmer than 80."

Recall that the relation between centigrade (C) and Fahrenheit (F) temperatures is

$$F = 32 + (9/5)\,C$$

Thus 40°C = 104°F and 80°C = 176°F. The zero point on both scales is arbitrary. Because of this, we can't express C or F temperatures as ratios. If we did, we would get the contradiction "80°C is twice as hot as 40°C, but 176°F is only 1.69 times as hot as 104°F."

Ratio Scale

A *ratio scale* has all the characteristics of an interval scale, and in addition it has a true (not arbitrary) zero point.

Companies may spend millions of dollars every year trying to develop new products. Many innovations never leave the laboratory. Even when they do, the chances of commercial success are low. The record of Company G is shown in Table 2–5.

TABLE 2–5

Year	*Successful Products Introduced*
1974	1
1975	0
1976	3
1977	2
1978	2

Based on Table 2–5, we can say that, relative to 1974, Company G introduced (1) more or (2) two more or (3) three times as many products in 1976.

Table 2–6 compares the four measurement scales. Imagine that a and b are two numerical observations (e.g., $a = 4$ and $b = 2$). The asterisks identify the ways in which we can validly compare a and b in the different scales.

TABLE 2–6

Scale	$a > b$ *or* $a < b$	$a - b$	a/b
Nominal			
Ordinal	*		
Interval	*	*	
Ratio	*	*	*

SIGMA NOTATION

Statisticians use the Greek capital letter *sigma,* Σ, when they want a "shorthand" way to indicate the addition of certain observations. Consider the following example.

Once a year, *Fortune* magazine publishes data on the 500 largest United States industrial companies, the so-called "Fortune 500." Table 2–7 identifies the four sales leaders in 1976.

TABLE 2–7

Position	*Company*	*1976 Sales (in $ billions)*
1	Exxon	49
2	General Motors	47
3	Ford Motor	29
4	Texaco	26

Source: *Fortune,* May 1977, p. 366.

Given the variable X = 1976 sales, the observation in the ith position is X_i; for example, $X_2 = 47$ and $X_4 = 26$. We note that

$$\sum_{i=1}^{4} X_i$$

is the *summation* of items in positions $i = 1$ through 4 inclusive; that is,

$$\sum_{i=1}^{4} X_i = X_1 + X_2 + X_3 + X_4$$
$$= 49 + 47 + 29 + 26 = 151$$

Moreover,

$$\sum_{i=2}^{3} X_i = X_2 + X_3 = 47 + 29 = 76$$

and

$$\sum_{i=1}^{3} X_i = X_1 + X_2 + X_3 = 49 + 47 + 29 = 125$$

Suppose we let N = the number of observations in a *population* and n = the number of observations in a *sample*. Then,

$$\sum_{i=1}^{N} X_i = \text{the summation of } all \text{ population items}$$

and

$$\sum_{i=1}^{n} X_i = \text{the summation of } all \text{ sample items}$$

In these circumstances, we delete the *index of summation, i,* and just write ΣX.

There are some important properties connected with the Σ operation:

1. If we multiply each observation of a variable, X, by a constant, k, the corresponding summation, ΣkX, is equal to $k\Sigma X$; that is,

$$\Sigma kX = k\Sigma X \tag{2–1}$$

2. If each of the N observations of a variable, X, has the same value, the corresponding summation, ΣX, is equal to the product Nk; that is,

$$\sum_{i=1}^{N} X_i = Nk \tag{2–2}$$

   when $X_1 = X_2 = \ldots = X_N = k$.

   Similarly,

$$\sum_{i=1}^{n} X_i = nk$$

   when $X_1 = X_2 = \ldots = X_n = k$.

3. If each variable X observation in the ith position is added to each variable Y observation in the ith position, the corresponding summation of the sum of X and Y,

$$\sum_{i=1} (X_i + Y_i) = (X_1 + Y_1) + (X_2 + Y_2) + \ldots,$$

is equal to the sum of the individual summations; that is,

$$\Sigma(X + Y) = \Sigma X + \Sigma Y \qquad (2\text{–}3)$$

Also,

$$\Sigma(X + Y + Z) = \Sigma X + \Sigma Y + \Sigma Z$$

4. If each variable Y observation in the ith position is subtracted from each variable X observation in the ith position, the corresponding summation of the difference between X and Y,

$$\sum_{i=1} (X_i - Y_i) = (X_1 - Y_1) + (X_2 - Y_2) + \ldots,$$

is equal to the difference between the individual summations; that is,

$$\Sigma(X - Y) = \Sigma X - \Sigma Y \qquad (2\text{–}4)$$

Also,

$$\Sigma(X - Y - Z) = \Sigma X - \Sigma Y - \Sigma Z$$

5. We can manipulate the terms to the right, or "inside," of Σ before we add; for example,

$$\Sigma(X + Y)^2 = \Sigma(X^2 + 2XY + Y^2)$$

The following example clarifies several of these properties.

States finance their budgets largely through income and sales taxes. Pennsylvania residents pay a 2.2 percent state personal income tax. Table 2–8 displays information on people with incomes consisting of X = wages and Y = interest.

TABLE 2–8

Taxpayer	*Position in List*	*W*	*Y*	*Tax* $[.022(X + Y)]$
A	1	$13,000	$ 700	$301.40
B	2	15,000	900	349.80
C	3	10,000	400	228.80
		$38,000	$2,000	$880.00

The total tax liability of A, B, and C is

$$\sum_{i=1}^{3} .022(X_i + Y_i)$$

We can derive this summation by any one of the following methods:

$\Sigma\, .022(X + Y) = \$880$	
$\Sigma(.022X + .022Y) = \$880$	Property #5
$\Sigma\, .022X + \Sigma\, .022Y = \$880$	Property #3
$.022\, \Sigma X + .022\, \Sigma Y = \$880$	Property #1

MEASURES OF CENTRAL TENDENCY

Summary measures identify specific characteristics of a collection, or *distribution,* of data. Before computing parameters and statistics, however, researchers often pair the different values of X with their *frequency, f*. The next example illustrates this procedure.

Retailers incur losses of more than $2 billion a year due to shoplifting. Company D has six clothing stores. Management wants to pick the best way to reduce theft. It will either (1) install closed-circuit TV cameras or (2) hire security guards. D's losses for 1978 are reproduced in Table 2–9.

TABLE 2–9

Store	X *Shoplifting Losses (in $ thousands)*
A	5
B	9
C	3
D	16
E	7
F	5

Since 3, 5, 7, 9, and 16 are the different values of X in Table 2–9, the *frequency distribution* of X would look like Table 2–10.

TABLE 2–10

X *(in $ thousands)*	f *Number of Stores*
3	1
5	2
7	1
9	1
16	1
	$N = 6$

Instead of showing the frequency of a value, we can determine its *relative frequency,* or f/N. Table 2–11 presents the *relative frequency distribution* of X in this example.

We will next calculate some parameters of the distribution of X = losses. Measures that yield "typical" or "average" or "representative" values are said to be *measures of central tendency.* These include (1) arithmetic means, (2) medians, and (3) modes.

TABLE 2–11

X (in $ thousands)	f/N Proportion of Stores
3	1/6
5	2/6
7	1/6
9	1/6
16	1/6
	6/6 = 1

Arithmetic Mean

Imagine a steel rod having a scale etched on it, as in Figure 2–1.

FIGURE 2–1

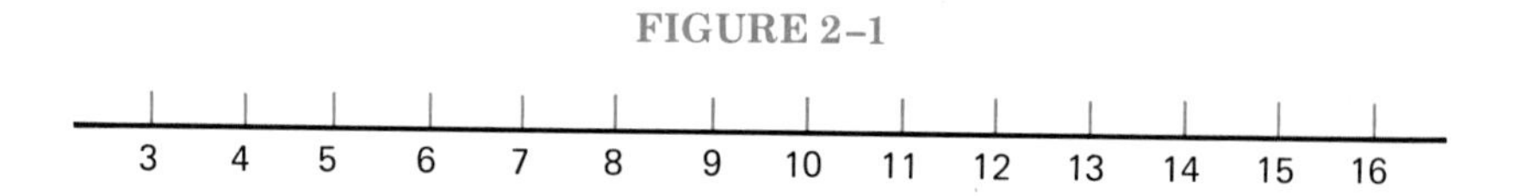

Adding a 1-pound weight at the value of X_i in our shoplifting example results in Figure 2–2. The *arithmetic mean* is the rod's "center of gravity." We could balance the rod by placing a fulcrum at the mean value.

FIGURE 2–2

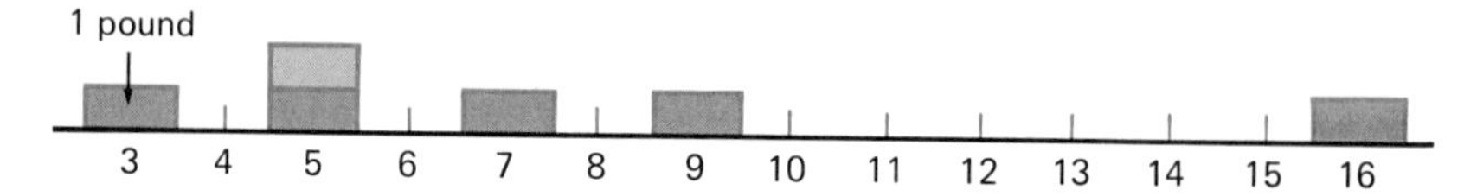

The balancing point is the sum of all N or n observations divided by the number of observations. Statisticians symbolize the population mean (a parameter) with the Greek letter *mu*, μ. We will refer to the sample mean (a statistic) as $\overline{X}$ (i.e., X-bar). Therefore,

$$\mu = \frac{\sum_{i=1}^{N} X_i}{N} \tag{2–5}$$

$$\overline{X} = \frac{\sum_{i=1}^{n} X_i}{n} \tag{2–6}$$

Since, in our example,

$$\mu = \frac{5 + 9 + 3 + 16 + 7 + 5}{6} = 7.5 \text{ (i.e., \$7,500)},$$

we can visualize where to locate the fulcrum (see Figure 2–3). Notice that the mean value isn't always an *actual* value of X.

FIGURE 2–3

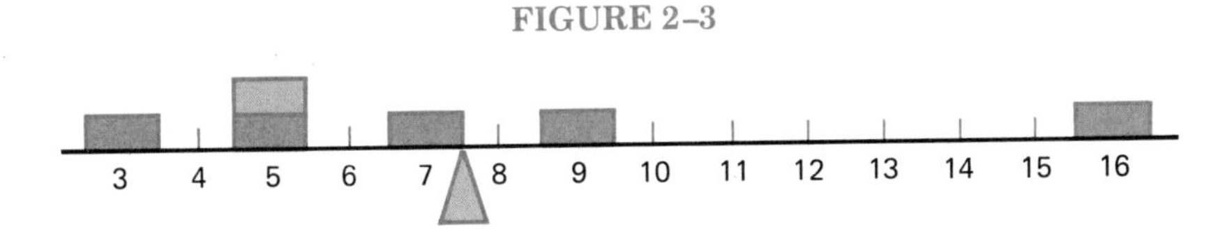

The mean has several virtues:

1. It is a popular concept.
2. It is easy to calculate.
3. We can perform algebraic operations with it; for example,
   A. If we know μ and N, we can solve for ΣX:

$$\mu = \Sigma X/N$$
$$N\mu = \Sigma X$$

   B. If we know the means ($\overline{X}_1$ and $\overline{X}_2$) of two samples, as well as the number of observations (n_1 and n_2) in these samples, we can solve for the mean of the combined data:

$$\text{Combined mean} = \frac{\text{sum of combined items}}{\text{number of combined items}}$$
$$= \frac{n_1\overline{X}_1 + n_2\overline{X}_2}{n_1 + n_2}$$

One disadvantage of μ, however, is its sensitivity to *extremes* (i.e., observations much smaller or much larger than the others). The mean is not very "representative" when extremes are present. Without the value 16 in our example, μ would equal

$$\frac{5 + 9 + 3 + 7 + 5}{5} = 5.8 \text{ (i.e., \$5,800)}$$

The extreme X_i "pulls" the mean in its direction (from 5.8 to 7.5).

Median

The *median* is a "central" value because it is the observation in the *middle* of a distribution after the observations have been listed in ascending or descending order; that is,

$$\text{Rank of the median} = \frac{N+1}{2} \text{ (population)} \quad (2\text{–}7)$$

$$\text{Rank of the median} = \frac{n+1}{2} \text{ (sample)} \quad (2\text{–}8)$$

The number of observations above and below the median are, therefore, identical. When N or n is an even number, two observations lie in the middle. Under these circumstances, the median is defined as the *mean* of the middle observations. Thus the median of

3
5
5 — Middle observations
7 — [Rank of median = (6 + 1)/2 = 3.5]
9
16

is (5 + 7)/2 = 6 (i.e., $6,000).

Extremes don't affect the median as much as they do the mean. The median of 3, 5, 5, 7, 9 is 5. If the extreme were 1,600 instead of 16, the median would still rise from 5 to 6. In contrast, the mean would skyrocket from 5.8 to 271.5. Inasmuch as the median is determined by the *position* of observations, we can't compute the median of two combined distributions unless we first rank the $N_1 + N_2$ or $n_1 + n_2$ items.

Mode

The *mode* is the value with the greatest frequency. In our example, the X mode is 5 (i.e., $5,000). Unlike the mean and the median, the mode of a set of numerical observations (e.g., 2, 4, 5, 7, 9, 10) may not exist, or the mode may have more than one value, for example,

3, (5, 5) 6, 7, (9, 9) 10, 11
mode mode

The mode is particularly appropriate when the data are categorical, as in the following example.

A textile manufacturer is reluctant to fire any of his employees. Nonetheless, 20 were dismissed between 1977 and 1979. The reasons for the dismissals are contained in Table 2–12, which shows that "chronic absenteeism" is the "typical," or *modal,* cause of dismissal.

The preferred measure of central tendency will depend on the context. Because of the mean's mathematical properties and its popularity, we will discuss it extensively.

TABLE 2–12

Cause	Number of Workers
Carelessness	6
Chronic absenteeism	12
Unauthorized use of equipment	2

MEASURES OF DISPERSION

If the data are numerical, we can *always* derive a mean. Each of the populations in Table 2–13 has a mean of 7.5. Although 7.5 is "typical" of the values in columns 1, 2, and 3, it "represents" column 1 best. The mean gives a very poor impression of a set of X_is whenever the observations aren't close together.

TABLE 2–13

X	X	X
7.5	5	0
7.5	9	0
7.5	3	0
7.5	16	15
7.5	7	15
7.5	5	15

We wish to develop parameters and statistics that indicate *how different* the X_is are. Measures that summarize the "spread" or variability of a distribution are called *measures of dispersion.* As long as $X_1 = X_2 = \ldots = k$ (see column 1), all measures of dispersion will equal zero. Departures from zero indicate that at least one observation is not like the others. Researchers must personally evaluate the *size* of such a measure (Is it "large" or "small"?).

TABLE 2–14

Measure of dispersion is "small"	*Measure of dispersion is "large"*
The items are very similar.	The items are not similar.
The items are very close to the mean.	The items are not close to the mean.
The mean provides an accurate impression of the data.	The mean doesn't provide an accurate impression of the data.

Since the differences between the X_is and the mean are large when the distribution is quite spread out, we will look for measures of dispersion that force the researcher to compute (X_i − mean). Table 2–14 reviews our line of reasoning thus far.

Range

Like the mean, the *range*,

$$\text{Range} = \text{Largest item} - \text{Smallest item}, \quad (2\text{–}9)$$

is familiar to many people. In the shoplifting example, X = losses, the range is $16 - 3 = 13$ (i.e., $13,000). The range is influenced by extremes. Since it reports nothing about the concentration of observations around the mean, we won't investigate this measure of dispersion further.

Total and Average Deviations

Intuitively, the *total deviation,* $\Sigma(X - \mu)$, and the *average deviation,* $\Sigma(X - \mu)/N$, seem like useful measures. Table 2–15 computes $\Sigma(X - \mu)$ for X = losses.

TABLE 2–15

X	μ	$X - \mu$
5	7.5	−2.5
9	7.5	1.5
3	7.5	−4.5
16	7.5	8.5
7	7.5	−0.5
5	7.5	−2.5
		$\Sigma(X - \mu) = 0$

Even though none of the items is equal to μ, the total deviation is zero. The explanation for this "paradox" is that four negative deviations "canceled out" two positive deviations.

Irrespective of the distribution, $\sum^{N}(X - \mu) = 0$. We can verify this by replacing the specific values in Table 2–15 with symbols in Table 2–16. Since $\sum^{N}(X - \mu)$ is the summation of a difference and μ is a constant, we know that

$$\sum^{N}(X - \mu) = \sum^{N} X - \sum^{N} \mu \qquad \text{Property \#4}$$

$$= \sum X - N\mu \qquad \text{Property \#2}$$

$$= N\mu - N\mu = 0$$

TABLE 2–16

X	μ	$X - \mu$
X_1	μ	$X_1 - \mu$
X_2	μ	$X_2 - \mu$
X_3	μ	$X_3 - \mu$
X_4	μ	$X_4 - \mu$
X_5	μ	$X_5 - \mu$
X_6	μ	$X_6 - \mu$

To include $X_i - \mu$ in the calculation of a measure of dispersion, we have to bypass the problem of $\Sigma(X_i - \mu) = 0$. Two common approaches are (1) working with absolute values and (2) squaring.

Mean Absolute Deviation

An absolute value is the "distance" between a number and zero, rather than the "distance" and "direction." Because -3 is a distance of 3 units from zero, its absolute value, written as $|3|$, is 3. The *mean absolute deviation* is the average difference between the observations and the mean when the signs of the deviations are dropped:

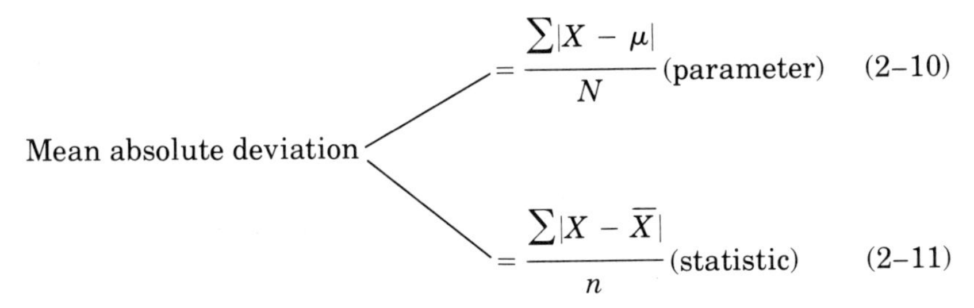

Mean absolute deviation

$$= \frac{\sum|X - \mu|}{N} \text{ (parameter)} \qquad (2\text{–}10)$$

$$= \frac{\sum|X - \overline{X}|}{n} \text{ (statistic)} \qquad (2\text{–}11)$$

Table 2–17 computes $|X - \mu|$ for X = losses, and we find that the mean absolute deviation is $20/6 = 3.33$ (i.e., $3,330).

TABLE 2–17

X	μ	$X - \mu$	$\|X - \mu\|$
5	7.5	−2.5	2.5
9	7.5	1.5	1.5
3	7.5	−4.5	4.5
16	7.5	8.5	8.5
7	7.5	−0.5	0.5
5	7.5	−2.5	2.5
			20.0

Inasmuch as the mean absolute deviation isn't important in developing the theory of statistical inference, we will emphasize squaring.

Variance

The population *mean squared deviation,* or *variance,* is symbolized by σ^2 (σ is the Greek lower case letter *sigma*):

$$\sigma^2 = \frac{\sum^{N}(X - \mu)^2}{N} \tag{2–12}$$

Given the computations in Table 2–18, we find that the variance in our example X = losses is $\sigma^2 = 107.5/6 = 17.9$ (i.e., 17.9 square \$ thousands).

TABLE 2–18

X *(in \$ thousands)*	μ *(in \$ thousands)*	$X - \mu$ *(in \$ thousands)*	$(X - \mu)^2$ *[in (\$ thousands)^2]*
5	7.5	−2.5	6.25
9	7.5	1.5	2.25
3	7.5	−4.5	20.25
16	7.5	8.5	72.25
7	7.5	−0.5	0.25
5	7.5	−2.5	6.25
			107.50

Computation of the sample variance, or s^2, does not parallel that of σ^2:

$$s^2 = \frac{\sum^{n}(X - \overline{X})^2}{n - 1} \tag{2–13}$$

We will substantiate dividing by $n - 1$, instead of n, in Chapter Eight. The reason for $n - 1$ is *not* obvious. It relates to our ultimate interest in deriving inferences from samples.

Standard Deviation

Because the variance is expressed in squared units, it doesn't possess a physical identity (e.g., What are square \$ thousands?). We can't easily relate this measure to the data and the mean. Therefore, we will find the *positive* square root of σ^2 and s^2. Such a root is referred to as the *standard deviation*:

$$\sigma = \sqrt{\sigma^2} = \sqrt{\Sigma(X - \mu)^2/N} \tag{2–14}$$

$$s = \sqrt{s^2} = \sqrt{\Sigma(X - \overline{X})^2/(n - 1)} \tag{2–15}$$

Returning to our example, we see that $\sigma = \sqrt{17.9} = 4.2$; that is,

$$\sigma = \sqrt{17.9 \text{ (\$ thousands)}^2} = 4.2 \text{ (\$ thousands)} = \$4{,}200$$

Notice that σ is expressed in the same units as X and μ.

The next question is whether the standard deviation is "large" or "small." If we interpret 4.2 as "large," we simultaneously imply that (1) the data are widely spread out (i.e., the items are not similar) and (2) 7.5 is not very representative of the data (i.e., the mean value doesn't offer an accurate description of the actual X_is).

When we don't want to judge the size of σ, we can still get information from it. The information is a consequence of *Chebyshev's theorem:*

> For $k \geq 1$, at least $(1 - 1/k^2)(100\%)$ of the observations in *any* distribution have a value between $\mu - k\sigma$ and $\mu + k\sigma$.

Suppose we (1) list the X observations in our example in ascending order and (2) move $k = 1.5$ standard deviations below and above the mean (see Figure 2–4). Fifty-six percent, $(1 - 1/1.5^2)(100\%)$, or more of the X_is must lie in the interval 1.2 to 13.8. In fact, 5/6, or 83 percent, of the observations do.

FIGURE 2–4

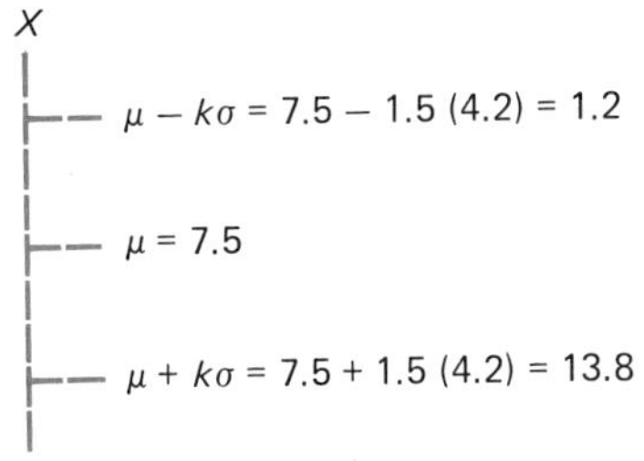

X

$\mu - k\sigma = 7.5 - 1.5\ (4.2) = 1.2$

$\mu = 7.5$

$\mu + k\sigma = 7.5 + 1.5\ (4.2) = 13.8$

Coefficient of Variation

Researchers may wish to study the standard deviations of two sets of data. Comparing σs when the variables are measured in different units (e.g., tons vs. gallons) or in different magnitudes (e.g., \$ billions vs. \$ millions) isn't valid. In such situations, statisticians compute the *coefficient of variation,* which states σ or s *relative* to the mean:

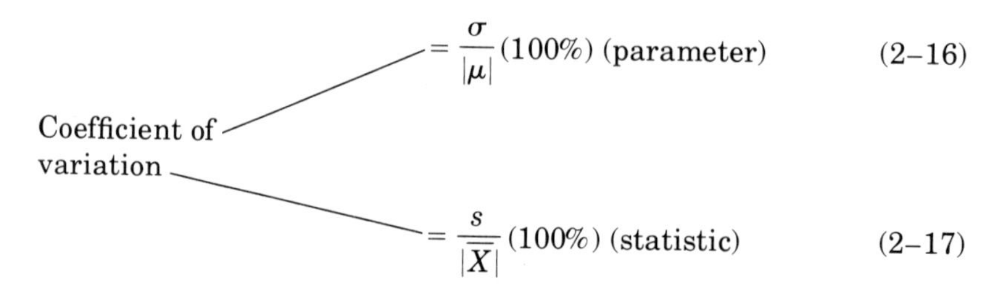

$$\text{Coefficient of variation} = \frac{\sigma}{|\mu|}(100\%) \text{ (parameter)} \qquad (2\text{–}16)$$

$$\text{Coefficient of variation} = \frac{s}{|\overline{X}|}(100\%) \text{ (statistic)} \qquad (2\text{–}17)$$

This removes the effect of different units or magnitudes. The following example shows the usefulness of the coefficient of variation.

An economist specializing in industrial organization is investigating the characteristics (e.g., pricing policy, reliance on external financing, etc.) of all companies in industry H. For X = assets and Y = capacity utilization rates (i.e., current output/capacity output), she finds

$$\mu_x = \$750{,}000 \qquad \mu_y = .86$$
$$\sigma_x = \$\ 27{,}000 \qquad \sigma_y = .05$$

Since X is in dollars and Y is a proportion, the economist can't compare σ_x and σ_y directly. Instead, she derives

$$\frac{\$27{,}000}{\$750{,}000}(100\%) = 3.6\%$$

and

$$\frac{.05}{.86}(100\%) = 5.8\%$$

Capacity utilization rates are, thus, more "erratic" than company asset levels.

CODING DATA

Calculating summary measures is usually tedious when the values of X are very large or small. Researchers frequently simplify their data by *coding* it. To do so, they transform the X_is through (1) adding or subtracting a constant, a, and/or (2) multiplying or dividing by a constant, b. The derived values are referred to as Y. Table 2–19 illustrates three coding processes.

TABLE 2–19

Divide X_i by 1,000,000		*Multiply X_i by 10,000*		*Add 10 to X_i*	
X	Y	X	Y	X	Y
35,000,000	35	.0058	58	−4	6
24,000,000	24	.0094	94	6	16
29,000,000	29	.0072	72	−8	2
43,000,000	43	.0018	18	5	15

The relationship $Y_i = a + bX_i$ establishes a particular connection between the summary measures of the X and Y distributions. We will confirm this in terms of a population of Xs. A proof developed with a sample would follow the same logic.

Regardless of how X is converted into Y, we know

$$\mu_y = \frac{\Sigma Y}{N}$$

Notice, however, that

$$\mu_y = \frac{\Sigma Y}{N} = \frac{\Sigma\,(a + bX)}{N}$$

$$= \frac{\Sigma a}{N} + \frac{\Sigma bX}{N}$$

$$= \frac{Na}{N} + \frac{b\Sigma X}{N} = a + b\mu_x \qquad (2\text{–}18)$$

We can draw the following conclusions from Equation 2–18:

1. Just adding or subtracting a will make $\mu_y = a + \mu_x$ since $b = 1$ under these circumstances.
2. Just multiplying or dividing by b will make $\mu_y = b\mu_x$ since $a = 0$ under these circumstances.

We can express the variance of Y as

$$\sigma_y^2 = \frac{\Sigma(Y - \mu_y)^2}{N}$$

$$= \frac{\Sigma[(a + bX) - (a + b\mu_x)]^2}{N}$$

$$= \frac{\Sigma(bX - b\mu_x)^2}{N} = \frac{\Sigma[b(X - \mu_x)]^2}{N}$$

$$= b^2\frac{\Sigma(X - \mu_x)^2}{N} = b^2\sigma_x^2 \qquad (2\text{–}19)$$

Then,

$$\sigma_y = \sqrt{b^2\sigma_x^2} = |b|\sigma_x \tag{2–20}$$

Note that $\sigma_y^2 = \sigma_x^2$ when we add or subtract a constant. The logic behind this is that such an operation doesn't change the *distance* from one X_i to another. On the other hand, when we multiply or divide by a constant, we affect the "spread" of the data. If $X = 10$ and $X = 20$ are multiplied by 5, the distance between the transformed data is $100 - 50 = 50$, instead of $20 - 10 = 10$. The variance will thus increase. The next example suggests the practicality of coding.

Gas pumps constructed during the 1970s can't indicate prices in excess of 99.9 cents a gallon. If prices were based on liters, the oil companies would not have to develop completely new gauges for prices of $1.00 or more a gallon.

One gallon of gas is equal to 3.785 liters. Given a price of $2.00 per gallon, the current "three-window" pump would read 5–2–8/10 cents per liter.

Station A pumps an average (μ_x) of 800 gallons per day, with a variance (σ_x^2) of 500 square gallons.

Suppose the owner would like to restate his gallons-per-day measures in terms of liters per day. Inasmuch as

$$\begin{aligned} Y_i\,(\text{liters}) &= a + bX_i\,(\text{gallons}) \\ &= 0 + 3.785\,X_i \end{aligned}$$

he can quickly find

$$\begin{aligned} \mu_y &= 3.785\,\mu_x = 3{,}028 \text{ liters} \\ \sigma_y^2 &= (3.785)^2\,\sigma_x^2 = 7{,}163 \text{ square liters} \\ \sigma_y &= 3.785\,\sigma_x = 84.6 \text{ liters} \end{aligned}$$

SUMMARY

Researchers compute parameters and statistics because these measures identify many characteristics of a set of data. The mean, median, and mode are "representative" or "typical" values. When the data are spread out, the mean is not very representative of the X_is. The variance and the standard deviation indicate how much the X_is differ from μ or $\overline{X}$. Researchers must interpret the size of this difference themselves. Computing parameters is often simplified when the data are coded.

We will discuss additional topics in descriptive statistics in the next chapter. Our interest will shift from ungrouped data to data organized into classes.

EXERCISES

1. Employees at Company G rarely quit. They indicated the company's major attractions:

Reason for Remaining	*Number of Employees*
Good pay	60
Socially-important work	20
Pleasant colleagues	10
	90

What type of scale do the data represent?

2. Promotions in a certain corporation are based on the recommendations of department heads. Ten members of the accounting department are being considered for an executive position overseas. These employees received the following ratings:

Rating	*Number of Employees*
Excellent	2
Good	7
Fair	1
	10

What type of scale do the data represent?

3. Observations for variables X and Y are:

X	Y
6	−2
−4	3
5	4

Compute the following summations:

a. ΣX
b. ΣY
c. $\Sigma(XY)$
d. $\Sigma(X + Y)^2$
e. $\Sigma(4X)$
f. $\Sigma(X/Y)$
g. $(\Sigma X)(\Sigma Y)$
h. $(\Sigma X)^2(\Sigma Y)$
i. $\Sigma(2X)(3Y)$
j. $\Sigma(X^2)(Y)$

4. Prove that

$$\sum_{i=1}^{4} (X_i + Y_i + Z_i) = \sum_{i=1}^{4} X_i + \sum_{i=1}^{4} Y_i + \sum_{i=1}^{4} Z_i$$

5. Prove that

$$\sum_{i=1}^{3} 5X_i = 5 \sum_{i=1}^{3} X_i$$

6. Write the following expression in terms of Σ:

$$X_2 - Y_2 + 6 + X_3 - Y_3 + 6$$

7. By using the data

X	Y
2	4
3	1
−1	5

substantiate the following:

a. $\Sigma XY \neq (\Sigma X)(\Sigma Y)$
b. $\Sigma X^2 \neq (\Sigma X)^2$
c. $\Sigma(X/Y) \neq \dfrac{\Sigma X}{\Sigma Y}$

8. Some advocates of welfare reform have proposed a "negative income tax" system. Suppose that this system has the following characteristics:

a. Families earning Y dollars or less are guaranteed a welfare check of X dollars.
b. Families earning more than Y dollars but less than $10,000 will get a check for under X dollars. The size of the check will be determined by subtracting from X an amount equal to a proportion, t, times the difference between family income, Z, and Y.

Three families have incomes greater than Y but less than $10,000. Use Σ to show the total amount of welfare assistance these families would receive.

9. The National Highway Safety Council was interested in finding out whether motorists knew that the braking distance for a car traveling at 30 miles an hour was over 130 feet. Eight motorists believed the braking distance was 50′, 78′, 29′, 300′, 67′, 94′, 148′, and 94′. Compute the mean, median, and modal values.

10. Financial analysts are often curious about the "liquidity" of a company. That is, can the company repay its short-term debt? One index of liquidity is the "current ratio," or current assets/current liabilities. The numerator includes cash, accounts receivable, in-

ventories, and several other items. Current liabilities are such things as accounts payable. The "quick ratio" is another index. It is equal to (current assets − inventories)/current liabilities. These indices were computed for each of the firms in industry U:

X *Current Ratio*	Y *Quick Ratio*
1.8	0.7
2.5	1.6
2.8	1.2
2.7	0.9
3.2	1.5
2.8	2.1

a. Calculate the mean, median, and modal values for X and Y.
b. With respect to which ratio are the companies in industry U most similar?

11. The National Football League Management Council reported the following information on the 1977 salaries of football players:

Position	*Number of Players*	*Mean Salary*
Quarterback	88	$89,354
Kicker	59	$41,506

What was the average salary for quarterbacks and kickers combined?

12. The variance can be calculated more quickly if an alternative to the formula in the text is applied. Given the summation properties, prove the following:

a. $\sigma^2 = \dfrac{N\Sigma X^2 - (\Sigma X)^2}{N^2}$

b. $s^2 = \dfrac{n\Sigma X^2 - (\Sigma X)^2}{n(n-1)}$

13. The largest component of gross national product is personal consumption. Data on this variable are presented below:

Year	*Personal Consumption Spending (in \$ billions)*
1972	733
1973	810
1974	888
1975	973

Source: *Statistical Abstract of the United States, 1977,* p. 429.

a. Calculate the variance in personal consumption spending using Equation 2–13.
b. Compute s^2 using the alternative derived in question 12(b).

14. The 1976 data for the five largest-selling brands of liquor are:

Brand	*Sales (in hundreds of thousands of cases)*
Seagram's 7 Crown	6.5
Smirnoff	5.8
Bacardi	4.2
Seagram's VO	3.7
Canadian Club	3.6

Source: *Business Week,* March 21, 1977, p. 143.

Viewing the observations as a population, calculate the mean, range, mean absolute deviation, variance, and standard deviation.

15. Whenever the President submits his budget to Congress, certain critics fight for a "balanced" budget, that is, one in which spending is equal to tax revenues. Unlike the person on the street, however, the federal government can always finance its purchases by selling bonds. The information below indicates that budget deficits were common in the 1970s:

Year	*Excess of Outlays over Receipts (in $ billions)*
1970	2.8
1971	23.0
1972	23.4
1973	14.8
1974	4.7
1975	45.1

Source: *Economic Indicators,* December 1977, p. 32.

Calculate the variance for this sample of deficits.

16. The actions of the Federal Reserve System influence the supply of money. Economists are presently debating which money supply concept should be relevant for guiding policy. The "narrowly-defined" money supply, or M_1, consists of cash and checking account deposits, whereas M_2 contains M_1 plus time deposits at commercial banks. Thirty-six economists indicated their preference:

Money Supply Concept	*Number of Economists*
M_1	20
M_2	12
Other	4

What was the modal preference?

17. National defense expenditures are often a target of people who feel that government spending should be limited. Based on the data below, find out whether X = national defense expenditures is more erratic than Y = national defense expenditures as a percentage of total outlays.

Year	*Total Outlays (in $ billions)*	X *(in $ billions)*
1971	211.4	76.8
1972	232.0	77.4
1973	247.1	75.1
1974	269.6	78.6
1975	326.1	86.6
1976	365.7	90.0

Source: *Economic Indicators,* December 1977, p. 33.

18. Government analysts rely on a variety of measurements to assess the state of the economy. One such "indicator" is the "capacity utilization rate." The Department of Commerce collected the following rates from a sample of ten manufacturing firms: 76%, 89%, 85%, 94%, 67%, 84%, 91%, 56%, 77%, and 83%. Compute the variance and the standard deviation.

19. When the financial troubles of the steel industry intensified in 1977, much national attention was directed at the trade policies of the United States. Steel executives were upset by the ability of foreign producers to sell steel in this country at below cost. Union officials noted that steel imports were responsible for thousands of layoffs. Congress has attempted to restrict steel imports through passage of a so-called "trigger-price" program. From the information on iron and steel imports displayed below, calculate the mean and median values of imports.

Year	*Value of Imports (in $ billions)*
1970	2.0
1971	2.6
1972	2.7
1973	2.8
1974	4.8
1975	4.0
1976	3.8

Source: *Statistical Abstract of the United States, 1977,* p. 875.

20. The following home run totals were achieved by the National League East teams in 1977:

Club	*X* *Number of Home Runs*
Philadelphia	186
Pittsburgh	133
St. Louis	96
Chicago	111
New York	88
Montreal	138

a. Assuming that the data constitute a population, compute the mean and variance.
b. Code the data by subtracting 88 from each observation. Calculate the mean and variance of the transformed data.
c. If only part (b) were done, how could μ_x and σ_x^2 be found?

21. The personnel department at Company R administered a test to seven employees. The scores were:

X *Score*
310
620
950
470
860
730
260

The observations represent a population.

a. How can the company code these scores so that their mean is 500 and their standard deviation is 100?
b. List the coded scores.

22. In the spirit of "social responsibility," Corporation Q has decided to provide its workers with a financial incentive to stop smoking. Before the program was started, Q learned that its 1,000 employees averaged 46 cigarettes during a 40-hour week. The standard deviation was eight cigarettes.

a. At least how many employees smoked between 30 and 62 cigarettes per workweek?
b. At least how many employees smoked between 34 and 58 cigarettes per workweek?

23. A certain TV producer currently has five programs on the commercial networks. A ratings service has identified the size of the audience for these shows:

Program	X *National Audience*
A	24,500,000
B	15,000,000
C	19,500,000
D	30,000,000
E	27,000,000

a. To code the data, divide each observation by 1,000,000. Calculate the mean and variance of the new population.
b. Use the answer in part (a) to derive the mean and variance of the original data.

24. The average cost of a retailer's inventory is $38. The standard deviation is $6. If the retailer operates on the basis of a 30 percent markup over cost, what would be the average price and the variance after the inventory was made available for sale?

25. Companies can finance their activities by issuing stocks and bonds or by using retained earnings. The latter funds are the profits which remain once taxes and dividends have been paid. A researcher has calculated X = retained earnings/total funds for all 300 companies in industry D. The average was 73 percent, with σ = 5 percent.

a. How many companies get between 63 and 83 percent of their funds from retained earnings?
b. How many companies get between 60 and 86 percent of their funds from retained earnings?

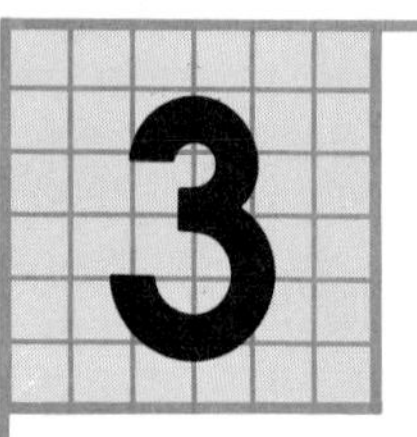

Descriptive Statistics: *Grouped Data*

CONSTRUCTING CLASSES

We have seen that summary measures help researchers analyze sets of data. We may also want to supply our audience with a list of the observations. Yet showing *every* item could be impractical. In this chapter, we'll investigate how to present data compactly by means of a *grouped frequency distribution.* Consider the following example.

The owner of a hardware store recently computerized his inventory records. He formerly had to close down every time his stock was counted. The latest inventory of lumber appears in Table 3–1.

TABLE 3–1

X *Lengths of 2″-by-4″ Lumber*	f *Number of Pieces*
24″	6,000
36″	1,000
48″	3,000
	10,000

Even though the table is small, we can, nonetheless, identify the specific length of every piece of wood. This is so because the variable, lengths of 2″-by-4″ lumber, has only three different values. Given 1,000 different lengths, however, a table showing the number of pieces of each would have 1,000 rows and would be unwieldy.

When a variable has many *distinct* values, the values are often combined into *groups,* or *classes,* before the observations are displayed in tabular form. Table 3–2 is a grouped frequency distribution of lumber sizes. It shows, for instance, that 2,000 pieces are *somewhere* between 4.0 and 12.9 inches. Table 3–2 results from a compromise. The data are reported efficiently, but we can't determine exact values.

TABLE 3–2

X *Lengths of 2"-by-4" Lumber (in inches)*	f *Number of Pieces*
4.0–12.9	2,000
13.0–21.9	3,000
22.0–30.9	1,000
31.0–39.9	3,000
40.0–48.9	1,000
	10,000

The process of grouping observations is largely a matter of (1) the researcher's discretion and (2) "trial and error." However, three guidelines exist:

1. Avoid either excessive or insufficient detail. (Five to fifteen classes are usually adequate.)
2. Every observation should belong to *one* and *only one* class.
3. No observation should fall in a "gap." (Refer to the classes 13.0–21.9 and 22.0–30.9 in Table 3–2. A "gap" occurs between 21.9 and 22.0.)

Let's look at another example.

Television has been criticized because of its impact on children. In 1977 a national commission blamed TV for part of the decline in high school reading and writing skills. Certain psychologists, moreover, believe that violent programs increase antisocial behavior.

The principal of J. T. High School is interested in the television-viewing habits of his 50 tenth graders. Table 3–3 contains the population of "TV viewing times." Note the gaps between classes (between 4 and 5 hours, 8 and 9 hours, etc.).

We shall assign half of the gap between two classes to each of them. Figure 3–1 shows the implications of this for Class #3. The "theoretical" end-points, or *boundaries*, serve as the basis for computing the *class interval,* g_i:

$$g_i = \begin{matrix}\text{upper boundary}\\ \text{of Class } i\end{matrix} - \begin{matrix}\text{lower boundary}\\ \text{of Class } i\end{matrix} \qquad (3\text{–}1)$$

Hence, $g_3 = 12.5 - 8.5 = 4$ hours.

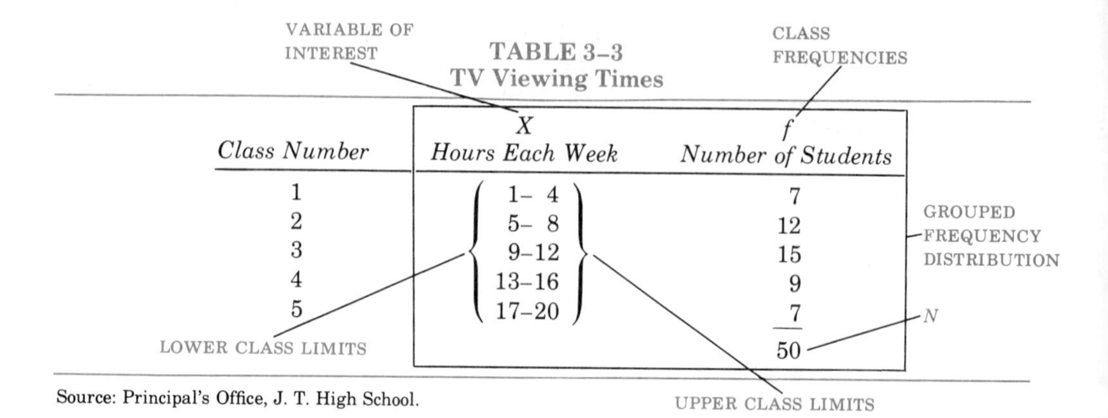

TABLE 3–3
TV Viewing Times

Class Number	X Hours Each Week	f Number of Students
1	1– 4	7
2	5– 8	12
3	9–12	15
4	13–16	9
5	17–20	7
		50

Source: Principal's Office, J. T. High School.

The *midpoint*, m_i, of a class is the average of the *class limits* (see Table 3–3):

$$m_i = \frac{\text{lower limit of Class } i + \text{upper limit of Class } i}{2} \tag{3–2}$$

Thus, $m_3 = (9 + 12)/2 = 10.5$ hours.

Knowledge of limits, boundaries, intervals, and midpoints is needed to (1) create a table such as 3–3 and (2) calculate the parameters of such a distribution.

Let's re-create Table 3–3. We first put the 50 observations in ascending order, as in Table 3–4. We want to cover the entire range of the data in $k = 5$ classes having equal intervals. In general, we solve

$$g^* = \text{estimated interval} = \frac{\text{range}}{k} \tag{3–3}$$

and, thereafter, raise g^* to a whole number. Thus, $g^* = (20 - 1)/5 = 3.8$, and $g = 4.0$.

FIGURE 3–1

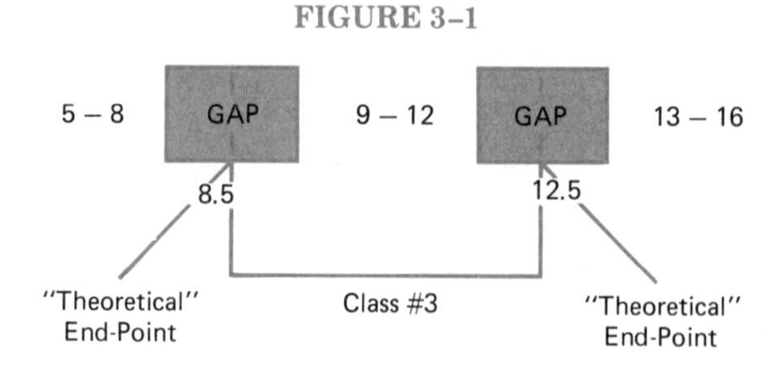

TABLE 3–4
TV Viewing Times (hours)

1	8	13
3	8	13
3	9	14
3	9	14
4	9	14
4	9	15
4	9	15
5	10	16
5	10	16
5	10	17
5	10	17
5	11	18
6	11	18
7	11	18
7	12	19
7	12	20
7	12	

Starting at the lowest value of X, we can write the lower limits (see Table 3–5). The upper limits must be selected so as to prevent

TABLE 3–5

	Lower Limits
	1–
Add $g = 4$	5–
Add $g = 4$	9–
Add $g = 4$	13–
Add $g = 4$	17–
	(21)

an observation from lying in a gap. If we had items like 4.3 or 16.2, the grouping scheme on the left in Table 3–6 (and used in Table 3–3) would be inadequate; we would need to use the scheme on the right.

TABLE 3–6

X	X
1– 4	1– 4.9
5– 8	5– 8.9
9–12	9–12.9
13–16	13–16.9
17–20	17–20.9

Having established the lower and upper limits for each class, we complete Table 3–3 by attaching the frequencies.

The next example highlights several additional concerns in constructing a grouped frequency distribution.

Among the many public assistance programs, Aid to Families with Dependent Children (AFDC) is most closely associated with the "welfare system." During his first year in office, President Jimmy Carter proposed major reforms in welfare. He hoped to eliminate the inequities, red tape, and work disincentives surrounding AFDC.

The number of people who receive AFDC funds has grown rapidly since 1960. Table 3–7 gives the state and District of Columbia totals for 1970.

TABLE 3–7
AFDC Recipients in 1970 (in thousands)

Wyoming	6	Arizona	61	Washington	150
Alaska	10	District of Columbia	61	Maryland	161
N. Dakota	12	Kansas	70	Alabama	164
New Hampshire	14	S. Carolina	71	Tennessee	169
Vermont	15	Iowa	75	Missouri	175
Nevada	16	Colorado	92	Louisiana	254
Montana	18	Connecticut	96	Massachusetts	256
Idaho	19	Wisconsin	97	Florida	259
S. Dakota	19	Minnesota	99	Georgia	260
Delaware	26	W. Virginia	100	Texas	331
Hawaii	32	Oregon	103	Ohio	339
Nebraska	37	Oklahoma	107	New Jersey	420
Utah	39	Indiana	114	Michigan	445
Rhode Island	46	Virginia	117	Illinois	484
Maine	52	Kentucky	139	Pennsylvania	565
New Mexico	58	Mississippi	139	New York	1,226
Arkansas	60	N. Carolina	147	California	1,542

Source: *Statistical Abstract of the United States, 1975,* p. 307.

If we want $k = 11$ classes, $g^* = (1{,}542 - 6)/11 = 139.6$. Suppose we choose an interval of $g = 150$ and begin at zero. This results in the lower limits shown in Table 3–8.

The distribution in Table 3–9 meets the three conditions that we established earlier. Nevertheless, this distribution can be improved. To account for two extremes (i.e., 1,226 and 1,542), we squeezed 34/51, or two-thirds, of the observations into only one class. There are three possible remedies:

TABLE 3–8

	Lower Limits
	0–
Add g : 150	150–
	300–
	450–
	600–
	750–
	900–
	1,050–
	1,200–
	1,350–
	1,500–
	(1,650)

TABLE 3–9

X *AFDC Recipients (in thousands)*	f *Number of States (including D.C.)*
0–149	34
150–299	9
300–449	4
450–599	2
600–749	0
750–899	0
900–1,049	0
1,050–1,199	0
1,200–1,349	1
1,350–1,499	0
1,500–1,649	1
	51

Source: *Statistical Abstract of the United States, 1975,* p. 307.

1. Eliminate the extremes and mention them in a footnote, as in Table 3–10. Despite the fact that Table 3–10 has one less class than Table 3–9, it provides more detail.
2. Use *unequal* class intervals. After 540–599 in Table 3–10, we could attach 600–1,559. Then the first ten classes would have intervals of 60, while the last interval would be 960.
3. Use an *open-ended* final class. Just one limit is stated for such classes (e.g., 600 and over).

TABLE 3–10

X *AFDC Recipients (in thousands)*	f *Number of States* (including D.C.)*
0–59	18
60–119	13
120–179	8
180–239	0
240–299	4
300–359	2
360–419	0
420–479	2
480–539	1
540–599	1
	49

*New York (1,226) and California (1,542) are excluded.

Source: *Statistical Abstract of the United States, 1975,* p. 307.

The latter two strategies often appear together. Table 3–11, a case in point, was prepared by the Census Bureau.

TABLE 3–11

	X *Money Income, 1971*	f *Number of Families (in thousands)*
	$ 0– 999	799.4
g = 1,000 ————	1,000– 1,999	1,385.7
	2,000– 2,999	2,238.4
	3,000– 3,999	2,558.2
	4,000– 4,999	2,878.0
g = 2,000 ————	5,000– 6,999	5,969.2
g = 3,000 ————	7,000– 9,999	9,859.8
g = 5,000 ————	10,000–14,999	14,336.6
g = 10,000 ————	15,000–24,999	10,392.7
Open-ended ————	25,000 or more	2,824.7
		$\approx$ 53,000.0

Source: *Statistical Abstract of the United States, 1973,* p. 329.

SUMMARY MEASURES FOR GROUPED DATA

When only a grouped distribution is available, and we want to calculate parameters or statistics, we have to make some assumptions about where the individual observations are located. Because of this, grouped summary measures are approximations.

Let's reconsider the TV-viewing distribution (see Table 3–3). We know that $\mu = \Sigma X/N = \Sigma X/50$. Imagine that the f_i items in Class i are either (1) spread evenly throughout i or (2) equal to the midpoint, m_i. In both situations, $m_i f_i$ will represent the sum of the f_i observations, and

$$\sum_{i=1}^{k} m_i f_i,$$

where k = the number of classes, will become the *assumed* total of all N observations. Hence,

$$\mu \approx \frac{\sum_{i=1}^{k} m_i f_i}{N} \qquad (3\text{–}4)$$

Table 3–12 shows the computation of mf values for the classes in Table 3–3. Thus we can compute $\mu \approx 513/50 = 10.3$ hours. (Note: The true μ, see Table 3–4, is $508/50 = 10.2$ hours.)

TABLE 3–12

Hours	*m*	*f*	*mf*
1–4	2.5	7	17.5
5–8	6.5	12	78.0
9–12	10.5	15	157.5
13–16	14.5	9	130.5
17–20	18.5	7	129.5
			513.0

The median observation is in the $(N + 1)/2 = 51/2 = 25.5$th position. Such an observation falls in Class #3. We'll assume that the items are "evenly spaced." Suppose we divide Class #3 into $f = 15$ subintervals of length $g/f = 4/15 = .267$ (see Figure 3–2). The

FIGURE 3–2

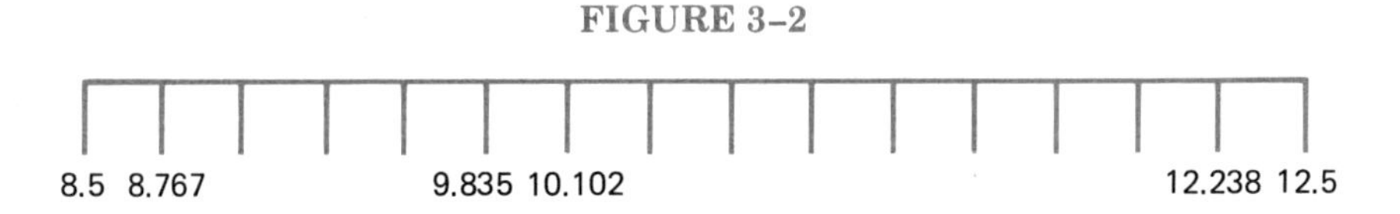

observations should lie halfway between these subintervals (see Figure 3–3). Inasmuch as 19 observations are below Class #3, the

25.5th item occupies rank 6.5 in that class. The median is, therefore, 10.102. (Note: The true median is 10.0.)

We derive the median much quicker if we apply the following:

$$\text{Median} \approx B + g\left[\frac{(N/2) - f_b}{f_a}\right] \tag{3–5}$$

where B = the lower boundary of the class containing the $(N/2)$th observation (i.e., the *median class*)
f_a = the frequency of the median class
f_b = the total number of observations below the median class

Using this formula, we get

$$\text{Median} \approx 8.5 + (4)\left[\frac{(50/2) - 19}{15}\right] = 10.1$$

Unlike the mean, the median can be derived even when open-ended classes are present. Although formulas exist for the mode, we generally refer to the *modal class* (e.g., Class #3) instead of the modal item.

If each observation were located at m_i, $(m_i - \mu)^2 f_i$ would equal the sum of the squared deviations for Class i. We could, then, compute the variance as follows:

$$\sigma^2 \approx \frac{\sum_{i=1}^{k} (m_i - \mu)^2 f_i}{N} \tag{3–6}$$

Based on the computations in Table 3–13, $\sigma^2 \approx 1{,}229.20/50 = 24.58$. (Note: The true variance is $1{,}198.8/50 = 23.98$. Moreover, $\sigma^2 = 24.58$ implies that the grouped standard deviation is $\sigma \approx 4.96$.)

FIGURE 3–3

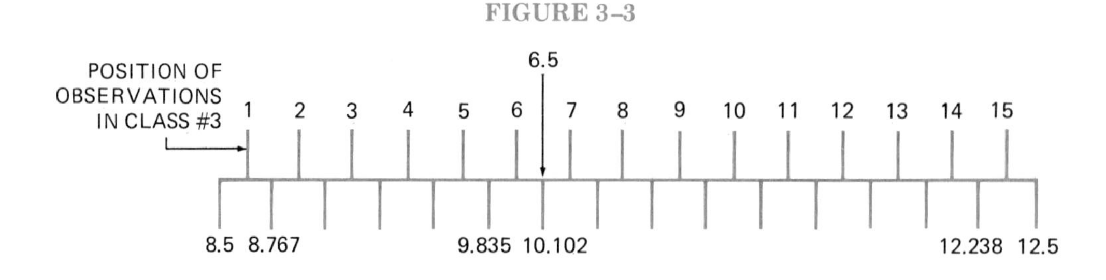

TABLE 3–13

Hours	m	f	μ	$m - \mu$	$(m - \mu)^2$	$(m - \mu)^2 f$
1–4	2.5	7	10.3	−7.8	60.84	425.88
5–8	6.5	12	10.3	−3.8	14.44	173.28
9–12	10.5	15	10.3	.2	.04	.60
13–16	14.5	9	10.3	4.2	17.64	158.76
17–20	18.5	7	10.3	8.2	67.24	470.68
						1,229.20

GRAPHICAL FORMS

Because descriptive procedures communicate quantitative information, *graphs* are logically included under this heading. We will restrict our attention to histograms and frequency polygons.

A *histogram* consists of a series of adjacent rectangles. The base of the ith rectangle runs from the lower to the upper boundary of Class i (a distance of g units), while the height is f_i. Figure 3–4 is a histogram of the grouped frequency distribution in Table 3–3.

If we converted Table 3–3 to a grouped relative frequency distribution, as in Table 3–14, the vertical scale of the histogram would

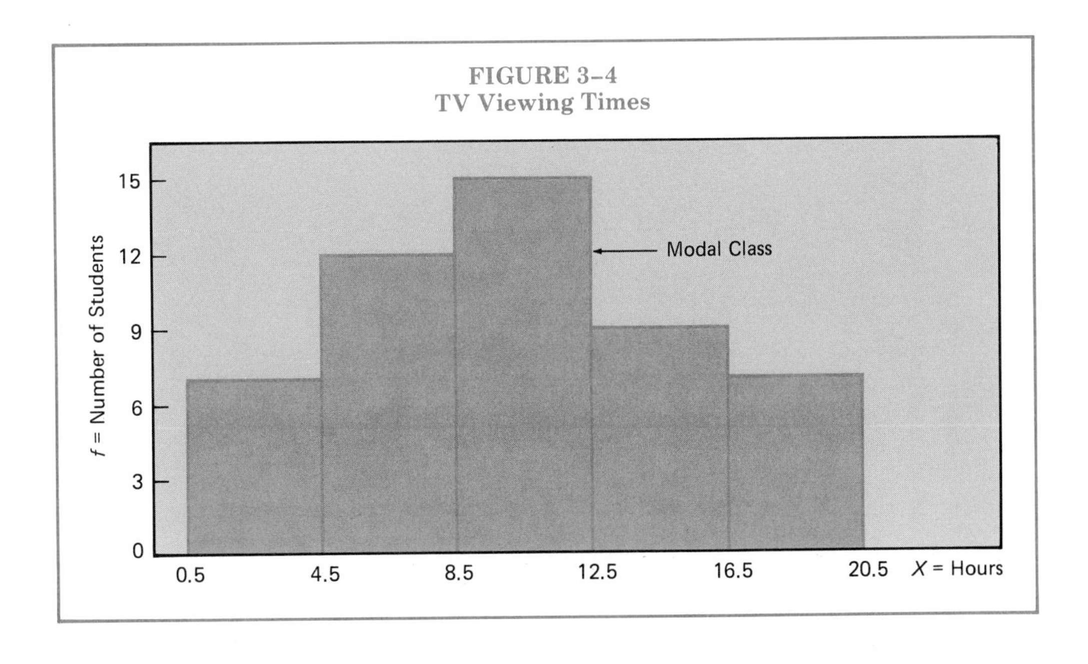

FIGURE 3–4
TV Viewing Times

FIGURE 3–5
TV Viewing Times

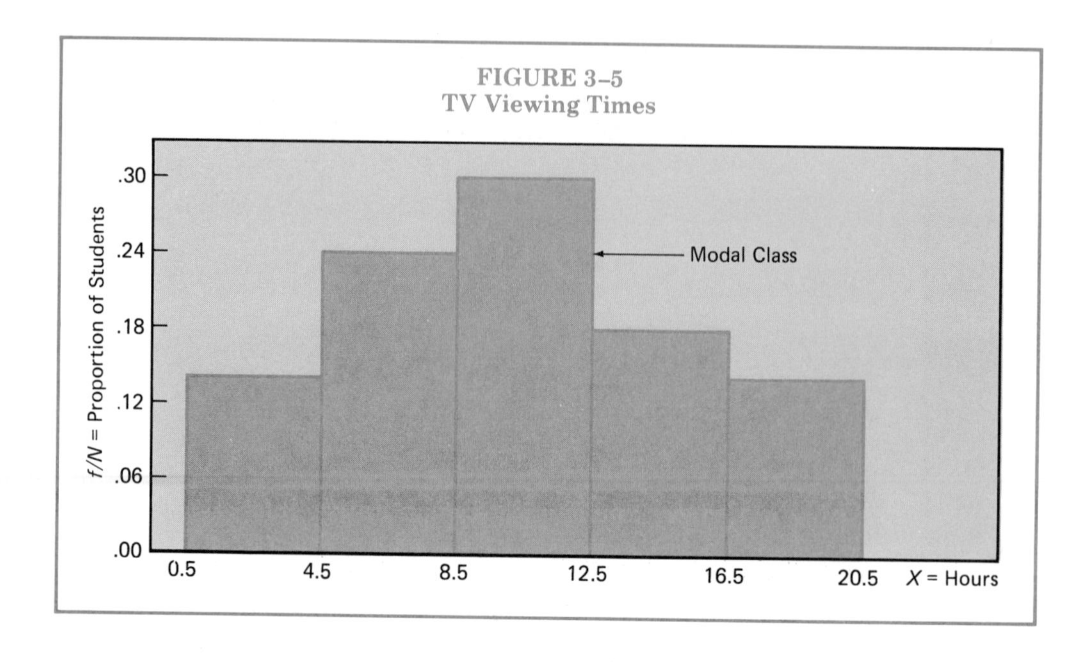

change (see Figure 3–5). Notice, however, that the *shape* of the histogram is unaffected.

TABLE 3–14

X *Hours*	f/N *Relative Frequency*
1–4	.14
5–8	.24
9–12	.30
13–16	.18
17–20	.14

To construct a *frequency polygon,* researchers plot classes as points having the coordinates $m_i f_i$. These points are then connected, and the polygon is "tied" to the X axis (see Figure 3–6). The "tying" is brought about by introducing a fictitious class at both ends of the distribution.

We can determine by looking at a histogram or frequency polygon whether a distribution is *symmetrical* and *unimodal* (i.e., single-peaked). A distribution is symmetrical if we are able to fold its graph at the median and find that one part coincides with the other, as in Figure 3–7. Under these circumstances, mean = median.

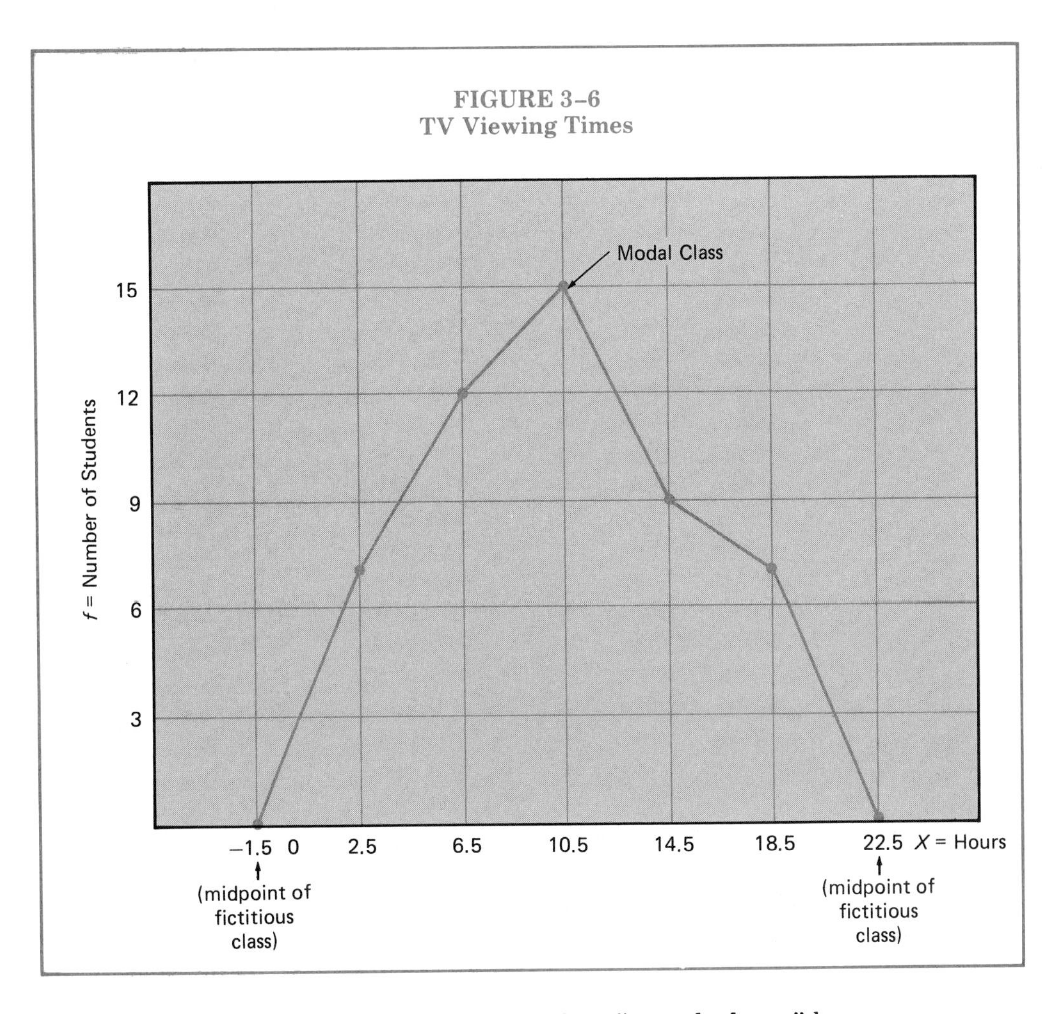
FIGURE 3–6
TV Viewing Times

Some unimodal distributions are *skewed,* or "stretched out," by extremes. Since extreme values influence the mean more than the median, mean > median when the extremes lie to the right of the median, and mean < median when the extremes lie to the left.

FIGURE 3–7

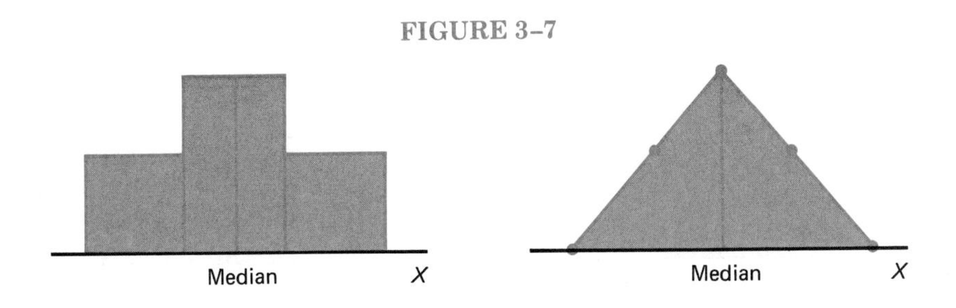

The graphs in Figure 3–8 indicate *positive skewness*; that is, the distributions are "pulled" to the right by extreme values. Consider the following example.

FIGURE 3–8

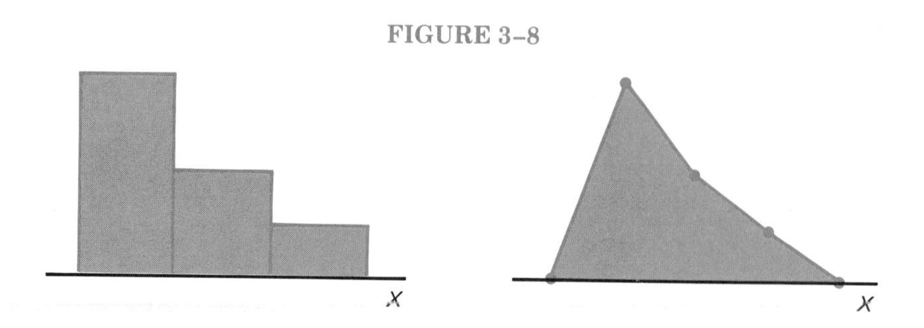

Over half of all manufacturing profits in the United States are made by several hundred large corporations. Yet there are hundreds of thousands of manufacturers. Most companies are small (e.g., assets < \$250,000). A grouped distribution for which X = assets of manufacturing firms and f = number of firms would be concentrated at the low (left) end of the horizontal axis. The relatively few very large corporations would skew the distribution to the right.

Distributions that are "pulled" to the left possess *negative skewness,* illustrated in Figure 3–9. The following is an example.

The before-tax yields of municipal bonds are not high. However, the wealthy are attracted to municipals because the interest on such bonds is exempt from federal taxes. People in the 50 percent tax bracket will benefit more than people in the 25 percent bracket. A grouped distribution for which X = income level of municipal bondholders and f = number of bondholders would be concentrated at the upper end of the horizontal axis.

FIGURE 3–9

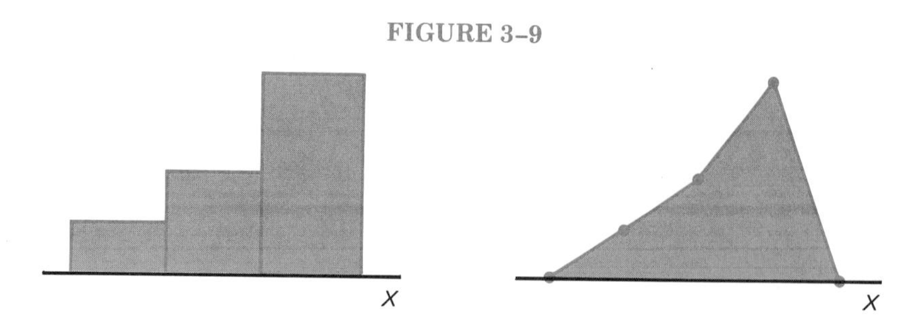

Skewness is commonly measured by the following formula:

$$\text{Pearson's coefficient of skewness} = \frac{3(\mu - \text{median})}{\sigma} \tag{3–7}$$

This coefficient is zero whenever a distribution is symmetrical (because $\mu - \text{median} = 0$). The product $(3/\sigma)(\mu - \text{median})$ gives the researcher a frame of reference since the coefficient will be less than -1 or greater than $+1$ only if a distribution is *very* skewed. Inasmuch as

$$\frac{3(10.3 - 10.1)}{4.96} = .12,$$

the distribution of X = TV-viewing times is *slightly* skewed to the right.

SUMMARY

We have expanded our discussion of descriptive statistics by examining situations in which grouping is necessary. For many published data sets, only grouped frequency distributions are presented. The researcher can, nonetheless, compute parameters and statistics for such observations. Graphing is another technique for displaying quantitative information. In later chapters, we will mention symmetry and skewness quite often.

Throughout Chapters Four, Five, and Six, we shall explore probability, the second of the major topics that lead into statistical inference.

EXERCISES

1. For each of the following collections of data, derive a grouped frequency distribution consisting of five classes:

1–1. Taxpayers across the country have begun to challenge the constitutionality of their state's method of funding public education.

The total amount of money spent by a school district is composed of local tax dollars plus state and federal aid. Despite this aid, the property tax is the major source of revenue. Many people object to the fact that the value of property in a district appears to determine the quality of the educational system.

a. Expenditures per pupil in the school districts of State W are as follows:

Expenditures per Pupil (in $)		
892	1,208	1,785
945	841	1,802
863	1,866	937
996	2,237	2,486
1,238	784	1,977
1,857	1,009	1,468
2,006	1,562	854
921	820	
1,449	916	

Source: Department of Education, State W.

b. Property value per student is also reported on a district basis:

Property Value per Pupil (in $)		
50,000	110,000	211,000
72,000	87,000	163,000
40,500	84,000	41,000
83,000	250,000	230,000
100,000	103,000	103,000
110,000	98,000	115,000
98,000	123,500	65,000
64,000	28,000	
123,500	43,000	

Source: Department of Education, State W.

1–2. Certain economists in the early 1960s urged the federal government to give a share of its income tax receipts back to the cities and states. This "revenue-sharing" arrangement did not become a reality until Congress passed the State and Local Fiscal Assistance Act in 1972. Under this legislation, the federal government uses a formula to determine how much each state will receive. The 1972 allocations were as follows:

1972 Allocations (in $ millions)					
Alabama	89.6	Kentucky	86.0	N. Dakota	21.4
Alaska	5.9	Louisiana	121.2	Ohio	211.3
Arizona	47.9	Maine	30.6	Oklahoma	58.2
Arkansas	53.9	Maryland	106.0	Oregon	52.4
California	554.3	Massachusetts	163.4	Pennsylvania	274.4
Colorado	53.9	Michigan	221.8	Rhode Island	23.9
Connecticut	66.5	Minnesota	104.8	S. Carolina	71.3
Delaware	15.9	Mississippi	87.5	S. Dakota	23.3
D.C.	23.7	Missouri	96.9	Tennessee	97.8
Florida	145.2	Montana	19.7	Texas	245.4
Georgia	108.5	Nebraska	38.3	Utah	30.1
Hawaii	23.5	Nevada	11.3	Vermont	14.4
Idaho	21.0	New Hampshire	16.3	Virginia	105.3
Illinois	270.4	New Jersey	165.0	Washington	76.9
Indiana	51.7	New Mexico	31.0	W. Virginia	51.4
Iowa	74.7	New York	582.6	Wisconsin	131.5
Kansas	51.6	N. Carolina	134.5	Wyoming	9.8

Source: U.S. Treasury Department, as quoted in "Congressional Quarterly Weekly Report," December 16, 1972, p. 3,136.

1–3. The OPEC countries have just called a meeting to discuss setting a new price for a barrel of oil. A sample of energy experts in the United States indicated the extent to which oil prices might be raised:

Expected Price Increase (in percent)			
5	12	23	6
20	35	21	10
32	42	10	19
18	8	4	5
14	15	7	3
30	36	27	18

Source: Energy News Service, September 1979.

1–4. One of the empirical tools available to members of the antitrust division of the Justice Department is the "concentration ratio." This ratio divides the value-of-shipments of the four largest firms in an industry by the shipments of all firms in the industry. The automobile concentration ratio is over 90 percent. Where the concentration ratio is high, companies may be tempted to fix prices or engage in other illegal market activity. The data following are concentration ratios for a sample of industries:

Industry Concentration Ratios (in percent)				
82	35	3	67	84
28	93	16	58	73
36	54	19	33	89
61	72	47	29	30
78	82	63	49	71
66	59	60	46	37
25	22	63	75	72

Source: U.S. Industry Reports, 1979.

1–5. In an effort to prevent a recession, the President wants to reduce both individual and corporate taxes. Although a small tax cut would be ineffective, a large one would have an unfortunate effect on inflation. A sample of members of Congress suggested the following aggregate tax cuts:

Aggregate Tax Cuts (in $ billions)			
20	14	29	16
9	7	34	50
35	41	36	42
27	36	18	34
5	21	30	18
15	17	28	38
21	13	31	27

Source: S. L. Survey Company, May 1978 Poll.

2. For each of the following distributions,

   a. Compute the mean, median, variance, and standard deviation.
   b. Construct a histogram and a frequency polygon.
   c. Compute the coefficient of skewness.

2–1. A significant element in the failure of the United States to recover swiftly from the 1974–1975 recession was inadequate business investment. Among other reasons, the corporate sector was hesitant to spend money because of its uncertainty about the cost and supply of energy. The 1980 investment plans of a sample of companies are:

Plant and Equipment Expenditures (in $ millions)	*Number of Companies*
1.0–2.4	10
2.5–3.9	7
4.0–5.4	6
5.5–6.9	3
7.0–8.4	1
	27

2–2. The Council of Economic Advisers during the Kennedy administration felt that "full employment" would be reached when the unemployment rate was at 4 percent. Inasmuch as Americans experienced high unemployment and inflation simultaneously during the 1970s, policymakers came to believe that inflation would skyrocket if the economy ever operated at 4 percent unemployment. The "full employment" concept is, therefore, being revised. Fifty-five economists identified the unemployment rate which they now consider "full employment." (Assume a population.)

Unemployment Rate (in percent)	*Number of Economists*
4.0–4.4	8
4.5–4.9	10
5.0–5.4	18
5.5–5.9	9
6.0–6.4	6
6.5–6.9	3
7.0–7.4	1
	55

2–3. Due to the huge hikes in gasoline prices within recent years, analysts looked for a decline in gasoline consumption. Economists compute "price elasticity of demand" to show the relation between the percentage change in the demand for a commodity and the percentage change in its price. Elasticities are generally expressed as positive values. The price elasticity of gasoline seems to be low. The following estimates of this elasticity have been selected from a sample of published studies:

Price Elasticity of Gasoline	*Number of Studies*
.03–.12	5
.13–.22	7
.23–.32	8
.33–.42	10
.43–.52	9
.53–.62	6
.63–.72	4
.73–.82	3
	52

2–4. "Minimum wage" legislation was created during the 1930s. Many economists have argued that these wages restrict the employment opportunities of the unskilled. Their rationale is that employers will not want to pay anyone more than his or her productivity is worth. Eighteen business executives revealed the number of additional people they would hire if minimum wages were abolished. (Assume a sample.)

Additional Employees	*Number of Executives*
0– 2	7
3– 5	5
6– 8	3
9–11	2
12–14	1
	18

3. For each of the following distributions, compute the mean and the standard deviation:

3–1. Two of the most well-known forecasting models are the Wharton and Brookings models. To evaluate the accuracy of a sample of forecasting procedures, a researcher measured the error in the 1979 predicted value of GNP, that is, |actual GNP − predicted GNP|:

Error in the GNP Forecast (in $ billions)	*Number of Models*
0– 2	4
3– 5	3
6– 8	5
9–11	1
12–14	3
	16

3–2. In many of the large cities throughout the United States, public transit lines incur deficits. Research shows that Americans are concerned with comfort, convenience, and speed when they travel. The cost of traveling is less important than these other factors. The pricing actions of 43 transit authorities were investigated:

Change in Fare Between 1977 and 1979 (in cents)	*Population Number of Transit Authorities*
−15– −6	3
− 5– 4	5
5–14	9
15–24	12
25–34	14
	43

3–3. Some economists, such as Milton Friedman of the University of Chicago, believe that the impact of monetary policy occurs with a long and unpredictable lag. Therefore, Friedman has encouraged the Federal Reserve Board to abandon its attempts at "fine tuning" the economy. He favors increasing the money supply by a fixed percentage every year. The following data are the percentage increases advocated by a sample of "fixed-rule" supporters:

Annual Increase in the Money Supply (in percent)	*Number of Economists*
2.0–2.4	5
2.5–2.9	3
3.0–3.4	2
3.5–3.9	4
4.0–4.4	6
4.5–4.9	7
	27

3–4. The manufacturer of an electronic component is interested in determining the temperature at which it stops functioning. The results of a test on a sample of components are as follows:

Temperature at Which Component Failed (in °F below zero)	*Number of Components*
50–40.1	6
40–30.1	5
30–20.1	3
20–10.1	2
	16

4. For each of the following distributions, compute the median and identify the modal class:

4–1. Firms examine a variety of things before they decide to invest. The "payback period" is the time between when an investment was made and when that project returned its cost. A researcher calculated the payback period for a sample of Company W's investments:

Payback Period (in years)	*Number of Investments*
0–1.9	5
2.0–3.9	6
4.0–5.9	3
6.0 and over	2
	16

4–2. When City A buys from a private company, it must first request sealed bids. The city is currently funding work at 15 construction sites. Data on the number of bidders for this construction are displayed below:

Number of Bidders	*Number of Construction Sites*
4 and under	2
5– 7	3
8–10	4
11–13	3
14–16	3
	15

Probability: *Basic Concepts*

INTERPRETING PROBABILITY

The business world is filled with uncertainty. For example, Company A hopes to increase its market share by either (1) reducing prices or (2) advertising more heavily. Before choosing a strategy, A must guess how B, a major rival, will react. A's management believes that B would quickly match a price cut. Thus A decides to engage in a "media blitz" instead. The following example illustrates the nature of chance.

The treasurer of Corporation Q drops key D and key E into his trouser pocket. D opens a filing cabinet, whereas E opens the executive washroom. However, the keys are the same size and shape. If the treasurer reaches into his pocket for one of the keys, whether he draws out D or E is a matter of chance.

A process that produces results which are not predetermined or certain is referred to as a *random experiment*. The distinct possibilities make up the potential outcomes, or *sample space* (see Table 4–1). These possibilities are occasionally displayed as points lying

TABLE 4–1

Random Experiment	*Sample Space*	*Result*
Select a key	key D, key E	1 key is selected

inside a rectangle, as in Figure 4–1, which is called a *Venn diagram*.

FIGURE 4–1

·Key D	·Key E

If we were to toss a well-balanced coin, we would not feel that the chances of "heads" were greater or less than the chances of "tails." The example of the two keys is another case where the outcomes are equally likely. We can state the following: When h_1, h_2, . . . , h_M are the M *equally likely* and *countable* outcomes in a sample space, the *probability* of h_i is $1/M$; that is,

$$P(h_i) = 1/M \tag{4-1}$$

The probability of selecting key D, P(key D), is, then, $1/M = 1/2$.

Although an experiment might be done only once, probability is a *long-run* phenomenon. To clarify this, suppose that the process of selecting a key is repeated many times. After each repetition, or *trial,* we can (1) record the outcome (D or E), (2) compute

$$\frac{f}{n} = \frac{\text{number of trials in which D has appeared}}{\text{number of trials}},$$

and (3) drop the key back in the pocket. Table 4–2 shows the results of eight such trials, and the data are graphed in Figure 4–2.

TABLE 4–2

Trial	*Outcome*	*Relative Frequency of Key D in All Completed Trials*
#1	D	1/1 = 1.00
#2	E	1/2 = .50
#3	E	1/3 = .33
#4	D	2/4 = .50
#5	D	3/5 = .60
#6	E	3/6 = .50
#7	E	3/7 = .42
#8	E	3/8 = .38

Between $n = 1$ and $n = 8$, the relative frequency of D jumps up and down. However, as n increases, f/n will get closer and closer to a specific value. That value is the probability of D:

$$P(h_i) = \lim_{n \to \infty} (f/n) \tag{4-2}$$

where the right side of the equation is "the limit of f/n as n approaches infinity." Over "a great many repetitions," the ratio

$$\frac{\text{number of Ds}}{\text{number of trials}}$$

should converge on, or move steadily toward, .50 (see Figure 4–3).

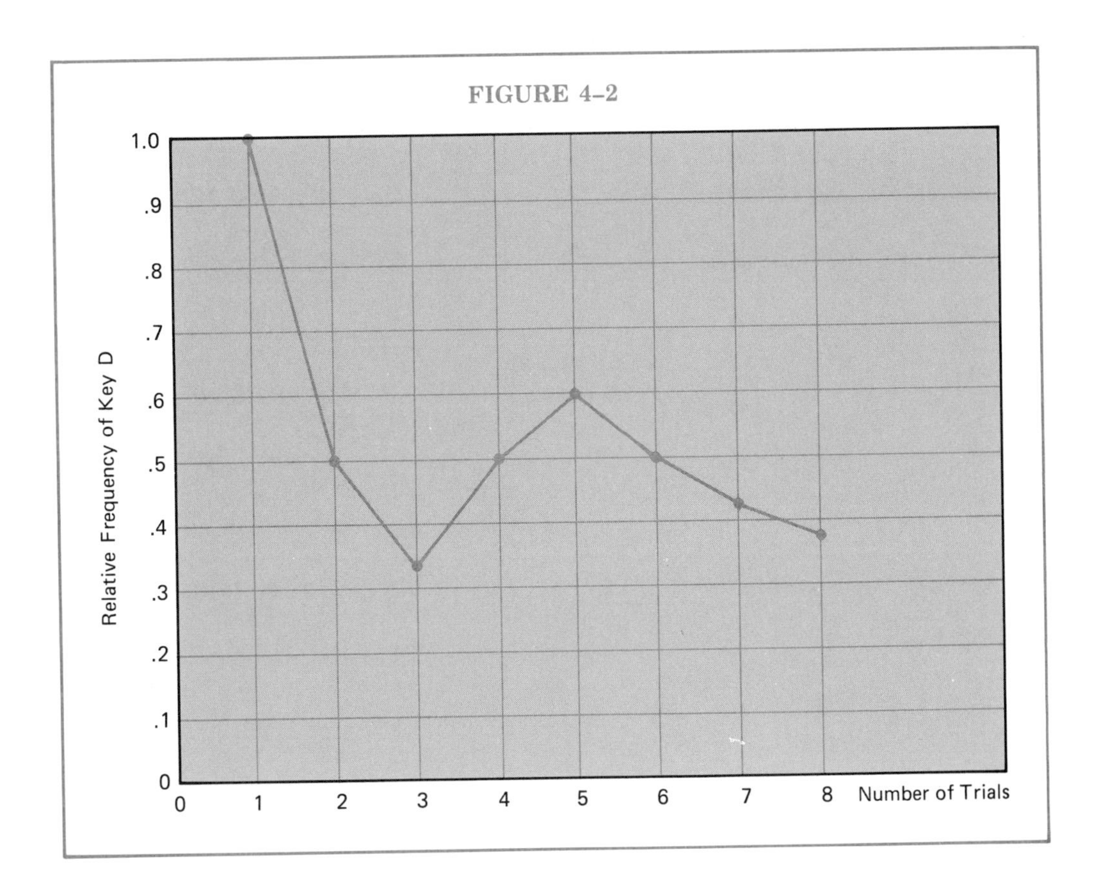

Note that Equation 4–1 is the basis for calculating probabilities when the outcomes are equally likely, but that Equation 4–2 is the *definition* of probability. Using the former, we are able to find $P(h_i)$ *a priori,* that is, without actually repeating an experiment "a great many times."

Since probabilities are long-run proportions, we can conclude (1) the probability of an outcome is never less than 0 nor greater than 1:

$$0 \leq P(h_i) \leq 1 \tag{4–3}$$

and (2) the sum of the probabilities of the M outcomes is 1:

$$P(h_1) + P(h_2) + \ldots + P(h_M) = 1 \tag{4–4}$$

Consider the following example.

Due to parking and fuel costs, Mr. Green takes the bus rather than driving to work. Labor and management are scheduled to negotiate a contract at 9:00 A.M. on Tuesday. Since Green is an im-

FIGURE 4–3

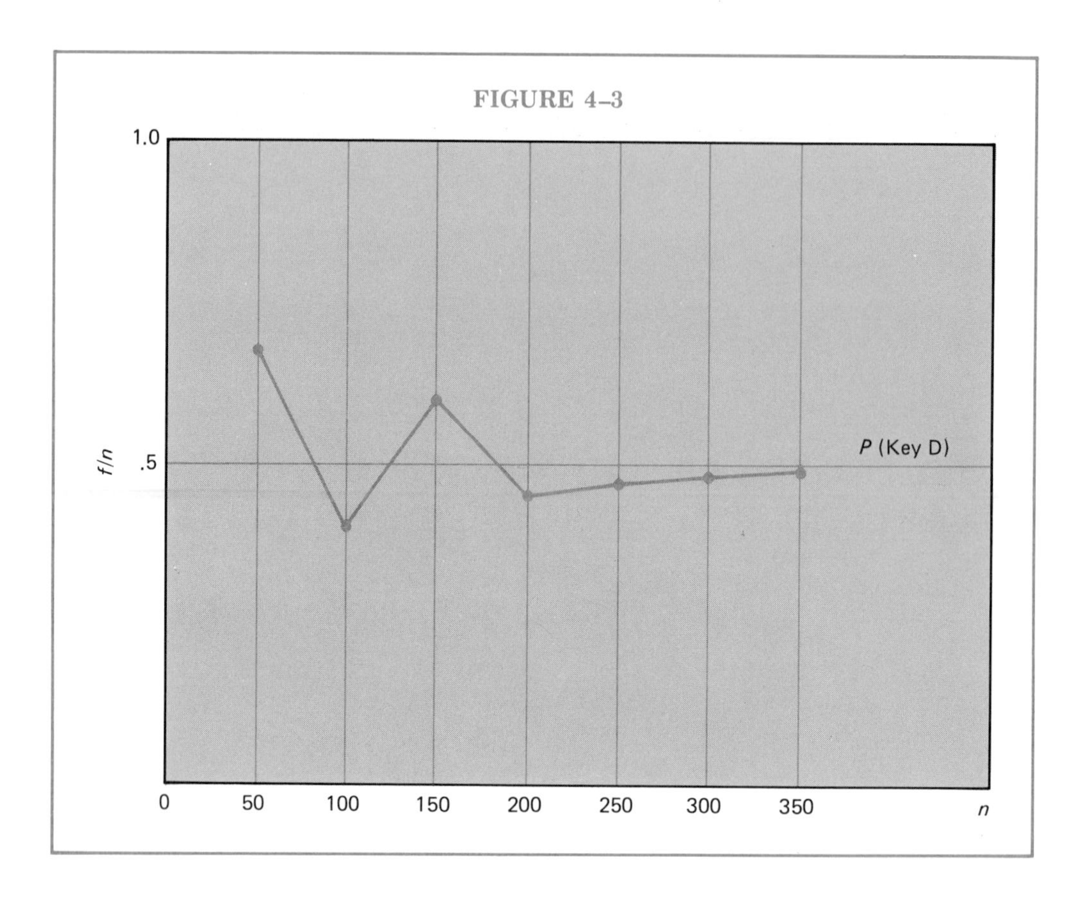

portant member of management's bargaining team, no agreements can be reached before he arrives. If his bus is late, the session will be delayed, and union leaders will become angered. He therefore wonders whether he should drive to work on that day. The question is: What are the chances his bus will be late on Tuesday?

Figure 4–4 diagrams the sample space. Like Figure 4–1 it contains two points. Yet we can't assume that "late" and "isn't late" are *equiprobable*.

FIGURE 4–4

·Bus is late	·Bus isn't late

We shall estimate P(late) empirically by viewing Green's past experiences as a set of trials. The bus was late 40 times in the past 200 days. Hence, estimated $P(\text{late}) = f/n = 40/200 = .20$. With a

20 percent chance that his bus will be late, Green decides to drive to work on Tuesday.

When an experiment is nonrepetitive and the outcomes are not equally likely, probabilities are derived *subjectively*. The next example is an illustration.

Some companies have lost trade secrets because their board rooms were "bugged" by competitors. The president of Studio R is currently reading a script involving industrial espionage. He intends to make a movie about it. Will the movie be a success or a failure? Figure 4–5 diagrams the sample space.

FIGURE 4–5

·Success	·Failure

The president has to treat P(success) as an *index*, or measure, of the confidence he has in "success." An outcome that an experimenter believes is very likely will be assigned a probability near 1. When probabilities are personal judgments, different people will attach different probabilities to the same outcome. The fact that Studio R will make the movie suggests that the president feels P(success) is "high," perhaps .80 or more.

Table 4–3 distinguishes the three "approaches" to probability.

TABLE 4–3

Approach to Probability	*Method of Computing Probability*	*Interpretation of* $P(h_i)$
A priori (also referred to as "classical")	$P(h_i) = 1/M$ (where each h_i is equally likely)	Long-run relative frequency of h_i
Empirical	Estimated $P(h_i) = f/n$	Long-run relative frequency of h_i
Subjective	$P(h_i)$ arises from a personal assessment	Degree of belief in h_i

Note: Regardless of the approach, Equations 4–3 and 4–4 apply.

THE PROBABILITY OF AN EVENT

Every "point" in the sample space is unique. Nonetheless, an outcome can possess *some* characteristics in common with other outcomes. Consider the following example.

The decline in birth rates over the past twenty years has forced college administrators to recognize the role of careful planning. Schools that construct new facilities in the face of shrinking enrollments or that cease to be innovative in their curricula will be burdened with financial problems. As a step in the direction of planning, the Dean of University L wants to put together a Committee on the Future. The committee will include one member from each academic department.

The faculty in business administration is listed in Figure 4–6. All 20 names are written on cards and thrown in a box. After shaking the box, the dean will draw one card. By virtue of this procedure, the possible outcomes are equiprobable; for example, $P(\text{SQ}) = 1/M = 1/20$. We say, then, that h_i is chosen *at random.*

FIGURE 4–6
Sample Space for "Select a Faculty Member"

· SQ	· DM	· CP	· RL
· AV	· BT	· RE	· NR
· CG	· PR	· ET	· SN
· LB	· VE	· TI	· GU
· KT	· CW	· DH	· JM

Suppose we identify two qualities connected with a faculty member: highest academic degree (B.A., M.A., or Ph.D.) and sex (female or male). Figure 4–7 is the detailed sample space.

FIGURE 4–7

· SQ (B.A., female)	· DM (B.A., male)	· CP (B.A., male)	· RL (B.A., male)
· AV (Ph.D., female)	· BT (Ph.D., male)	· RE (Ph.D., male)	· NR (Ph.D., male)
· CG (Ph.D., female)	· PR (Ph.D., male)	· ET (Ph.D., male)	· SN (Ph.D., male)
· LB (Ph.D., female)	· VE (Ph.D., male)	· TI (M.A., male)	· GU (M.A., male)
· KT (M.A., female)	· CW (M.A., male)	· DH (M.A., male)	· JM (M.A., male)

We will define event A as all outcomes having characteristic A. If A = female, then event A is represented by SQ, AV, CG, LB, and KT. Researchers may be interested in the probability of an event, rather than in the probability of a particular point, for instance, $P(\text{female})$ instead of $P(\text{SQ})$.

If $h_1, h_2, \ldots, h_j$ are the outcomes in A,

$$P(A) = P(h_1) + P(h_2) + \ldots + P(h_j) \qquad (4\text{–}5)$$

Whenever the sample space has M equiprobable and countable outcomes, the following also generates $P(A)$:

$$P(A) = \frac{\text{number of outcomes in } A}{M} \qquad (4\text{–}6)$$

In the "long run," we would choose SQ, AV, CG, LB, and KT each 5 percent of the time. Because of this, a female event would occur 25 percent of the time. We see that $P(\text{female}) = .05 + .05 + .05 + .05 + .05 = .25$ or $P(\text{female}) = 5/20 = .25$. The B.A. and M.A. events are indicated in Figure 4–8.

FIGURE 4–8

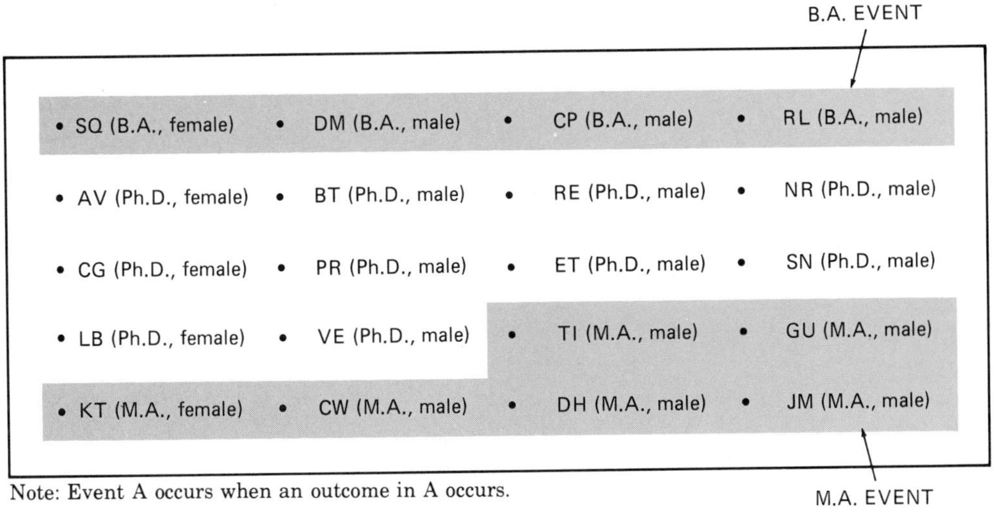

B.A. EVENT

• SQ (B.A., female)	• DM (B.A., male)	• CP (B.A., male)	• RL (B.A., male)
• AV (Ph.D., female)	• BT (Ph.D., male)	• RE (Ph.D., male)	• NR (Ph.D., male)
• CG (Ph.D., female)	• PR (Ph.D., male)	• ET (Ph.D., male)	• SN (Ph.D., male)
• LB (Ph.D., female)	• VE (Ph.D., male)	• TI (M.A., male)	• GU (M.A., male)
• KT (M.A., female)	• CW (M.A., male)	• DH (M.A., male)	• JM (M.A., male)

Note: Event A occurs when an outcome in A occurs.

M.A. EVENT

Let's look at another example. The courts have granted an injunction to stop an illegal strike at Company W. The leader of the strike, Bill Jones, is convinced that W will do one of the following things:

1. Fire him.
2. Suspend him for two weeks.
3. Suspend him for three weeks.
4. Ignore his role in the strike.

Jones gives these outcomes probabilities of .40, .20, .35, and .05, respectively.

For the event "suspension," we can compute

$$\begin{aligned} P(\text{suspension}) &= P(\text{2-week suspension}) + P(\text{3-week suspension}) \\ &= .20 + .35 = .55 \end{aligned}$$

THE MULTIPLICATION LAW

In our business faculty example, every faculty member belongs to one of six different degree and sex categories. Table 4–4 indicates the number of outcomes in each category. Note that two events can happen simultaneously (e.g., an h_i can be both a B.A. and a male). This is referred to as a *joint occurrence*. Thus DM is a joint occurrence of B.A. and male, while AV is a joint occurrence of Ph.D. and female (see Figure 4–8).

TABLE 4–4

		Degree			
		B.A.	M.A.	Ph.D.	*Row Total*
Sex	Female	1	1	3	5
	Male	3	5	7	15
	Column Total	4	6	10	20

If A and B are two events, where $h_1, h_2, \ldots, h_j$ are in both A and B simultaneously,

$$P(A \text{ and } B) = P(h_1) + P(h_2) + \ldots + P(h_j) \qquad (4\text{–}7)$$

Moreover, if the h_i are equiprobable and countable,

$$P(A \text{ and } B) = \frac{\text{number of outcomes in both } A \text{ and } B \text{ simultaneously}}{M} \qquad (4\text{–}8)$$

Let's now divide the entries in Table 4–4 by $M = 20$ (see Table 4–5). The ratios inside the lines are *joint probabilities*. We see that 3/20 is the probability of choosing a professor who is a Ph.D. and a female (i.e., a female Ph.D.). The probabilities of single events (M.A., male, etc.) are reproduced in the "margins" (i.e., the column

TABLE 4–5

		Degree			
		B.A.	M.A.	Ph.D.	*Row Total*
Sex	Female	1/20	1/20	3/20	5/20
	Male	3/20	5/20	7/20	15/20
	Column Total	4/20	6/20	10/20	20/20

and row totals). Hence, we usually label them as *marginal probabilities*; for example, $P(\text{M.A.}) = 6/20$ and $P(\text{male}) = 15/20$.

Suppose that a Ph.D. is picked. To find the probability that the Ph.D. is a female, we can act as though the sample space were restricted to the Ph.D. outcomes. The probability of a female *given* that the Ph.D. event has occurred, or $P(\text{female}|\text{Ph.D.})$, is called a *conditional probability*.

If the outcomes in a sample space are equiprobable and countable, the conditional probability of event A given event B is

$$P(A|B) = \frac{\begin{array}{c}\text{number of outcomes}\\ \text{in both } A \text{ and } B \text{ simultaneously}\end{array}}{\begin{array}{c}\text{number of}\\ \text{outcomes in } B\end{array}} \tag{4–9}$$

$P(\text{female}|\text{Ph.D.})$ is, thus, equal to 3/10.

The following example isolates marginal, joint, and conditional probabilities.

Company G would like to build a nuclear reactor in a region with 500 adults. When polled in 1978, 300 of these adults felt that nuclear plants were safe. One year later, following the Three Mile Island crisis, a new survey was conducted. The results are reported in Table 4–6.

TABLE 4–6

		1979 Response		
		Safe	Unsafe	*Row Total*
1978 Response	Safe	180	120	300
	Unsafe	0	200	200
	Column Total	180	320	500

Assuming that an adult is selected at random, we can calculate the following:

1. The probability that the adult viewed nuclear plants as unsafe in 1979:

$$\text{Marginal} \rightarrow P(\text{1979 unsafe}) = 320/500 = .64$$

2. The probability that the adult viewed nuclear plants as safe in 1978 and unsafe in 1979:

$$\text{Joint} \rightarrow P(\text{1978 safe and 1979 unsafe}) = 120/500 = .24$$

3. If the adult selected believed nuclear plants were safe in 1978, the probability that he or she viewed them as unsafe in 1979:

$$\text{Conditional} \rightarrow P(\text{1979 unsafe}|\text{1978 safe}) = 120/300 = .40$$

By manipulating Equation 4–8, we are able to express the joint probability of A and B as the *product* of two other probabilities:

$$P(A \text{ and } B) = \frac{\begin{array}{c}\text{number of}\\ \text{outcomes in both}\\ A \text{ and } B \text{ simultaneously}\end{array}}{M}$$

$$= \left(\frac{\begin{array}{c}\text{number of}\\ \text{outcomes in } B\end{array}}{M}\right)\left(\frac{\begin{array}{c}\text{number of}\\ \text{outcomes in both}\\ A \text{ and } B \text{ simultaneously}\end{array}}{\begin{array}{c}\text{number of}\\ \text{outcomes in } B\end{array}}\right)$$

$$= P(B)P(A|B) \tag{4–10}$$

or

$$P(A \text{ and } B) = \left(\frac{\begin{array}{c}\text{number of}\\ \text{outcomes in } A\end{array}}{M}\right)\left(\frac{\begin{array}{c}\text{number of}\\ \text{outcomes in both}\\ A \text{ and } B \text{ simultaneously}\end{array}}{\begin{array}{c}\text{number of}\\ \text{outcomes in } A\end{array}}\right)$$

$$= P(A)P(B|A) \tag{4–11}$$

Equations 4–10 and 4–11 state the *multiplication law of probability.*

When the occurrence of event A has no effect on the occurrence or nonoccurrence of event B, we say that A and B are *independent.* Formally, independence implies that the marginal probability of B is the same as the conditional probability of B given A:

$$P(B|A) = P(B) \tag{4–12}$$

or

$$\frac{\begin{array}{c}\text{number of}\\ \text{outcomes in both}\\ A \text{ and } B \text{ simultaneously}\end{array}}{\begin{array}{c}\text{number of}\\ \text{outcomes in } A\end{array}} = \frac{\begin{array}{c}\text{number of}\\ \text{outcomes in } B\end{array}}{M}$$

We can rearrange the last line so that

$$\frac{\begin{array}{c}\text{number of}\\ \text{outcomes in both}\\ A \text{ and } B \text{ simultaneously}\end{array}}{\begin{array}{c}\text{number of}\\ \text{outcomes in } B\end{array}} = \frac{\begin{array}{c}\text{number of}\\ \text{outcomes in } A\end{array}}{M}$$

and

$$P(A|B) = P(A) \tag{4-13}$$

Therefore, if B is independent of A, A is independent of B.

Inasmuch as $P(\text{female}|\text{B.A.}) = 1/4$ and $P(\text{female}) = 5/20$ in our example, female and B.A. are independent events. On the other hand, female and Ph.D. are *dependent*; that is, $P(\text{female}|\text{Ph.D.}) = 3/10 \neq P(\text{female})$. Although dependence implies an *association* between A and B, it doesn't indicate that A *causes* B or B *causes* A. Combining Equations 4–10, 4–11, 4–12, and 4–13, we may conclude that the multiplication law reduces to

$$P(A \text{ and } B) = P(A)P(B) \tag{4-14}$$

provided A and B are indeoendent. [Note: By extension, $P(A$ and B and $C) = P(A)P(B)P(C)$ when A, B, and C are independent events.] Let's look at another example.

In a recent study of its "returns" policy, Department Store D established the following facts:

1. 20 percent of all returned items were originally gifts.
2. 30 percent of all returned items are clothes.
3. Of the clothes that are returned, 40 percent were originally gifts.

Since $P(\text{gift}|\text{clothing}) = .40 \neq P(\text{gift}) = .20$, gift and clothing are dependent. The probability that a returned item will be both clothing and a gift is

$$\begin{aligned} P(\text{clothing and gift}) &= P(\text{clothing})P(\text{gift}|\text{clothing}) \\ &= (.30)(.40) = .12 \end{aligned}$$

THE ADDITION LAW

Returning to our business faculty example, suppose we wish to derive the probability of choosing *either* a Ph.D. *or* a female. Because some outcomes are both Ph.D. and female, we *can't* write

$$\begin{aligned} P(\text{Ph.D. or female}) &= \frac{\text{number of outcomes in either Ph.D. or female}}{M} \\ &= \frac{\text{number of outcomes in Ph.D.}}{M} + \frac{\text{number of outcomes in female}}{M} \\ &= P(\text{Ph.D.}) + P(\text{female}) \\ &= 10/20 + 5/20 = 15/20 = .75 \end{aligned}$$

This would count each female Ph.D. twice, once as a member of the Ph.D. event and once as a member of the female event (see Table 4–7). Let's subtract the joint outcomes. Then,

$$P(\text{Ph.D. or female}) = \frac{\begin{matrix}\text{number of} \\ \text{outcomes in either} \\ \text{Ph.D. or female}\end{matrix} - \begin{matrix}\text{number of} \\ \text{outcomes in both} \\ \text{Ph.D. and female}\end{matrix}}{M}$$

$$= P(\text{Ph.D.}) + P(\text{female}) - P(\text{Ph.D. and female})$$

$$= 10|20 + 5|20 - 3|20 = 12|20 = .60$$

TABLE 4–7

		Degree		
		B.A.	M.A.	Ph.D.
Sex	Female	1	1	3
	Male	3	5	7

OUTCOMES IN BOTH PH.D.
AND FEMALE SIMULTANEOUSLY

The following is the *addition law of probability*:

$$P(A \text{ or } B) = P(A) + P(B) - P(A \text{ and } B) \qquad (4\text{–}15)$$

$P(A \text{ or } B)$ is the probability that an h_i is either an A or a B or both an A and a B.

Events A and B are *mutually exclusive* if there are no joint A and B outcomes. Under these circumstances, Equation 4–15 appears as

$$P(A \text{ or } B) = P(A) + P(B) - 0$$
$$= P(A) + P(B) \qquad (4\text{–}16)$$

Female and male are mutually exclusive. We see that

$$P(\text{female and male}) = P(\text{female})\left(\frac{\begin{matrix}\text{number of} \\ \text{outcomes in both} \\ \text{female and male simultaneously}\end{matrix}}{\begin{matrix}\text{number of} \\ \text{outcomes in female}\end{matrix}}\right)$$

$$= P(\text{female})(0/5) = 0$$

Mutually exclusive outcomes are always dependent; for example, $P(\text{female}) = 5/20 \neq P(\text{female}|\text{male}) = 0$.

The addition law when $P(A \text{ and } B) = 0$ is one of the three *axioms of probability*:

AXIOM 1: Since $P(A)$ is the sum of the probabilities of the outcomes in A, and the individual probabilities are never negative,

$$P(A) \geq 0 \qquad (4\text{–}17)$$

AXIOM 2: If S is the event "sample space," S is *certain* to occur because every possible outcome is included in S; that is,

$$P(S) = 1 \qquad (4\text{–}18)$$

AXIOM 3: If A and B are mutually exclusive,

$$P(A \text{ or } B) = P(A) + P(B)$$

Moreover, if A, B, and C are mutually exclusive,

$$P(A \text{ or } B \text{ or } C) = P(A) + P(B) + P(C)$$

Using the axioms and Figure 4–9, we shall develop a proof of Equation 4–15. Each circle in Figure 4–9 encloses an event. Inasmuch as A and B are not mutually exclusive, the circles overlap.

FIGURE 4–9

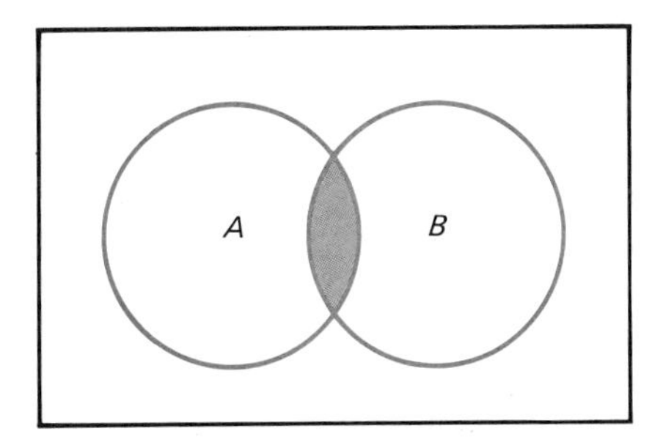

We'll symbolize the event "not A" as A' (i.e., A prime) and the event "not B" as B'. Figure 4–10 identifies four types of outcomes. Imagine that $C = A$ and B', $D = A$ and B, and $E = A'$ and B. Thus,

$$P(A \text{ or } B) = P(C \text{ or } D \text{ or } E)$$

Due to the fact that C, D, and E are mutually exclusive,

$$P(A \text{ or } B) = P(C) + P(D) + P(E) \qquad \text{Axiom \#3}$$

$$= P(A \text{ and } B') + P(A \text{ and } B) + P(A' \text{ and } B)$$

FIGURE 4–10

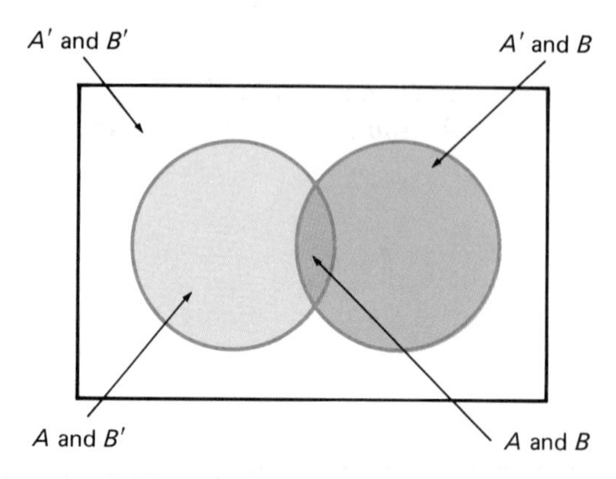

When we add $[P(A \text{ and } B) - P(A \text{ and } B)]$ to the right side, we get

$$P(A \text{ or } B) = \overbrace{P(A \text{ and } B') + P(A \text{ and } B)}^{P(A)}$$

$$+ \overbrace{P(A' \text{ and } B) + P(A \text{ and } B)}^{P(B)}$$

$$- P(A \text{ and } B)$$

The following example illustrates the addition law.

One hundred large companies operate in County T. All finance a health plan for their workers. Besides providing standard medical benefits, (1) 20 of the plans cover dental appointments, (2) 8 cover eye appointments, and (3) 5 cover both dental and eye appointments.

Suppose a company is randomly picked. The probability that its workers are covered for either dental or eye appointments or both dental and eye appointments is

$$P(\text{dental or eye}) = P(\text{dental}) + P(\text{eye}) - P(\text{dental and eye})$$
$$= .20 + .08 - .05 = .23$$

COUNTING METHODS

We now want to introduce some shortcuts for finding

$$M = \text{number of possible outcomes}$$

when a *series* of actions is required to complete *one* trial of an experiment. Consider the following example.

To remove himself from City V's crime and pollution, Mr. Smith built a home in a rural community. Smith's office, however, is still in the city. He can travel on one of three routes (A, B, or C) between his home and the highway leading to V. Smith can also choose one of four routes (D, E, F, or G) between the highway exit and his office.

Thus going to work involves two actions: (1) pick one route from among A, B, and C, and (2) pick one route from among D, E, F, and G. The sample space is reproduced in Figure 4–11.

FIGURE 4–11

· A,D	· B,D	· C,D
· A,E	· B,E	· C,E
· A,F	· B,F	· C,F
· A,G	· B,G	· C,G

If an experiment consists of k parts, such that

n_1 = the number of ways of performing the first part,
n_2 = the number of ways of performing the second part,
·
·
·
n_k = the number of ways of performing the kth part,

$$M = (n_1)(n_2) \dots (n_k) \qquad (4\text{–}19)$$

We see that M in our example is, indeed, (3)(4) = 12 pairs of routes.

Imagine an experiment in which three items are drawn from a hat containing $n = 5$ items:

1. When we select *with* replacement (i.e., an item is picked, recorded, and returned to the hat before the next pick), the sample space will have (5)(5)(5) = 125 points.
2. When we select *without* replacement (i.e., an item is picked but not returned to the hat), the sample space will have (5)(4)(3) = 60 points.

The next example illustrates the second case.

The W. X. Company has five vice-presidents. Based on selections from a hat, three will address a stockholders' meeting. Executive #1 must detail the cost of installing antipollution devices in W. X.'s plants. The second executive will explain the civil antitrust suit W. X. started against Company Q. The last executive picked will comment on the firm's foreign subsidiaries.

The 60 possible selections are displayed in Figure 4–12. The six groups in a particular column are identical in terms of people but different in terms of the *order* of the people. Each of the 60 points is a specific arrangement, or *permutation,* of $n = 5$ objects chosen $r = 3$ at a time.

FIGURE 4–12

Sample Space for "Select Three Vice-Presidents"

· *ABC*	· *ACD*	· *ADE*	· *BCD*	· *BDE*	· *CDE*	· *ABD*	· *ABE*	· *ACE*	· *BCE*
· *ACB*	· *ADC*	· *AED*	· *BDC*	· *BED*	· *CED*	· *ADB*	· *AEB*	· *AEC*	· *BEC*
· *BAC*	· *CAD*	· *DAE*	· *CBD*	· *DBE*	· *DCE*	· *BAD*	· *BAE*	· *CAE*	· *CBE*
· *BCA*	· *CDA*	· *DEA*	· *CDB*	· *DEB*	· *DEC*	· *BDA*	· *BEA*	· *CEA*	· *CEB*
· *CAB*	· *DAC*	· *EAD*	· *DBC*	· *EBD*	· *ECD*	· *DAB*	· *EAB*	· *EAC*	· *EBC*
· *CBA*	· *DCA*	· *EDA*	· *DCB*	· *EDB*	· *EDC*	· *DBA*	· *EBA*	· *ECA*	· *ECB*

The order of selection, however, may not be relevant to the experimenter. Consider the following situation.

The W. X. Company intends to pick three out of five vice-presidents to participate in a stockholders' meeting. The three will conduct a "Question and Answer" session.

Since only the composition of the panel, not the order in which the names are drawn, is important, we can view the sample space as in Figure 4–13. Each of these ten outcomes is said to be a *combination* of $n = 5$ objects chosen $r = 3$ at a time. (Note: *ABC, ACB, BAC, BCA, CAB,* and *CBA* form *one* combination.)

FIGURE 4–13

· *A,B,C*	· *A,C,D*	· *A,D,E*	· *B,C,D*	· *B,D,E*
· *C,D,E*	· *A,B,D*	· *A,B,E*	· *A,C,E*	· *B,C,E*

The *factorial* sign, !, is an instruction to compute the product

$$n! = (n)(n-1)(n-2)\ldots(1)$$

For example,

$$5! = (5)(4)(3)(2)(1) = 120$$

We define 0! to be 1. If $r = 7$ items are selected from $n = 10$ *different* items,

$$\begin{aligned}\text{Number of permutations} &= (10)(9)(8)(7)(6)(5)(4) = 604{,}800\\ &= (n)(n-1)(n-2)\ldots(n-r+1)\end{aligned}$$

Let's multiply the last line by 1, that is, $(n - r)!/(n - r)!$:

$$(n)(n-1)(n-2)\ldots(n-r+1)\left[\frac{(n-r)!}{(n-r)!}\right]$$

$$= \frac{\overbrace{(n)(n-1)(n-2)\ldots(n-r+1)[(n-r)(n-r-1)\ldots]}^{n!}}{(n-r)!}$$

Thus,

$$\text{Number of permutations} = \frac{n!}{(n-r)!} \qquad (4\text{–}20)$$

When $r = n$,

$$\text{Number of permutations} = \frac{n!}{(n-n)!} = \frac{n!}{0!} = \frac{n!}{1} = n! \qquad (4\text{–}21)$$

Observe that Figure 4–12 contains 60 permutations and 1/6 of 60, or 10, combinations. Inasmuch as $1/6 = 1/r!$, we can write

$$\text{Number of combinations} = (1/r!)\,(\text{number of permutations}) = \frac{1}{r!}\left[\frac{n!}{(n-r)!}\right] = \frac{n!}{r!(n-r)!} \qquad (4\text{–}22)$$

SUMMARY

Various activities that are associated with the business environment have uncertain outcomes. Probability theory offers a way to quantify the degree of uncertainty. Even though there are three approaches to computing $P(h_i)$ (i.e., the *a priori*, etc.), probabilities are always non-negative numbers between 0 and 1. We have identified marginal, joint, and conditional probabilities. The multiplication law and the addition law yield $P(A \text{ and } B)$ and $P(A \text{ or } B)$, respectively. When an experiment consists of more than one action, we can often determine M by formula, instead of by creating the sample space. We will study probability further in Chapter Five.

EXERCISES

1. Identify the probability approach used in each of the following situations:

   a. A person shuffles a deck of 52 cards and determines that the probability of selecting a three of diamonds is 1/52.
   b. A company develops a new product. Management feels that the probability the product will be a top seller is .7.
   c. An insurance investigator finds that ten of the last forty businesses to burn down in City A were the victims of arson. He estimates that .25 is the probability the next business fire will be due to arson.

2. Suppose a box contains five balls numbered 1 through 5.

   a. Determine the sample space for the experiment "select a ball at random."
   b. What is the probability of picking the #2 ball?
   c. What is the probability of picking either the #3 or #5 ball?
   d. What is the probability of picking an odd-numbered ball?
   e. What is the probability of picking a ball with a number of 3 or less?

3. a. Toss a quarter 50 times and record the relative frequency of heads after each toss.
   b. Following "a great many tosses," what number would the ratio of heads to tosses converge toward?

4. If a card is selected from a well-shuffled deck, what are the following probabilities?

   a. P(ace)
   b. P(two of Clubs)
   c. P(Diamond|red)
   d. P(three or five)
   e. P(five and black)

5. Suppose $P(A) = .4$ and $P(B) = .7$. If A and B are independent, what is $P(A \text{ and } B)$?

6. $P(A) = .5$ and $P(A \text{ and } B) = .2$. If A and B are independent, what is $P(B)$?

7. $P(A) = .2$, $P(B) = .6$, and $P(A \text{ and } B) = .1$.

   a. Are A and B independent?
   b. What is $P(B|A)$?
   c. What is $P(A|B)$?

8. Using the probability axioms, show that $P(A') = 1 - P(A)$, where A' is the event "not A."

9. If n items are selected from among n items, how many combinations are possible?

10. A corporate vice-president can ask any of four secretaries to type a memo. The secretary can photocopy the memo on any of six machines. In how many different ways can the memo be typed and copied?

11. A television station makes air-time available for responses to its editorials. The station has invited five viewers to present their rebuttals on Tuesday. Each viewer will speak for thirty seconds. In how many different ways can the station order the appearances of the viewers?

12. Suppose six items are chosen from among twelve different items.

a. How many permutations are possible?
b. How many combinations are possible?

13. Advocates of the federal Department of Education believe that it will reduce costs and red tape. Opponents, however, feel that the department may interfere extensively with local control of schools. One hundred educators and non-educators indicated whether they were pleased with the creation of this department:

	Educators A	*Non-Educators* C
For B	36	13
Against D	22	29

If one of these individuals is chosen at random, what are the following probabilities?

a. $P(A)$
b. $P(A \text{ or } B)$
c. $P(B \text{ and } C)$
d. $P(A|D)$
e. $P(C|B)$
f. $P(A \text{ and } B)$
g. $P(B|A)$

14. When New York City experienced a potential bankruptcy in 1975, Congress was hesitant to give the city loans. Some people believed that such loans would set a bad precedent, causing other municipalities to feel that the federal government would always

"bail them out" of financial difficulties. A researcher recorded the attitudes of 200 adults from the East and West coasts:

	East A	*West* C
For Aid B	60	30
Against Aid D	20	90

If one of these individuals is randomly selected, what are the following probabilities?

a. $P(C)$
b. $P(B \text{ or } C)$
c. $P(A \text{ and } D)$
d. $P(C|B)$
e. $P(D|A)$

15. After Chrysler appealed to Congress for aid in 1979, the researcher in question 14 interviewed the same people once again. He found that two-thirds of those "for" New York City loans were "for" Chrysler loans. If one of the individuals interviewed is picked at random, what is the probability he or she is "for" both types of loans?

16. During the 1970s, more and more women were appointed as directors of major companies. The number of women board members in 50 particular companies appears below:

Number of Women	*Number of Companies*
0	20
1	15
2	10
3	5

Suppose one of these companies is randomly chosen. Calculate the following probabilities:

a. P(2 or more women)
b. P(1 or 2 women)
c. P(0 or 3 women)

17. The Susan B. Anthony coin was introduced to reduce the cost of printing bills. Because many Americans have avoided using this coin, an investigator spoke to 100 men and 75 women. Ten of the men and 30 of the women liked the Anthony dollar. If one of these 175 people is picked at random, what are the following probabilities?

a. P(individual is a male and likes the coin)
b. P(individual is either a female or doesn't like the coin)
c. P(individual is a female|individual likes the coin)

18. There has been much debate over the "double taxation" of cor-

porate dividends. Companies pay taxes on their earnings. Then dividends, which come from after-tax earnings, are taxed along with the other income of stockholders. The table below indicates how 50 business leaders and 50 economists believe the dividend situation should be changed.

	Business Leaders A	*Economists* C
Tax Only the Company's Earnings B	25	30
Tax Only Dividends D	25	20

Suppose one of these individuals is chosen at random. Calculate the following probabilities:

a. $P(B|C)$
b. $P(A \text{ or } D)$
c. $P(C \text{ and } B)$
d. $P(B \text{ or } C)$

19. Corporation G's new emblem will portray the company as bold and dynamic. The emblem will contain four colors. How many different color combinations are possible if management chooses from a list of nine colors?

20. The probability that Mrs. Jones will invest in real estate is .3, whereas the probability that she will invest in stocks is .4. The probability that she will invest in both real estate and stocks is .25.

a. Calculate P(invest in neither real estate nor stocks).
b. Calculate P(invest in either real estate or stocks).
c. Are "invest in real estate" and "invest in stocks" independent?

21. John selects three cards from a deck. The cards are drawn with replacement.

a. What is the probability that all three cards are diamonds?
b. What is the probability that all three cards are red?
c. What is the probability that the second card is the jack of clubs?

22. The financial department at Company R includes seven employees, whereas the personnel department includes eight. The company wishes to send three people from finance and four people from personnel to an industry convention. How many different groups of seven employees are possible?

23. The probability that salesman A will meet his quota is .8. The probability that both salesmen A and B will meet their quotas is .7. If the sales of A and B are independent, what is the probability that B will meet his quota?

Discrete Random Variables

PROBABILITY DISTRIBUTIONS

Whether we estimate a parameter by a statistic or test a hypothesis about an unknown parameter, the logic of our procedures is a consequence of the relationship that exists between a population and its corresponding samples. Before we can identify this relationship, we must introduce the concept of a *random variable.*

We have already indicated the highest degree held by members of the business administration department at University L. A list of these degrees (i.e., B.A., B.A., M.A., Ph.D., M.A., etc.) would represent a distribution of categorical data where X = business faculty degrees. The data are also points in the sample space of the experiment "select a faculty member at random."

Because the outcomes are categorical, we can arbitrarily assign a number to each classification, for example, 0 = B.A., 1 = M.A., and 2 = Ph.D. (see Figure 5–1). P(M.A.) is, therefore, the same as $P(X = 1)$. The actual value of X is determined by chance. We will say that a variable of interest is a random variable whenever we view its possible values as outcomes of a random experiment.

In this chapter, we will concentrate on *discrete* random variables. The values of such variables are counted rather than measured. Consider the following example.

FIGURE 5–1

· SQ (0)	· DM (0)	· CP (0)	· RL (0)
· AV (2)	· BT (2)	· RE (2)	· NR (2)
· CG (2)	· PR (2)	· ET (2)	· SN (2)
· LB (2)	· VE (2)	· TI (1)	· GU (1)
· KT (1)	· CW (1)	· DH (1)	· JM (1)

The Board of Governors of the Federal Reserve System consists of seven people who meet periodically to establish monetary policy. If X = the number of board members who are absent from a scheduled meeting, there are eight possible values of X (i.e., 0, 1, 2, 3, 4, 5, 6, and 7). By contrast, if X = the length of the meeting, many values are possible (e.g., 57.89965 minutes, 200.45687 minutes, etc.).

The first X is discrete, while the second is *continuous*. Given the interval $X = a$ to $X = b$, we can write down *every* value of X from a to b when X is discrete. We cannot do so when X is continuous. The only values of X between $X = 2$ board members and $X = 5$ board members are 2, 3, 4, and 5. The values of X between $X = 10$ minutes and $X = 20$ minutes, however, are infinite. Figure 5–2 outlines our objective in this chapter.

FIGURE 5–2

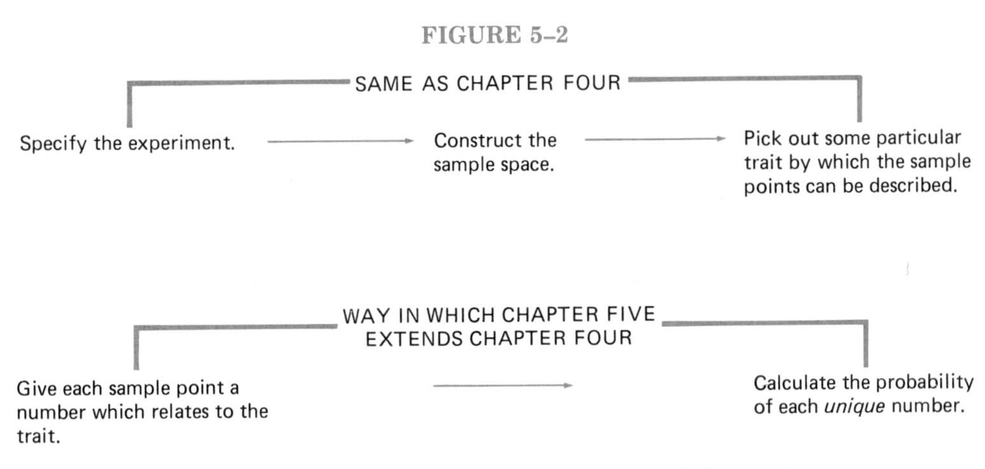

In our business faculty example, we can compute $P(X = 0) = .2$, $P(X = 1) = .3$, and $P(X = 2) = .5$. The pairing of *every* value of X with its probability creates the *probability distribution* of X. We will display such distributions in tables (see Table 5–1). Because every value of X appears in the table, $\Sigma P(X)$ must equal 1.0. Figure 5–3 is the histogram of the probability distribution of X in Table 5–1. The discrete outcomes are presented as class midpoints.

TABLE 5–1

X *Business Faculty Degrees*	$P(X)$
0	.2
1	.3
2	.5
	1.0

FIGURE 5–3

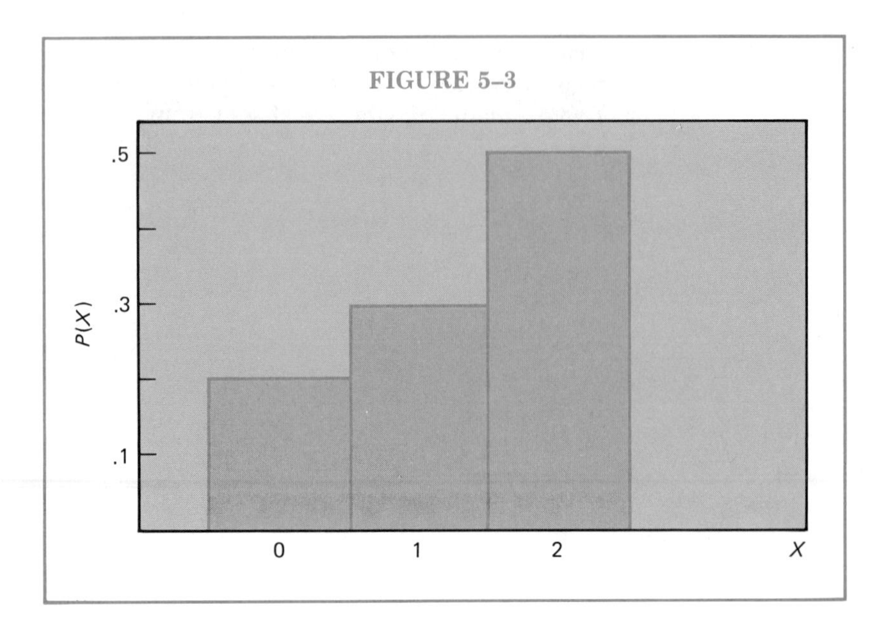

Calculating Parameters

When "all" trials of an experiment have been performed, the *population* relative frequency distribution of X is identical to the probability distribution of X. To establish the parameters of these distributions, suppose that 1,000 trials of "select a faculty member at random" are sufficient for the relative frequency of X to approach $P(X)$. Then, $X = 0$ should occur about $.2(1{,}000) = 200$ times; $X = 1$ should occur about $.3(1{,}000) = 300$ times; and $X = 2$ should occur about $.5(1{,}000) = 500$ times (see Table 5–2).

TABLE 5–2

X	*Number of Trials*
0	200
1	300
2	500
	1,000

We would determine μ_x (the average value of X) according to the following process:

1. Add 200 0s, 300 1s, and 500 2s.
2. Divide 1,000 into the total.

Thus,

$$\mu_x = \frac{(0)(200) + (1)(300) + (2)(500)}{1{,}000}$$

$$= (0)(.2) + (1)(.3) + (2)(.5) = 1.3$$

Since .2, .3, and .5 are probabilities as well as relative frequencies, we can write

$$\mu_x = \Sigma XP(X) = 1.3$$

This average is called the *expected value* of the random variable X. We will often symbolize it as $E(X)$; that is,

$$E(X) = \mu_x = \Sigma XP(X) \qquad (5\text{–}1)$$

One trial of the experiment will generate a 0, a 1, or a 2. The average outcome in the "long run" is 1.3. Like any average, the value of $E(X)$ is not confined to an actual value of X.

In calculating the variance, σ_x^2, of the relative frequency (or probability) distribution in our example, we must:

1. Subtract μ_x from 200 0s, 300 1s, and 500 2s.
2. Square the differences.
3. Add the squares.
4. Divide 1,000 into the squares.

This procedure yields

$$\sigma_x^2 = \frac{(0 - 1.3)^2(200) + (1 - 1.3)^2(300) + (2 - 1.3)^2(500)}{1{,}000}$$

$$= (0 - 1.3)^2(.2) + (1 - 1.3)^2(.3) + (2 - 1.3)^2(.5)$$

$$= .61$$

We will often symbolize the variance of the random variable X as $V(X)$, and thus we see that

$$V(X) = \sigma_x^2 = \Sigma(X - \mu_x)^2P(X) \qquad (5\text{–}2)$$

$E(X)$ and $V(X)$ indicate the central tendency and dispersion of outcomes following an infinite number of repetitions of an experiment. We will study these parameters extensively. Both Σ and $E(\ )$ are *instructions*. Given $E(\ )$, we have to *add* all of the products that are formed when the value of the expression to the right of E is multiplied by its probability. $V(X)$ is, then, $\Sigma(X - \mu_x)^2P(X)$ or $E[(X - \mu_x)^2]$ or $E[X - E(X)]^2$.

Besides academic degree, a member of the business faculty at University L can be described by his or her sex. If 0 = female and

1 = male, the probability distribution of Y = sex is as shown in Table 5–3. The parameters of this distribution are

$$E(Y) = \Sigma YP(Y) = (0)(.25) + (1)(.75) = .75$$

and

$$V(Y) = \Sigma(Y - \mu_y)^2P(Y) = (0 - .75)^2(.25) + (1 - .75)^2(.75) = .1875$$

Over a "great many" trials, Y would have an average value of .75. We again find that the expected value is something other than one of the discrete outcomes (i.e., 0 or 1) of the random variable.

TABLE 5–3

Y	$P(Y)$
0	5/20 = .25
1	15/20 = .75
	20/20 = 1.00

Joint and Conditional Probability

Because X and Y arise from the same experiment, each sample point has an X value and a Y value (see Figure 5–4). These joint occurrences are reproduced in Table 5–4.

FIGURE 5–4

· SQ ($X = 0, Y = 0$)	· DM($X = 0, Y = 1$)	· CP ($X = 0, Y = 1$)	· RL ($X = 0, Y = 1$)
· AV($X = 2, Y = 0$)	· BT ($X = 2, Y = 1$)	· RE ($X = 2, Y = 1$)	· NR($X = 2, Y = 1$)
· CG($X = 2, Y = 0$)	· PR ($X = 2, Y = 1$)	· ET ($X = 2, Y = 1$)	· SN ($X = 2, Y = 1$)
· LD ($X = 2, Y = 0$)	· VE ($X = 2, Y = 1$)	· TI ($X = 1, Y = 1$)	· GU($X = 1, Y = 1$)
· KT($X = 1, Y = 0$)	· CW($X = 1, Y = 1$)	· DH($X = 1, Y = 1$)	· JM($X = 1, Y = 1$)

TABLE 5–4

		X			
		0	1	2	*Total*
Y	0	1	1	3	5
	1	3	5	7	15
	Total	4	6	10	20

The *joint probability distribution* of X and Y is the complete listing of all joint events and their respective probabilities. The joint probabilities in Table 5–5 are derived from the information in Table 5–4. The marginal probabilities appear in the individual probability distributions of X and Y. The joint probability distribution is given in Table 5–6. Inasmuch as every sample point is accounted for by (0,0), (0,1), (1,0), (1,1), (2,0), and (2,1), the sum of the joint probabilities is 1.0. We see, for example, that the outcome "female Ph.D." or ($X = 2$, $Y = 0$) will take place 3/20 of the time if the experiment "select a faculty member" is repeated many times.

TABLE 5–5

		X			*Marginal Probability of Y*
		0	1	2	
Y	0	1/20	1/20	3/20	5/20
	1	3/20	5/20	7/20	15/20
	Marginal Probability of X	4/20	6/20	10/20	

TABLE 5–6

X,Y	P(X,Y)
0,0	1/20
0,1	3/20
1,0	1/20
1,1	5/20
2,0	3/20
2,1	7/20
	20/20 = 1.0

Equation 4–14 establishes the fact that the joint probability of two independent events is equal to the product of their marginal probabilities. We now wish to apply the notion of independence to random variables. Whenever two random variables are independent, the probability of *every* joint occurrence, $P(X_i, Y_j)$, is equal to the product of the marginal probabilities, $P(X_i)$ and $P(Y_j)$; that is,

$$P(X_i, Y_j) = P(X_i)P(Y_j) \tag{5–3}$$

for all combinations of X and Y.

In our business faculty example, are X and Y (degree and sex) independent? We see in Table 5–7 that they are not since the second and third columns of the table are different. The association between X and Y doesn't prove, however, that "sex *causes* degree."

TABLE 5–7

X,Y	$P(X,Y)$		$P(X)P(Y)$	
0,0	.05	=	(4/20)(5/20)	= .05
0,1	.15	=	(4/20)(15/20)	= .15
1,0	.05	≠	(6/20)(5/20)	= .075
1,1	.25	≠	(6/20)(15/20)	= .225
2,0	.15	≠	(10/20)(5/20)	= .125
2,1	.35	≠	(10/20)(15/20)	= .375

We can also determine *conditional probability distributions.* Suppose that $Y = 0$ has occurred. $P(X|Y = 0)$ is the conditional probability of X given $Y = 0$. By referring to the $Y = 0$ row of Table 5–4, we can construct the conditional probability distribution of X given $Y = 0$ (see Table 5–8). If we were to isolate all of the $Y = 0$ outcomes in the "long run," we would observe that $X = 0$ is paired with $Y = 0$ twenty percent of the time, $X = 1$ is paired with $Y = 0$ twenty percent of the time, and $X = 2$ is paired with $Y = 0$ sixty percent of the time. Conditional probability distributions are an important part of the material on regression and correlation.

TABLE 5–8

$X|Y = 0$	$P(X|Y = 0)$
0	1/5
1	1/5
2	3/5
	5/5 = 1.0

FUNCTIONS OF ONE RANDOM VARIABLE

Any theoretical discussion that we have in the remainder of the text will be an outgrowth of the behavior of random variables. In this and the next section, we will explore such behavior. Although we are going to develop proofs for discrete random variables, the *results* of the proofs apply even when random variables are continuous.

Variable U is *functionally* related to variable X if we can identify a rule by which a value of X is paired with *one* and *only one* value of U. $U = 3X$, for instance, is a rule which "multiplies X by 3" in order to get U. When $X = 4$, $U = 3(4) = 12$. Alternatively,

$U = 5 + 2X$ is a rule which "adds 5 to twice the value of X." Because of the rule, $X = 4$ is paired with $U = 5 + 2(4) = 13$. We want to compare the probability distribution of U with the probability distribution of X. The following example will serve as an illustration.

The yearly increase in a professor's salary at University L is determined by the function

$$U = 500 + 250X$$

where U = the yearly increase in salary and X = the highest degree. A faculty member with an M.A. (i.e., $X = 1$) would get a raise of $U = 500 + 250(1) = \$750$.

The mixture of degrees in Figure 5–1 implies that the business administration department will receive four \$500 raises, six \$750 raises, and ten \$1,000 raises. The probability distribution of U is presented in Table 5–9. Thus,

$$E(U) = (500)(.2) + (750)(.3) + (1{,}000)(.5) = \$825$$

and

$$V(U) = (500 - 825)^2(.2) + (750 - 825)^2(.3) + (1{,}000 - 825)^2(.5)$$
$$= 38{,}125$$

TABLE 5–9

U	$P(U)$
500	4/20 = .2
750	6/20 = .3
1,000	10/20 = .5
	20/20 = 1.0

We will show that the parameters of the X distribution can be used to calculate the parameters of the U distribution, or, specifically,

$$E(U) = \Sigma UP(U) = \Sigma(\text{function of } X)P(X)$$
$$= \Sigma(500 + 250X)P(X)$$

and

$$V(U) = \Sigma[U - E(U)]^2P(U)$$
$$= \Sigma[(\text{function of } X) - E(\text{function of } X)]^2P(X)$$
$$= \Sigma[(500 + 250X) - E(500 + 250X)]^2P(X)$$

We must now identify certain properties of random variables.

PROPERTY 1: The expected value of some constant, a, is equal to that constant; that is,

$$E(a) = a \tag{5-4}$$

If an experiment always leads to the same outcome, a, the distribution of X is

X	$P(X)$
a	1.0

Thus, $E(X) = aP(a) = a(1.0) = a$.

PROPERTY 2: Where $U = X + a$, the expected value of U is equal to the expected value of X plus the expected value of a; that is,

$$E(X + a) = E(X) + E(a) = E(X) + a \tag{5-5}$$

Equation 5–5 is derived as follows:

$$\begin{aligned} E(X + a) &= \Sigma(X + a)P(X) = \Sigma[XP(X) + aP(X)] \\ &= \Sigma XP(X) + \Sigma aP(X) \\ &= \Sigma XP(X) + a\Sigma P(X) \end{aligned}$$

The first term, $\Sigma XP(X)$, is $E(X)$. Since $\Sigma P(X) = 1.0$ for any probability distribution, $a\Sigma P(X) = a(1.0) = a$. Hence, $E(X + a) = E(X) + a$.

PROPERTY 3: Where $U = X - a$, the expected value of U is equal to the expected value of X minus the expected value of a; that is,

$$E(X - a) = E(X) - E(a) = E(X) - a \tag{5-6}$$

The derivation is

$$\begin{aligned} E(X - a) &= \Sigma(X - a)P(X) = \Sigma[XP(X) - aP(X)] \\ &= \Sigma XP(X) - \Sigma aP(X) \\ &= E(X) - a\Sigma P(X) = E(X) - a \end{aligned}$$

PROPERTY 4: Where $U = aX$, the expected value of U is equal to a times the expected value of X; that is,

$$E(aX) = \Sigma aXP(X) = a\Sigma XP(X) = aE(X) \tag{5-7}$$

Returning to our example of business faculty raises, we know according to Equation 5–5 that $E(U) = E(500 + 250X) = E(500) + E(250X)$. Since 500 and 250 are constants, $E(500 + 250X) = 500 + 250E(X)$. Inasmuch as $E(X) = 1.3$, $E(U) = 500 + 250(1.3) =$ \$825. This value agrees with $\Sigma UP(U) =$ \$825.

We turn next to the variance. Instead of writing it as a summation, we will express the variance as $E[(\ ) - E(\ )]^2$. The mathematical operations suggested by the symbols "inside" E can be performed before we compute $E[(\ ) - E(\ )]^2$.

PROPERTY 5: The variance of some constant, a, is equal to 0; that is,

$$V(a) = 0 \tag{5–8}$$

Earlier, we saw that $E(a) = a$. Then, $V(a) = E[a - E(a)]^2 = E(a - a)^2 = E(0) = 0$.

PROPERTY 6: Where $U = X + a$, the variance of U is equal to the variance of X; that is,

$$V(X + a) = V(X) \tag{5–9}$$

The derivation is

$$\begin{aligned} V(X + a) &= E[U - E(U)]^2 = E[(X + a) - E(X + a)]^2 \\ &= E[(X + a) - E(X) - a]^2 \qquad \text{By virtue of Equation 5–5} \\ &= E[X - E(X)]^2 = V(X) \end{aligned}$$

PROPERTY 7: Where $U = X - a$, the variance of U is equal to the variance of X; that is,

$$V(X - a) = V(X) \tag{5–10}$$

The derivation is

$$\begin{aligned} V(X - a) &= E[(X - a) - E(X - a)]^2 \\ &= E[(X - a) - E(X) + a]^2 \qquad \text{By virtue of Equation 5–6} \\ &= E[X - E(X)]^2 = V(X) \end{aligned}$$

PROPERTY 8: Where $U = aX$, the variance of U is equal to a^2 times the variance of X; that is,

$$V(aX) = a^2V(X) \tag{5–11}$$

The derivation is

$$\begin{aligned} V(aX) &= E[(aX) - E(aX)]^2 \\ &= E[(aX) - aE(X)]^2 \qquad \text{By virtue of Equation 5–7} \\ &= E\{a[X - E(X)]\}^2 = E\{a^2[X - E(X)]^2\} \\ &= a^2E[X - E(X)]^2 \qquad \text{By virtue of Equation 5–7} \\ &= a^2V(X) \end{aligned}$$

Going back to our example, we see that $V(500 + 250X) = V(500)$

$+ V(250X) = 0 + (250)^2V(X)$. Since $V(X) = .61$, $V(U) = (250)^2(.61) = 38{,}125$. Thus, both $E(U)$ and $V(U)$ are quickly derived from $E(X)$ and $V(X)$ as long as U and X are *linearly* related; that is, $U =$ (some constant) + (another constant)X.

FUNCTIONS OF TWO RANDOM VARIABLES

The values of some random variables are jointly determined by two or more random variables. This condition exists in the following example.

More and more women are suing their employers because of the latter's failure to provide "equal pay for equal work." The dean of University L wants to remedy past male-female salary differentials. He has decided to change the procedure for awarding raises. The new formula,

$$U = 600 + 100X - 200Y$$

where Y is sex (i.e., 0 = female and 1 = male), favors women. A female Ph.D. will receive a raise of $U = 600 + 100(2) - 200(0) = \$800$, but a male Ph.D. will receive only $U = 600 + 100(2) - 200(1) = \$600$.

The number of joint occurrences of X and Y, and the Us associated with these occurrences, are supplied in Table 5–10. The probability distribution of U is given in Table 5–11. We see that

$$E(U) = (400)(.15) + (500)(.25) + (600)(.40) + (700)(.05) + (800)(.15) = \$580$$

and

$$V(U) = (400 - 580)^2(.15) + (500 - 580)^2(.25) + (600 - 580)^2(.40) + (700 - 580)^2(.05) + (800 - 580)^2(.15) = 14{,}600$$

TABLE 5–10

		X		
		$X_0 = 0$	$X_1 = 1$	$X_2 = 2$
Y	$Y_0 = 0$	1 ($U = 600$)	1 ($U = 700$)	3 ($U = 800$)
	$Y_1 = 1$	3 ($U = 400$)	5 ($U = 500$)	7 ($U = 600$)

TABLE 5–11

U	$P(U)$
400	3/20 = .15
500	5/20 = .25
600	8/20 = .40
700	1/20 = .05
800	3/20 = .15

Calculating Parameters

Instead of calculating $E(U)$ and $V(U)$ from the marginal distribution of U, we will determine these parameters explicitly in terms of the joint probability distribution of X and Y (see Table 5–12).

TABLE 5–12

Joint Occurrence	*600 + 100X − 200Y*	*P(X,Y)*
X_0, Y_0	$600 + 100X_0 - 200Y_0$	1/20
X_1, Y_0	$600 + 100X_1 - 200Y_0$	1/20
X_2, Y_0	$600 + 100X_2 - 200Y_0$	3/20
X_0, Y_1	$600 + 100X_0 - 200Y_1$	3/20
X_1, Y_1	$600 + 100X_1 - 200Y_1$	5/20
X_2, Y_1	$600 + 100X_2 - 200Y_1$	7/20

The expected value of this distribution is the sum of all of the $(600 + 100X - 200Y)P(X,Y)$ products. We *cannot* indicate such a summation as

$$\sum_{i=0}^{2} (600 + 100X_i - 200Y_i)P(X_i, Y_i)$$

The summation doesn't account for the instances in which the subscripts of X and Y differ, for example, $(600 + 100X_0 - 200Y_1)P(X_0, Y_1)$. Moreover, Y_2 is not a value of Y.

The appropriate symbol is the *double summation*

$$\sum_{j=0}^{1} \sum_{i=0}^{2}$$

where the X index is $i = 0,1,2$ and the Y index is $j = 0,1$. We must add as follows:

1. Start with $i = 0$ and $j = 0$.
2. Proceed through all i, holding j at 0.
3. Proceed through all i, holding j at 1.

Moves to another j happen only after we have completed a series of is; for example,

$$\sum_{j=0}^{1}\sum_{i=0}^{2}(X_i + Y_j) = (X_0 + Y_0) + (X_1 + Y_0) + (X_2 + Y_0) + (X_0 + Y_1) + (X_1 + Y_1) + (X_2 + Y_1)$$

Therefore,

$$E(600 + 100X - 200Y) = \sum_{j=0}^{1}\sum_{i=0}^{2}(600 + 100X_i - 200Y_j)P(X_i, Y_j)$$

and

$$V(600 + 100X - 200Y) = \sum_{j=0}^{1}\sum_{i=0}^{2}[(600 + 100X_i - 200Y_j) - E(600 + 100X_i - 200Y_j)]^2 P(X_i, Y_j)$$

We will list the properties that establish the connection between Tables 5–11 and 5–12. For simplicity, the proofs will evolve from Table 5–13.

PROPERTY 9: The expected value of the sum of two random variables is equal to the sum of their individual expected values; that is,

$$E(X + Y) = E(X) + E(Y) \qquad (5\text{–}12)$$

Similarly,

$$E(X + Y + Z) = E(X) + E(Y) + E(Z)$$

We know that $E(X) = X_1P(X_1) + X_2P(X_2)$ and $E(Y) = Y_1P(Y_1) + Y_2P(Y_2)$. Then,

$$\begin{aligned} E(X + Y) &= \sum_j\sum_i(X_i + Y_j)P(X_i, Y_j) \\ &= [(X_1 + Y_1)p_{11} + (X_2 + Y_1)p_{21} + (X_1 + Y_2)p_{12} + (X_2 + Y_2)p_{22}] \\ &= [X_1(p_{11} + p_{12}) + X_2(p_{21} + p_{22})] + [Y_1(p_{11} + p_{21}) + Y_2(p_{12} + p_{22})] \\ &= E(X) + E(Y) \end{aligned}$$

PROPERTY 10: The expected value of the difference between two random variables is equal to the difference between their individual expected values; that is,

$$E(X - Y) = E(X) - E(Y) \qquad (5\text{–}13)$$

TABLE 5–13

	X_1	X_2	*Marginal Probability of Y*
Y_1	p_{11}	p_{21}	$P(Y_1) = p_{11} + p_{21}$
Y_2	p_{12}	p_{22}	$P(Y_2) = p_{12} + p_{22}$
Marginal Probability of X	$P(X_1) = p_{11} + p_{12}$	$P(X_2) = p_{21} + p_{22}$	

This equality is derived as follows:

$$\begin{aligned} E(X - Y) &= \sum\sum(X - Y)P(X,Y) \\ &= [(X_1 - Y_1)p_{11} + (X_2 - Y_1)p_{21} + (X_1 - Y_2)p_{12} \\ &\quad + (X_2 - Y_2)p_{22}] \\ &= [X_1(p_{11} + p_{12}) + X_2(p_{21} + p_{22})] - [Y_1(p_{11} + p_{21}) \\ &\quad + Y_2(p_{12} + p_{22})] \\ &= E(X) - E(Y) \end{aligned}$$

An expression such as $(X + Y)$ or $(X - Y)$ is said to be a *linear combination* of two random variables. The general form of a combination is $aX + bY$, where a and b are constants. If the combination is $(X + Y)$, $a = 1$ and $b = 1$. When the combination is $(X - Y)$, however, $a = 1$ and $b = -1$.

PROPERTY 11: The expected value of $aX + bY$ is equal to a times the expected value of X plus b times the expected value of Y; that is,

$$E(aX + bY) = aE(X) + bE(Y) \tag{5–14}$$

The derivation is

$$\begin{aligned} E(aX + bY) &= \sum\sum(aX + bY)P(X,Y) \\ &= [(aX_1 + bY_1)p_{11} + (aX_2 + bY_1)p_{21} \\ &\quad + (aX_1 + bY_2)p_{12} + (aX_2 + bY_2)p_{22}] \\ &= [aX_1(p_{11} + p_{12}) + aX_2(p_{21} + p_{22})] + [bY_1(p_{11} + p_{21}) \\ &\quad + bY_2(p_{12} + p_{22})] \\ &= a[X_1(p_{11} + p_{12}) + X_2(p_{21} + p_{22})] + b[Y_1(p_{11} + p_{21}) \\ &\quad + Y_2(p_{12} + p_{22})] \\ &= aE(X) + bE(Y) \end{aligned}$$

Although Equations 5–12 and 5–13 are just special cases of Equation 5–14, we have given them separate identities because of their importance in later chapters.

We are now in a position to evaluate $E(600 + 100X - 200Y)$:

$$E(600 + 100X - 200Y) = E(600) + E(100X - 200Y)$$

By virtue of Equation 5–5

$$= 600 + E(100X - 200Y)$$

By virtue of Equation 5–4

$$= 600 + 100E(X) - 200E(Y)$$

By virtue of Equation 5–14

Because $E(X) = 1.3$ and $E(Y) = .75$, $E(U) = 600 + (100)(1.3) - (200)(.75) = \$580$. This is the same expected value that we computed as $\Sigma UP(U)$.

The variance of a linear combination, $V(aX + bY)$, is expanded below:

$$V(aX + bY) = E[(aX + bY) - E(aX + bY)]^2$$

$$= E[aX + bY - aE(X) - bE(Y)]^2$$

By virtue of Equation 5–14

$$= E\{a[X - E(X)] + b[Y - E(Y)]\}^2$$

$$= E\{a^2[X - E(X)]^2 + 2ab[X - E(X)][Y - E(Y)] + b^2[Y - E(Y)]^2\}$$

$$= a^2E[X - E(X)]^2 + 2abE[X - E(X)][Y - E(Y)] + b^2E[Y - E(Y)]^2$$ By virtue of Equation 5–14

$$= a^2V(X) + 2abE[X - E(X)][Y - E(Y)] + b^2V(Y)$$

Covariance

Part of the middle term in the last line is called the *covariance* of X and Y, or $Cov(X,Y)$:

$$\text{Cov}(X,Y) = E[X - E(X)][Y - E(Y)]$$

$$= \sum\sum [X - E(X)][Y - E(Y)]P(X,Y) \qquad (5\text{–}15)$$

A detailed discussion of Equation 5–15 won't appear until the correlation section of Chapter Eighteen.

The covariance is a measure of the way in which X and Y are related. If high values of X are usually paired with high values of

Y, Cov(X,Y) will be *positive.* On the other hand, if high values of X are usually paired with low values of Y, Cov(X,Y) will be *negative.* When X and Y are independent, there is *no* relation between them and Cov(X,Y) = 0. [In Chapter Eighteen, we will show that Cov(X,Y) = 0 doesn't automatically imply independence, even though independence does imply Cov(X,Y) = 0.]

To substantiate why independence produces Cov(X,Y) = 0, let's rewrite Cov(X,Y) as $E(XY) - \mu_x\mu_y$; that is,

$$\begin{aligned} E[X - E(X)][Y - E(Y)] &= E[XY - XE(Y) - YE(X) + E(X)E(Y)] \\ &= E[XY - X\mu_y - Y\mu_x + \mu_x\mu_y] \\ &= E(XY) - \mu_y E(X) - \mu_x E(Y) + E(\mu_x\mu_y) \quad \text{By virtue of Equation 5–14} \\ &= E(XY) - \mu_y E(X) - \mu_x E(Y) + \mu_x\mu_y \quad \text{By virtue of Equation 5–4} \\ &= E(XY) - \mu_y\mu_x - \mu_x\mu_y + \mu_x\mu_y \\ &= E(XY) - \mu_x\mu_y \end{aligned}$$

In the light of Table 5–13,

$$\begin{aligned} E(XY) &= \sum\sum(XY)P(X,Y) \\ &= (X_1Y_1)p_{11} + (X_2Y_1)p_{21} + (X_1Y_2)p_{12} + (X_2Y_2)p_{22} \end{aligned}$$

For independence, $P(X_i, Y_j)$ has to equal $P(X_i)P(Y_j)$:

$$p_{11} = P(X_1)P(Y_1)$$

$$p_{21} = P(X_2)P(Y_1)$$

$$p_{12} = P(X_1)P(Y_2)$$

$$p_{22} = P(X_2)P(Y_2)$$

We can, thus, state that

$$\begin{aligned} E(XY) &= (X_1Y_1)P(X_1)P(Y_1) + (X_2Y_1)P(X_2)P(Y_1) + (X_1Y_2)P(X_1)P(Y_2) + (X_2Y_2)P(X_2)P(Y_2) \\ &= [X_1P(X_1) + X_2P(X_2)][Y_1P(Y_1)] + [X_1P(X_1) + X_2P(X_2)][Y_2P(Y_2)] \\ &= [X_1P(X_1) + X_2P(X_2)][Y_1P(Y_1) + Y_2P(Y_2)] \\ &= E(X)E(Y) \end{aligned}$$

Moreover,

$$\text{Cov}(X,Y) = E(XY) - E(X)E(Y) = E(X)E(Y) - E(X)E(Y) = 0$$

The past few results lead to the following properties:

PROPERTY 12: If X and Y are two *dependent* random variables,

$$V(aX + bY) = a^2V(X) + 2ab\,\text{Cov}(X,Y) + b^2V(Y) \quad (5\text{–}16)$$

PROPERTY 13: If X and Y are two *independent* random variables,

$$V(aX + bY) = a^2V(X) + b^2V(Y) \quad (5\text{–}17)$$

We shall also isolate two special cases of Equation 5–17:

PROPERTY 14: If X and Y are independent, the variance of their sum is equal to the sum of their individual variances; that is,

$$V(X + Y) = V(X) + V(Y) \quad (5\text{–}18)$$

Similarly,

$$V(X + Y + Z) = V(X) + V(Y) + V(Z)$$

PROPERTY 15: If X and Y are independent, the variance of their difference is equal to the *sum* of their individual variances; that is,

$$V(X - Y) = V(X) + V(Y) \quad (5\text{–}19)$$

Returning to our example, we find that

$$\begin{aligned} V(600 + 100X - 200Y) &= V(600) + V(100X - 200Y) \\ &= 0 + V(100X - 200Y) \\ &\quad \text{By virtue of Equation 5–9} \\ &= (100)^2V(X) + (2)(100)(-200)\text{Cov}(X,Y) + (-200)^2V(Y) \\ &\quad \text{By virtue of Equation 5–16} \end{aligned}$$

Inasmuch as $E(X) = 1.3$ and $E(Y) = .75$,

$$\begin{aligned} \text{Cov}(X,Y) &= (0 - 1.3)(0 - .75)(.05) + (1 - 1.3)(0 - .75)(.05) + \\ &\quad (2 - 1.3)(0 - .75)(.15) + (0 - 1.3)(1 - .75)(.15) + \\ &\quad (1 - 1.3)(1 - .75)(.25) + (2 - 1.3)(1 - .75)(.35) \\ &= -.025 \end{aligned}$$

Recall that $V(X) = .61$ and $V(Y) = .1875$. Upon substitution, we get

$$\begin{aligned} V(600 + 100X - 200Y) &= 0 + (100)^2(.61) + (2)(100)(-200)(-.025) \\ &\quad + (-200)^2(.1875) \\ &= 14{,}600 = V(U) \\ &= \text{the variance of a linear combination of } X \text{ and } Y \end{aligned}$$

Table 5–14 reviews all 15 properties.

TABLE 5–14

Equation	*Property*
(5–4)	$E(a) = a$
(5–5)	$E(X + a) = E(X) + a$
(5–6)	$E(X - a) = E(X) - a$
(5–7)	$E(aX) = aE(X)$
(5–8)	$V(a) = 0$
(5–9)	$V(X + a) = V(X)$
(5–10)	$V(X - a) = V(X)$
(5–11)	$V(aX) = a^2V(X)$
(5–12)	$E(X + Y) = E(X) + E(Y)$
(5–13)	$E(X - Y) = E(X) - E(Y)$
(5–14)	$E(aX + bY) = aE(X) + bE(Y)$
(5–16) (dependence)	$V(aX + bY) = a^2V(X) + 2ab\text{Cov}(X,Y) + b^2V(Y)$
(5–17) (independence)	$V(aX + bY) = a^2V(X) + b^2V(Y)$
(5–18) (independence)	$V(X + Y) = V(X) + V(Y)$
(5–19) (independence)	$V(X - Y) = V(X) + V(Y)$

THE BINOMIAL PROBABILITY DISTRIBUTION

Creating the probability distribution of X = degrees was time consuming since we had to identify every outcome in the sample space. Given a certain set of conditions, however, statisticians can quickly derive an important probability distribution called the *binomial.*

When the outcomes of an experiment can be sorted into two and only two mutually exclusive classes (e.g., male and female or B.A. and non-B.A.), each repetition of it is known as a *Bernoulli trial.* We'll label these outcomes as "successes" and "failures." (We are not making a judgment. A "success" could be something that is "bad.")

If a "success" is a B.A. and a "failure" is a non-B.A. (i.e., an M.A. or a Ph.D.) in our business faculty example, we can write P(B.A.) = P(success) = 4/20 = .2 and P(non-B.A.) = P(failure) = 16/20 = .8. The random variable X = degrees is now restricted to two values, say, 1 = success and 0 = failure. Table 5–15 contains the probability distribution of X. The parameters are

$$E(X) = (1)(.2) + (0)(.8) = .2$$

and

$$V(X) = (1 - .2)^2(.2) + (0 - .2)^2(.8) = .16$$

TABLE 5–15

X	$P(X)$
1	.2
0	.8

Suppose we generalize the above by letting the Greek letter *pi*, π, represent P(success). With P(failure) $= 1 - \pi$, Table 5–15 would turn into Table 5–16, and the parameters become

$$E(X) = (1)(\pi) + (0)(1 - \pi) = \pi \tag{5–20}$$

and

$$\begin{aligned} V(X) &= (1 - \pi)^2(\pi) + (0 - \pi)^2(1 - \pi) \\ &= \pi - 2\pi^2 + \pi^3 + \pi^2 - \pi^3 \\ &= \pi - \pi^2 = \pi(1 - \pi) \end{aligned} \tag{5–21}$$

TABLE 5–16

X	$P(X)$
1	π
0	$1 - \pi$

Researchers are often interested in the *cumulative* number of "successes" in a series of trials. Imagine a situation in which the following conditions apply:

1. The outcome of any trial is *either* a success *or* a failure.
2. The probability of a success, π, is *constant* from trial to trial.
3. The outcomes are *independent.*

Table 5–17 shows that these conditions characterize the series "six tosses of a well-balanced coin." They also approximate the description of many "real-world" series. Consider the following example.

For a long time, the federal budget has not been balanced. A

TABLE 5–17

Conditions	*Coin-Toss Series*
1. Bernoulli trials	The outcome of any trial will be a head = success or a tail = failure.
2. π is constant	P(head) = .50 on every toss.
3. Independence	The appearance of a head on one trial will have no effect on the outcomes of other trials; for example, P(head|5 previous heads) = P(head) = .50.

growing body of Americans dislikes deficits, irrespective of the state of the economy. Some people have even urged the passage of a constitutional amendment placing ceilings on the budget.

An investigator puts the names of all United States representatives and senators into a hat, shuffles the names, and selects three congressmen. He wants to discuss the budget amendment with the individuals in his sample. Assume that 20 percent of the 535 representatives and senators are for the amendment.

Each congressman is a "for" or "against" outcome. The probability of a "for" on the first selection is

$$P(\text{for}) = P(\text{success}) = \pi = \frac{\text{number of for's}}{\text{number of congressmen}} = 107/535 = .20$$

Since $N = 535$ is large relative to the $n = 3$ trials of "select a congressman," the conditional probability of a success and π will never differ by much; for example, $P(\text{for}|2 \text{ previous for's}) = 105/533 = .197$. Hence, we can view the outcomes as though they were independent.

The researcher will observe a particular sequence of successes and failures. Suppose we define a random variable for each trial, where X_1 = attitude of the congressman selected on trial #1
X_2 = attitude of the congressman selected on trial #2
X_3 = attitude of the congressman selected on trial #3

If 0 = against and 1 = for, the sample space for a series of three trials will be as shown in Figure 5–5.

FIGURE 5–5

· $(X_1 = 0, X_2 = 0, X_3 = 0)$	· $(X_1 = 1, X_2 = 0, X_3 = 0)$
· $(X_1 = 0, X_2 = 1, X_3 = 0)$	· $(X_1 = 0, X_2 = 0, X_3 = 1)$
· $(X_1 = 1, X_2 = 1, X_3 = 0)$	· $(X_1 = 1, X_2 = 0, X_3 = 1)$
· $(X_1 = 0, X_2 = 1, X_3 = 1)$	· $(X_1 = 1, X_2 = 1, X_3 = 1)$

Due to the fact that X_1, X_2, and X_3 are virtually independent, $P(X_1, X_2, X_3) = P(X_1)P(X_2)P(X_3)$. Table 5–18 is the joint probability distribution of X_1, X_2, and X_3. When r is the number of successes in any sequence, that is,

$$r = X_1 + X_2 + X_3,$$

the probability of a *specific* sequence is $\pi^r(1 - \pi)^{n-r}$.

TABLE 5–18

X_1,X_2,X_3	$P(X_1,X_2,X_3)$
0,0,0	$(.8)(.8)(.8) = .512 = \pi^0(1-\pi)^3$
1,0,0	$(.2)(.8)(.8) = .128 = \pi^1(1-\pi)^2$
0,1,0	$(.8)(.2)(.8) = .128 = \pi^1(1-\pi)^2$
0,0,1	$(.8)(.8)(.2) = .128 = \pi^1(1-\pi)^2$
1,1,0	$(.2)(.2)(.8) = .032 = \pi^2(1-\pi)^1$
1,0,1	$(.2)(.8)(.2) = .032 = \pi^2(1-\pi)^1$
0,1,1	$(.8)(.2)(.2) = .032 = \pi^2(1-\pi)^1$
1,1,1	$(.2)(.2)(.2) = .008 = \pi^3(1-\pi)^0$
	1.000

Table 5–18 is useful only insofar as we can derive the probability distribution of r from it. This distribution is given in Table 5–19.

TABLE 5–19

$r = X_1 + X_2 + X_3$	$P(r)$
0	$P(0,0,0) = .512$
1	$P(1,0,0) + P(0,1,0) + P(0,0,1) = .384$
2	$P(1,1,0) + P(1,0,1) + P(0,1,1) = .096$
3	$P(1,1,1) = .008$
	1.000

The parameters are

$$E(r) = (0)(.512) + (1)(.384) + (2)(.096) + (3)(.008) = .6$$

and

$$V(r) = (0 - .6)^2(.512) + (1 - .6)^2(.384) + (2 - .6)^2(.096) + (3 - .6)^2(.008) = .48$$

To interpret Table 5–19, remember that r is the result of three Bernoulli trials; for example, $P(r = 1)$ is the probability of choosing one "for" in three draws. In 38.4 percent of a "great many" 3-trial series, we would find only one success.

The random variable r is called a *binomial variable,* while the distribution in Table 5–19 is referred to as a *binomial distribution.* A histogram of that distribution is presented in Figure 5–6.

Because π is constant, the individual probability distributions of X_1, X_2, and X_3 are *identical* (see Table 5–20). We note that

$$E(r) = E(X_1 + X_2 + X_3)$$

$$= E(X_1) + E(X_2) + E(X_3) \quad \text{By virtue of Equation 5–12}$$

$$= \pi + \pi + \pi = n\pi \quad (5\text{–}22)$$

FIGURE 5–6

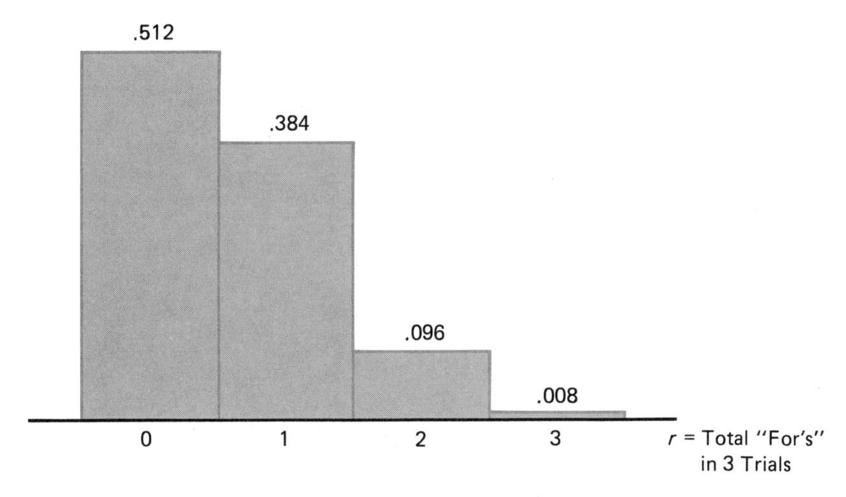

TABLE 5–20

X_1	$P(X_1)$	X_2	$P(X_2)$	X_3	$P(X_3)$
1	π	1	π	1	π
0	$1 - \pi$	0	$1 - \pi$	0	$1 - \pi$

Thus, in our example, $E(r) = n\pi = (3)(.2) = .6$.

Furthermore, as long as the outcomes are independent,

$$\begin{aligned} V(r) &= V(X_1 + X_2 + X_3) \\ &= V(X_1) + V(X_2) + V(X_3) \qquad \text{By virtue of Equation 5–18} \\ &= \pi(1 - \pi) + \pi(1 - \pi) + \pi(1 - \pi) \\ &= n\pi(1 - \pi) \end{aligned} \tag{5–23}$$

Thus, $V(r) = n\pi(1 - \pi) = (3)(.2)(.8) = .48$.

The binomial distribution is an example of a probability distribution that can be represented by a mathematical function. Such a *probability mass function,* or *discrete probability function,* pairs each value of a random variable with its probability. The *binomial probability function* is

$$P(r) = \frac{n!}{r!(n - r)!}\pi^r(1 - \pi)^{n - r} \tag{5–24}$$

The computations in Table 5–21 confirm that Equation 5–24 produces the probability distribution of r faster than our initial approach.

TABLE 5–21

r	$P(r)$
0	$\frac{3!}{0!3!}(.2)^0(.8)^3 = .512$
1	$\frac{3!}{1!2!}(.2)^1(.8)^2 = .384$
2	$\frac{3!}{2!1!}(.2)^2(.8)^1 = .096$
3	$\frac{3!}{3!0!}(.2)^3(.8)^0 = .008$
	1.000

In general,

$$P(r) = \begin{pmatrix}\text{number of}\\ \text{sequences}\\ \text{containing}\\ r\text{ successes}\end{pmatrix}\begin{pmatrix}\text{probability}\\ \text{of a specific}\\ \text{sequence of}\\ r\text{ successes}\end{pmatrix}$$

For example,

$$P(r = 2) = (3)[\pi^r(1 - \pi)^{n - r}] = (3)[(.2)^2(.8)^1]$$
$$= (3)(.032) = .096$$

The first quantity in parentheses is the same as the number of *permutations* of n things picked n at a time, where r things are alike and $n - r$ things are alike. Consider the three *different* letters A, B, C. If we choose these letters three at a time, we can form $n! = 3! = 6$ permutations. By contrast, there are only three permutations of A, A, C:

$A\ A\ C$
$C\ A\ A$
$A\ C\ A$

Since we have two As and one C, $r = 2$ and $(n - r) = 1$. Under these circumstances, the permutation formula is changed to

$$\frac{n!}{r!(n - r)!}$$

Therefore,

$$P(r = 2) = \frac{3!}{2!1!}[(.2)^2(.8)^1]$$
$$= \frac{n!}{r!(n - r)!}\pi^r(1 - \pi)^{n - r}$$

Whenever we perform a series of n independent Bernoulli trials, and $P(\text{success})$ is constant, we will know three things:

FIGURE 5–7

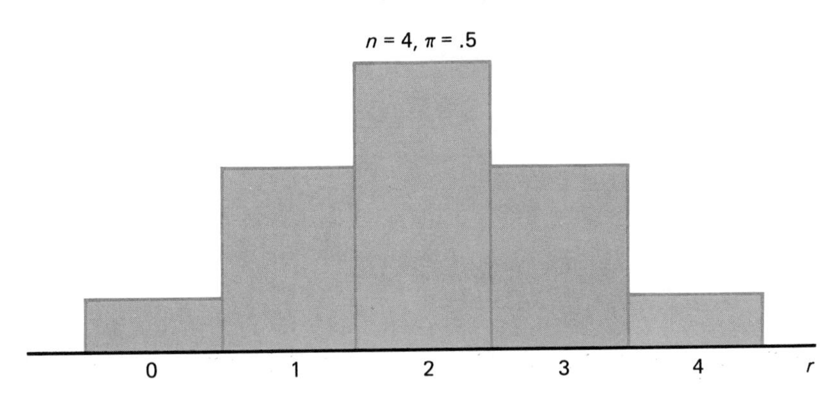

FIGURE 5–8

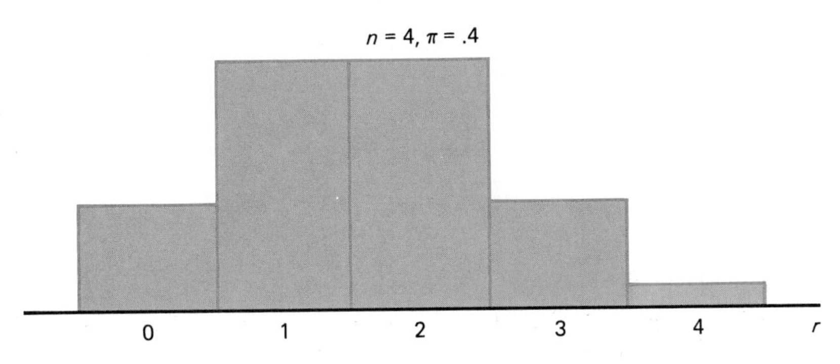

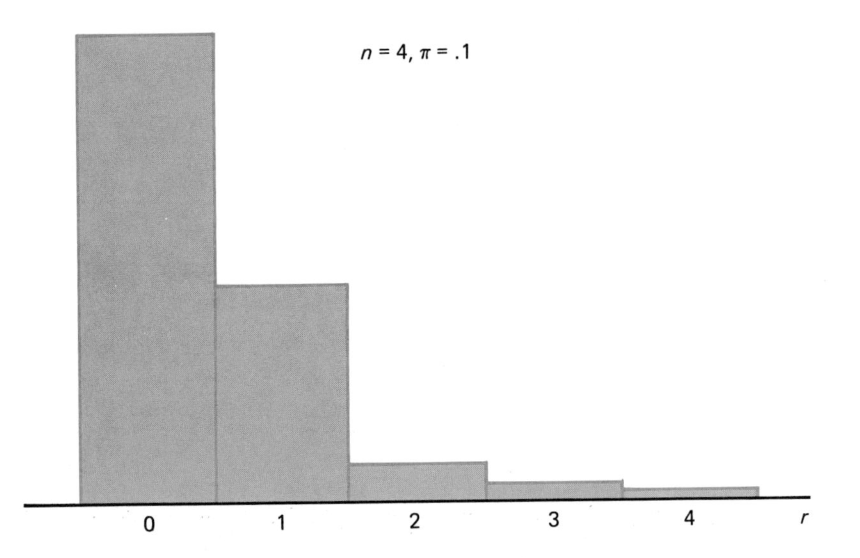

1. The probability of any number of successes is represented by Equation 5–24.
2. The expected value of r is equal to $n\pi$.
3. The variance of r is equal to $n\pi(1 - \pi)$.

Returning to "six tosses of a well-balanced coin," we can now compute

$$P(r = 2 \text{ heads}) = \frac{6!}{2!(6-2)!}(.50)^2(1 - .50)^4 = .234$$

$$E(r) = (6)(.5) = 3.0$$

$$V(r) = (6)(.5)(.5) = 1.5$$

In Chapters Six and Eleven, we will explain how the *shape* of the binomial distribution frequently justifies approximating the binomial with a symmetrical *normal curve*. Let's identify the characteristics of the binomial distribution:

1. The binomial is symmetrical for any n if $\pi = .5$ (see Figure 5–7).
2. The binomial becomes more skewed as π moves farther from .5 and n is held constant (see Figure 5–8).
3. The binomial becomes less skewed as n increases and $\pi \neq .5$ is held constant (see Figure 5–9).

FIGURE 5–9

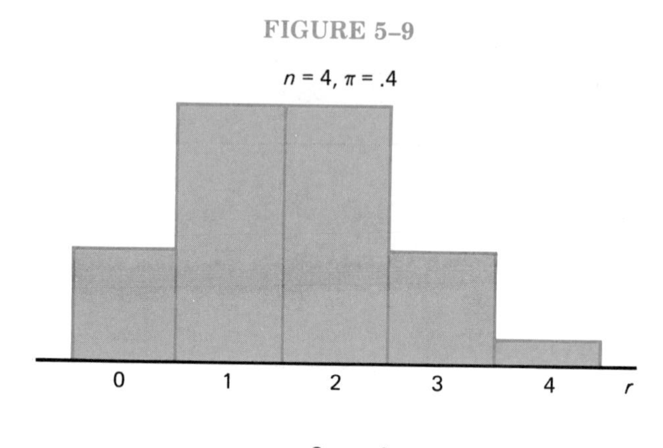

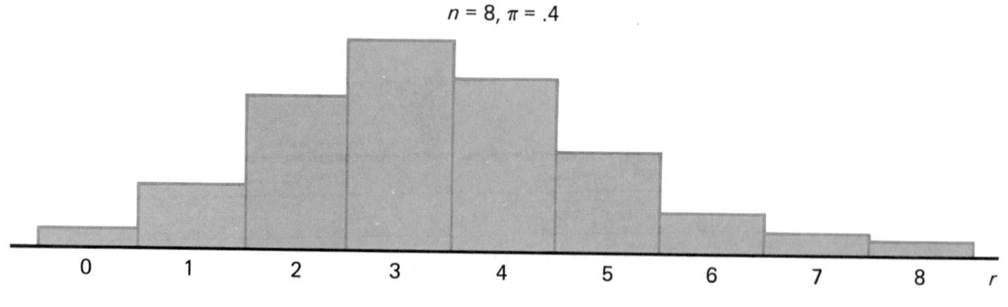

SUMMARY

Random variables and their probability distributions are important to the development of statistical inference theory. The properties of random variables substantiate many of the statements that we are going to make in later chapters. In fact, we have already used these properties to determine the parameters of (1) variables restricted to the values 1 and 0 and (2) binomial variables. The binomial distribution has a variety of applications. Given the three conditions that we mentioned (i.e., Bernoulli trials, constant π, and independence), the binomial probability function will yield the probability of r successes in n trials. In Chapter Six, we will investigate continuous, rather than discrete, random variables.

EXERCISES

1. Compute $E(X)$ and $V(X)$ if the probability distribution of X is

X	$P(X)$
0	.2
1	.3
2	.2
3	.2
4	.1

2. Compute $E(Y)$ and $V(Y)$ if the probability distribution of Y is

Y	$P(Y)$
3	.3
4	.1
5	.4
6	.2

3. Suppose $U = 5 + 2X$. Use the probability distribution in question 1 to determine $E(U)$ and $V(U)$.

4. Suppose $U = 3Y$. Use the probability distribution in question 2 to determine $E(U)$ and $V(U)$.

5. Show that $V(X) = E[X - E(X)]^2$ can be written as $V(X) = E(X^2) - [E(X)]^2$.

6. Express $V(X) = E(X^2) - [E(X)]^2$ in terms of summations.

7. Use the answer to question 6 to compute the variance of the probability distribution in question 1.

8. The joint probability distribution of X and Y is

		X	
		1	2
Y	2	.2	.2
	3	.2	.4

Calculate the following:

a. $E(X)$
b. $V(X)$
c. $E(Y)$
d. $V(Y)$
e. $\mathrm{Cov}(X,Y)$

9. Suppose $U = 2X + 3Y$. Calculate $E(U)$ and $V(U)$ if the joint probability distribution of X and Y is

		X	
		3	4
Y	6	.1	.3
	7	.1	.5

10. Suppose $U = 4X - 2Y$. Use the joint probability distribution in question 9 to compute $E(U)$ and $V(U)$.

11. The joint probability distribution of X and Y is

		X			
		0	1	2	3
Y	0	.01	.05	.10	.06
	1	.11	.03	.05	.20
	2	.04	.13	.16	.06

a. Construct the conditional probability distribution of $X|Y = 1$.
b. Construct the conditional probability distribution of $Y|X = 2$.

12. Complete the following table under the assumption that X and Y are independent.

		X: 0	X: 1	X: 2	*Marginal Probability of Y*
Y	0	.10	–	–	–
	1	.05	–	–	–
	2	.25	–	–	–
Marginal Probability of X		–	.30	.30	

13. The sample space for the experiment "select an object (i.e., A, B, C, etc.) at random" appears as

· $A(X = 1, Y = 1)$	· $B(X = 2, Y = 2)$
· $C(X = 3, Y = 1)$	· $D(X = 3, Y = 2)$
· $E(X = 1, Y = 1)$	· $F(X = 3, Y = 1)$
· $G(X = 2, Y = 1)$	· $H(X = 1, Y = 2)$
· $I(X = 1, Y = 2)$	· $J(X = 3, Y = 2)$

a. Determine the joint probability distribution of X and Y.
b. Compute Cov(X,Y).

14. Given

$$X_1 = 2 \qquad Y_1 = 3$$
$$X_2 = 3 \qquad Y_2 = 4$$
$$X_3 = 4 \qquad Y_3 = 5$$
$$Y_4 = 6$$

calculate the following summations:

a. $\sum_{j=1}^{4} \sum_{i=1}^{3} X_i Y_j$

b. $\sum_{j=3}^{4} \sum_{i=2}^{3} X_i Y_j$

c. $\sum_{j=1}^{3} \sum_{i=1}^{2} (X_i + Y_j)$

15. The riskiness of an investment is often measured by the standard deviation of the probability distribution of returns. Compute the expected value and the standard deviation of returns on Project A if the probability distribution is

X Return on A	$P(X)$
\$ 50	.1
100	.3
150	.3
200	.2
250	.1

16. Compute the expected value and the standard deviation of returns on Project B if the probability distribution is

Y Return on B	$P(Y)$
\$100	.6
200	.2
300	.1
400	.1

17. Mr. Evans invests G dollars in stock C and stock D. Suppose

X = return on C
Y = return on D
a = the proportion of G invested in C
b = the proportion of G invested in D

The expected value and variance of returns for such a portfolio of stocks is $E(aX + bY)$ and $V(aX + bY)$. Calculate these parameters if $a = .6$, $b = .4$, and the joint probability distribution of X and Y is

		X	
		\$50	\$100
Y	\$100	.1	.4
	\$200	.3	.2

18. Janet is going to toss a well-balanced coin seven times.

a. Determine the probability distribution of heads.
b. Compute the expected value and variance of this distribution.
c. Construct a histogram of the probability distribution.

19. Bob places four red balls and six purple balls into a hat. He will select three balls with replacement.

a. Determine the probability distribution of purple balls.
b. Compute the expected value and variance of this distribution.
c. Construct a histogram of the probability distribution.

20. For each of the following situations, assume that the conditions (e.g., independent trials) associated with the binomial distribution are fulfilled.

20–1. The U.S. Olympic Team is financed through voluntary contributions rather than by subsidies from the federal government. Suppose that 30 percent of all Americans would be willing to have $1 of their taxes allocated to a special Olympics Fund. If six Americans are picked,

a. What is the probability that three will be for this proposal?
b. What is the probability that at least two will be against it?

20–2. A variety of congressional bills have attempted to reduce the size of the oil companies. Suppose that 60 percent of all Americans are in favor of reorganizing the oil industry. If ten Americans are selected,

a. What is the probability that no more than three will be in favor of reorganization?
b. What is the probability that five will be in favor of reorganization?

20–3. The largest employer in a particular community is Company E. Because of current EPA standards, E will have to pay a substantial fine if it doesn't reduce its pollution. Since the company has been losing money for the last several years, E's management may decide to close the factory permanently instead of buying expensive antipollution devices. Suppose that 60 percent of the people in the community would like the EPA to exempt Company E from the pollution rules. If twelve people are picked,

a. What is the probability that eight will be in favor of an exemption?
b. What is the probability that between four and nine will be in favor of an exemption?

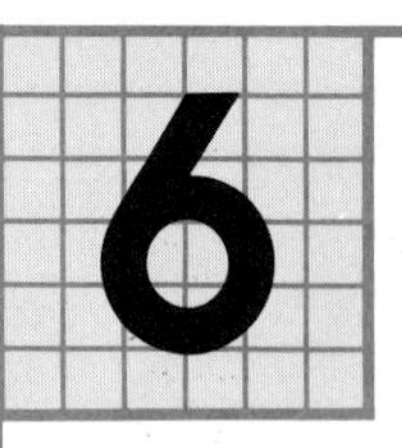

6 The Normal Distribution

CONTINUOUS RANDOM VARIABLES

Many of the inference procedures that we will discuss are connected with the *normal distribution.* Before we can analyze this particular probability distribution, however, we must first clarify the distinction between a *discrete* and a *continuous* random variable.

Based on Figure 6–1, let's define an experiment as "spin a dial." If pegs are attached to the wheel, the dial will stop only at position 0, 1, 2, or 3. (The relevance of labeling the uppermost spot as both 0 and 4 will be evident once the pegs are eliminated.) None of the four possible positions has a special influence over the dial. Because of this, each position is equally likely, and the probability associated with any outcome in the sample space of Figure 6–2 is $1/M = 1/4$.

FIGURE 6–1

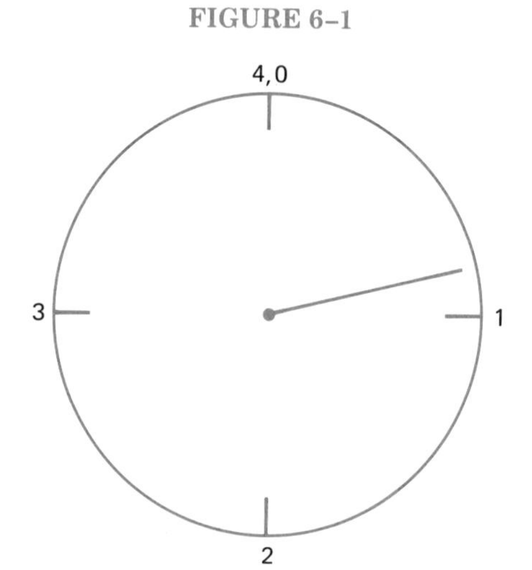

FIGURE 6–2

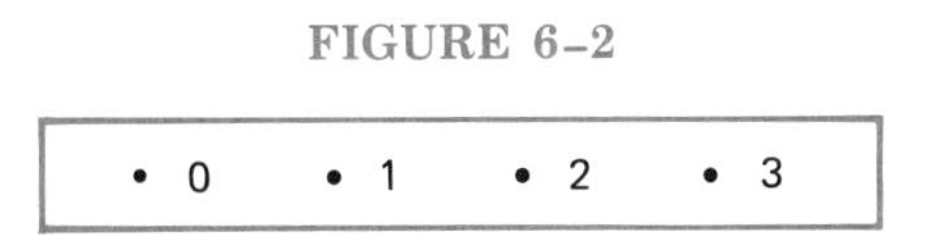

The probability distribution of the random variable X = resting place of the dial is detailed in Table 6–1. We can represent such a distribution by a histogram (see Figure 6–3). Based on Chapter Five, we should recognize that X is a discrete random variable. Its values are restricted to 0, 1, 2, and 3.

TABLE 6–1

X	$P(X)$
0	1/4
1	1/4
2	1/4
3	1/4

FIGURE 6–3

Suppose that we now change the nature of the experiment by removing the pegs, as in Figure 6–4. Under these circumstances, the dial can stop *anywhere* on the wheel. Positions 0, 1, 2, and 3 are still feasible, but 0.001847, 2.1335905, 3.20850, and so on are potential outcomes as well. Every location is again equally likely. The random variable X, however, has an infinite number of values. The probability of a *specific* value is $1/M = 1/\infty = 0$. Since an infinite number of outcomes lie between any two values of X, say $X = a$ and $X = b$, we refer to X as a *continuous* random variable.

Notice that $P(X) = 0$ is one implication of the latter experiment. If, between any two points on the wheel, there are an infinite number of resting places, we would hardly have any confidence that the dial would stop *exactly* at 2.300000003065874. On the other hand,

FIGURE 6–4

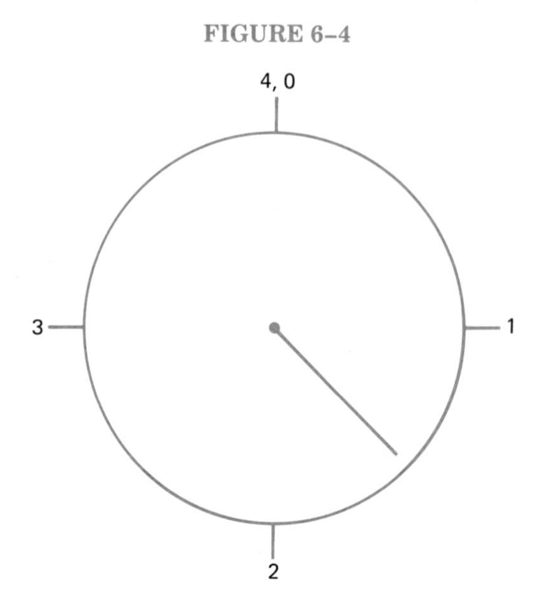

we could be somewhat confident that the dial would stop *between* 2.0 and 2.5 (see Figure 6–5).

Inasmuch as 1/8 [i.e., (2.5 − 2.0)/4] of all of the infinite outcomes fall in the interval 2.0 to 2.5, the probability of the event $2.0 \leq X \leq 2.5$ is 1/8. Although the probability of some *individual* value of a continuous random variable is zero, we see that non-zero probabilities can be computed for *intervals* of values.

We have previously mentioned that *probability mass functions*

FIGURE 6–5

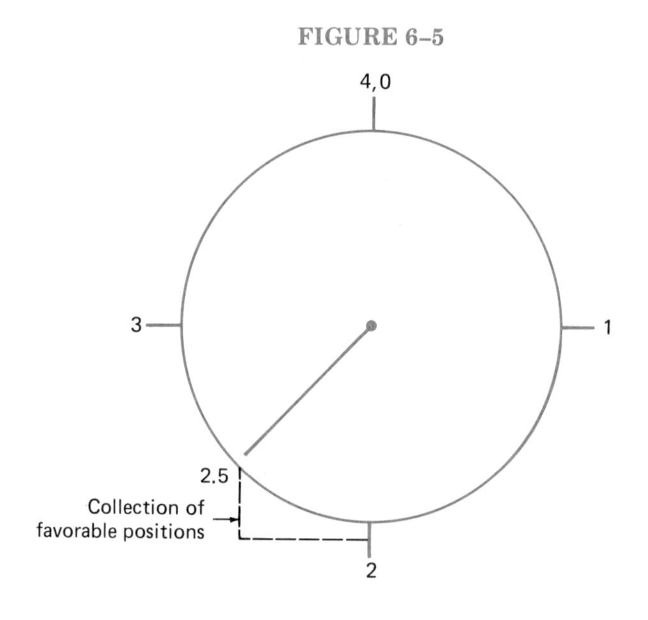

can be used to describe discrete probability distributions. These functions relate every value of the random variable X to $P(X)$. The sum of all the probabilities is 1.0. For the first experiment (i.e., with pegs), we can write

$$Y = P(X) = 1/4 \qquad \text{for } X = 0,1,2,3$$

On the other hand, continuous probability distributions are described by *probability density functions.* Given a density function, we can identify the probability that X will lie in the interval $X = a$ to $X = b$, that is, $P(a \leq X \leq b)$. The function that will satisfy this condition for the second experiment is

$$Y = 1/4 \qquad \text{for each of the infinite values of } X \text{ between 0 and 4}$$

From this last function, we can construct a curve. Some of the points on the curve are (1, 1/4), (2.1876, 1/4), and (3.54, 1/4). Notice that Y, or the *probability density,* is the height of the curve. Probability density is *not* similar to probability. The graph of the second function looks like Figure 6–6.

FIGURE 6–6

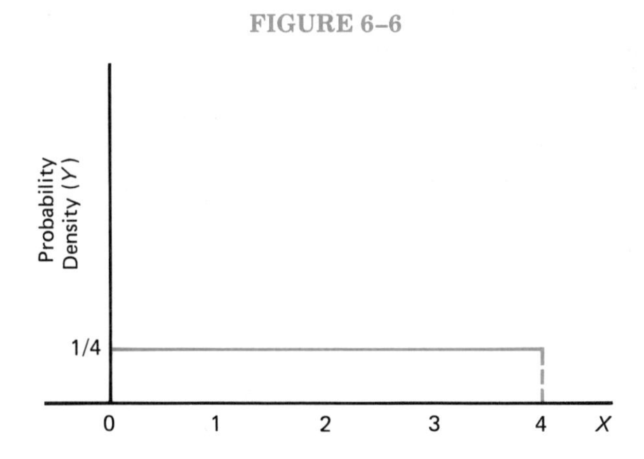

With continuous random variables, $P(a \leq X \leq b)$ is identical to the *area under the curve* between a and b. Hence, the probability that the dial will stop between 1.4 and 2.7 corresponds to the area enclosed between 1.4 and 2.7 (see Figure 6–7). This area is the area of a rectangle having a height of 1/4 and a length of $2.7 - 1.4 = 1.3$. As a result,

$$\begin{aligned} P(1.4 \leq X \leq 2.7) &= (\text{height})(\text{length}) \\ &= (1/4)(1.3) = .325 \end{aligned}$$

A probability of .325 is consistent with the fact that 1.3/4, or .325, of all the infinite points occur between 1.4 and 2.7 (see Figure 6–8).

FIGURE 6–7

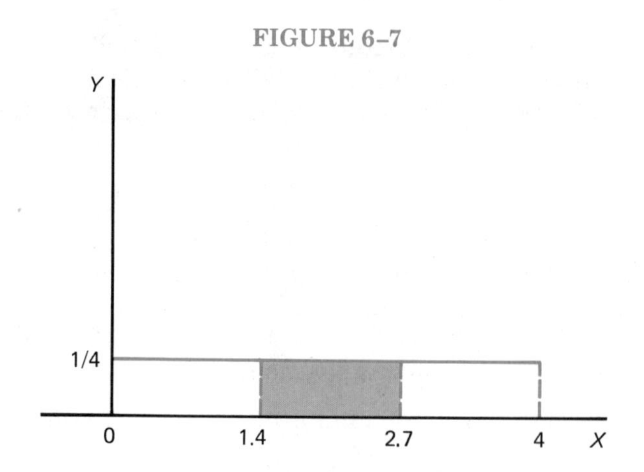

We know that the dial is certain to stop at some location on the wheel. Therefore, $P(0 \leq X \leq 4) = 1.0$. If the density function is to generate this probability, the entire area under the rectangle in Figure 6–9 must be 1.0. The total area, in fact, is $(1/4)(4) = 1.0$. Generalizing, we can note that a probability density function (1) yields a curve which encompasses an area of 1.0 and (2) establishes a correspondence between $P(a \leq X \leq b)$ and the area under the curve bordered by $X = a$ and $X = b$.

As with discrete distributions, continuous distributions have both expected values and standard deviations. These parameters are calculated differently in the continuous case, but their inter-

FIGURE 6–8

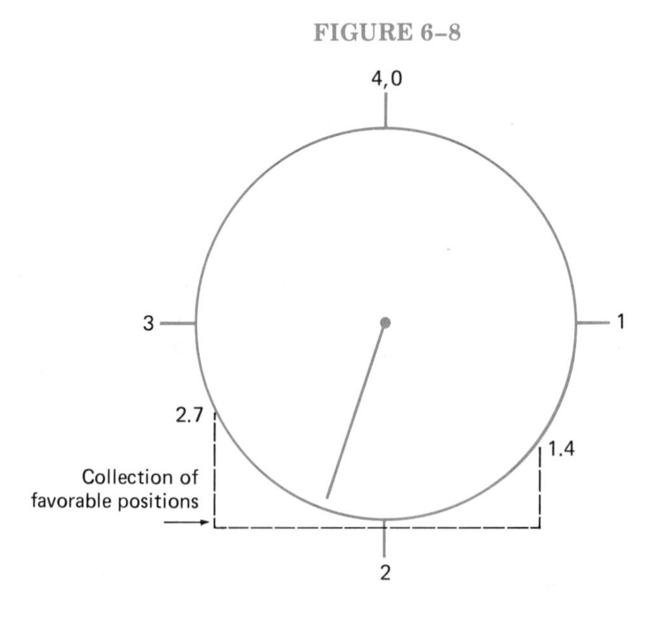

FIGURE 6–9

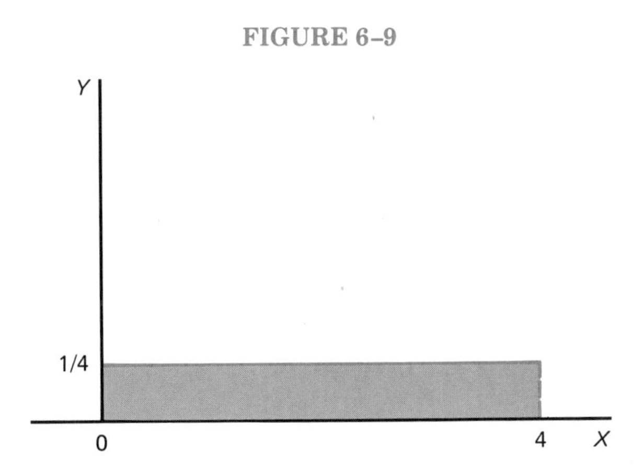

pretation remains the same; that is, μ is the "typical" outcome in the "long run" and σ measures the "spread" of the outcomes.

In reality, continuous random variables are measured as though they were discrete. To reinforce this fact, imagine that we have a ruler that is divided into ¼″ segments. Although an infinite number of measurements lie between 1″ and 2″, we can record only the discrete outcomes 1″, 1¼″, 1½″, 1¾″, and 2″. Yet, within the context of statistical research, probability histograms belonging to random variables that are naturally discrete or that become discrete because of measurement problems are often approximated by density curves. In many instances, the probabilities computed from the continuous distributions are quite accurate. We will show, moreover, that the continuous normal distribution is even the theoretically proper distribution under certain circumstances.

THE NORMAL DISTRIBUTION

As noted in Chapter Five, the probability distribution of X when the experiment is to randomly select one item from a population is identical to the relative frequency distribution of that population. We often would like to make probability statements about some distribution, but either (1) we don't have enough details or (2) the distribution is extremely complex. Under these circumstances, one alternative is to *approximate*. The *normal distribution* can be a good "stand-in" for roughly symmetrical and unimodal distributions.

Characteristics of the Normal Distribution

The density curve that represents a normal distribution is "bell-shaped," as in Figure 6–10. We should emphasize that "normal"

FIGURE 6–10

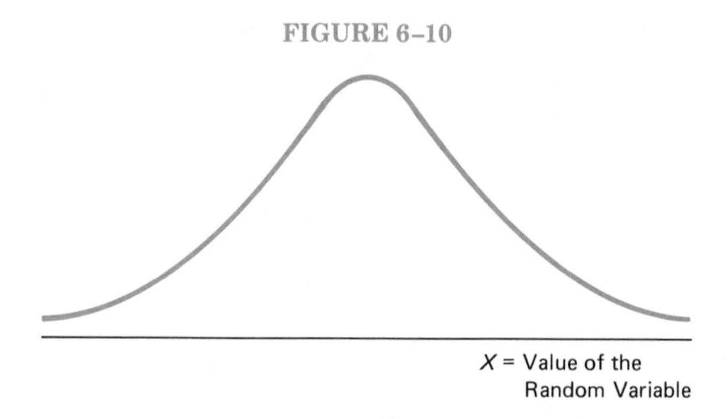

is just a name. None of its familiar connotations (e.g., usual) is implied. Thus we can't claim that a nonnormal distribution is "abnormal."

The following relationship is the density function associated with the normal distribution:

$$Y = \frac{1}{\sigma\sqrt{2\pi}} e^{-(1/2)[(X - \mu)/\sigma]^2} \qquad (6\text{–}1)$$

for any value of X between $-\infty$ and ∞

As before, Y is the probability density, μ and σ are parameters, and X is a value of the random variable. Both e and π are constants, where $e \approx 2.72$ and $\pi \approx 3.14$. Consequently, π no longer indicates what it did in Chapter Five.

Given a normal distribution with a mean of μ and a standard deviation of σ, Equation 6–1 would pair one value of Y with a particular value of X. Two X,Y pairs are identified in Figure 6–11. Only those bell-shaped curves that derive from Equation 6–1 belong to normally distributed random variables.

The highest point on the normal curve in Figure 6–12 is directly above $X = \mu$. Notice that the curve is symmetrical. Moreover, while

FIGURE 6–11

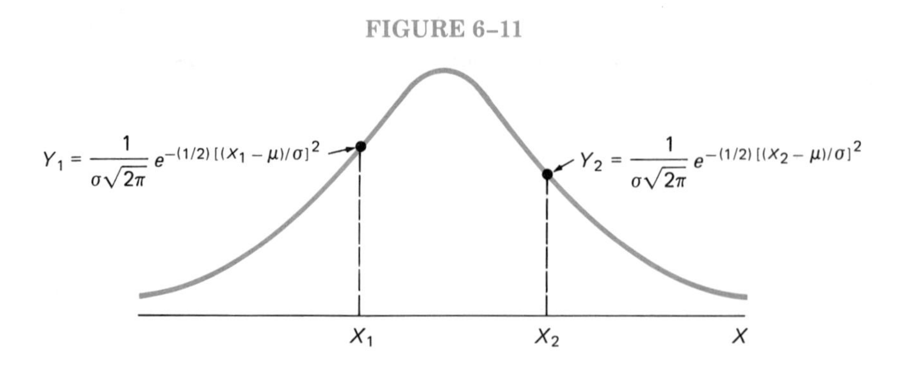

FIGURE 6–12

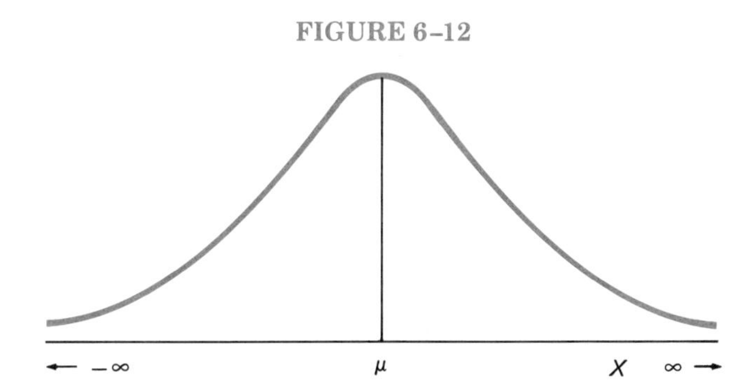

the value of Y progressively falls as we move farther to the right or to the left of μ, there is *no* value of X at which the curve intersects the horizontal axis (i.e., the normal curve approaches the X axis *asymptotically*).

To substantiate these properties, let's write X as $\mu + k\sigma$ or $\mu - k\sigma$, where the positive number k establishes how many standard deviations X exceeds or is short of μ. If $\mu = 10$ and $\sigma = 5$, for example, $X = 20$ is the same as $\mu + 2\sigma$ and $X = 3$ is $\mu - 1.4\sigma$.

Since $e^{-a} = 1/e^a$, Equation 6–1 is the same as

$$Y = \left[\frac{1}{\sigma\sqrt{2\pi}}\right]\left[\frac{1}{e^{(1/2)[(X - \mu)/\sigma]^2}}\right] \qquad (6\text{–}2)$$

Because σ and π are positive constants, the first bracketed quantity is a positive constant. In the second bracketed quantity, e, μ, and σ are constants. Hence, any change in Y is attributable solely to a change in X.

When $X = \mu + 0\sigma = \mu$,

$$Y = \left[\frac{1}{\sigma\sqrt{2\pi}}\right]\left[\frac{1}{e^{(1/2)[(\mu - \mu)/\sigma]^2}}\right]$$

$$= \left[\frac{1}{\sigma\sqrt{2\pi}}\right]\left[\frac{1}{e^{(1/2)(0/\sigma)^2}}\right]$$

$$= \frac{1}{\sigma\sqrt{2\pi}} \qquad (6\text{–}3)$$

inasmuch as $e^{(1/2)(0)^2} = e^0 = 1$.

By contrast, when X is a value greater than μ (i.e., $X = \mu + k\sigma$),

$$Y = \left[\frac{1}{\sigma\sqrt{2\pi}}\right]\left[\frac{1}{e^{(1/2)[(\mu + k\sigma - \mu)/\sigma]^2}}\right]$$

$$= \frac{1}{(\sigma\sqrt{2\pi})[e^{(1/2)k^2}]} \qquad (6\text{–}4)$$

If X is a value less than μ (i.e., $X = \mu - k\sigma$),

$$Y = \left[\frac{1}{\sigma\sqrt{2\pi}}\right]\left[\frac{1}{e^{(1/2)[(\mu - k\sigma - \mu)/\sigma]^2}}\right]$$

$$= \frac{1}{(\sigma\sqrt{2\pi})[e^{(1/2)k^2}]} \tag{6–5}$$

Due to the fact that

$$\frac{1}{\sigma\sqrt{2\pi}} > \frac{1}{(\sigma\sqrt{2\pi})[e^{(1/2)k^2}]}$$

for any value of k other than zero, we can conclude that Y reaches its maximum value at $k = 0$ or $X = \mu$.

Also, as X moves farther away from μ in either direction, k gets larger and the value of Y in Equations 6–4 and 6–5 gets smaller. Y, therefore, declines beyond $X = \mu$. Since the denominator of Equations 6–4 and 6–5 is always positive, however, Y itself is always positive. This condition prevents Y from ever coming into contact with the horizontal axis. For such an occurrence, Y would have to be zero at some value of X.

Because of symmetry, values of X *equidistant* from μ must be paired with the same Ys. Imagine that two of these values are $X_1 = \mu + k_1\sigma$ and $X_2 = \mu - k_1\sigma$. After substituting X_1 and X_2 into Equation 6–2, we find that

$$Y_1 = Y_2 = \frac{1}{(\sigma\sqrt{2\pi})[e^{(1/2)k_1^2}]}$$

As a result of Equation 6 1, normal curves become more "pinched in" around μ when σ decreases. Consider two distributions, one with parameters μ_1, σ_1 and the other with parameters μ_2, σ_2. Figure 6–13 might illustrate their relationship if $\sigma_1 > \sigma_2$ but $\mu_1 = \mu_2$. If the standard deviations are the same but $\mu_1 < \mu_2$, we might have Figure 6–14.

Since an infinite number of μ, σ combinations are possible, infinitely many normal curves exist. The basis for each curve is a density function. Because the total area under a density curve is always 1.0, any curve which derives from Equation 6–1 must enclose an area of 1.0 (see Figure 6–15).

We will be interested in some interval of Xs. The expression $P(X_1 \leq X \leq X_2)$ indicates the probability that a random variable X will have a value of X_1 *or* X_2 *or* anything in between, as Figure 6–16 illustrates. In other words, $P(X_1 \leq X \leq X_2) = P(X_1) + P(X_2)$

FIGURE 6–13

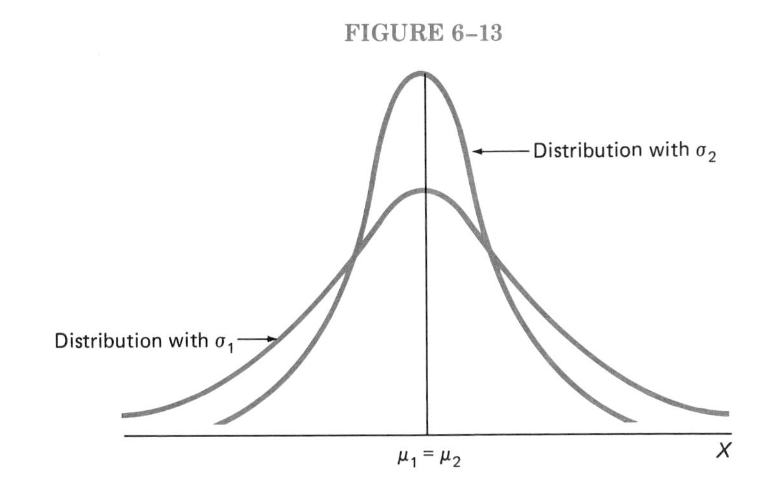

FIGURE 6–14

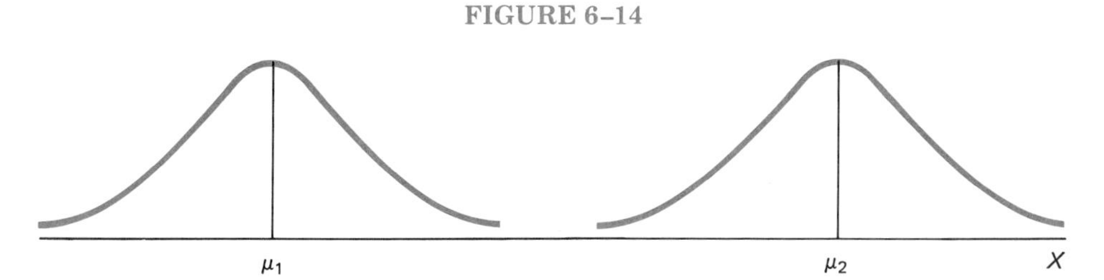

FIGURE 6–15

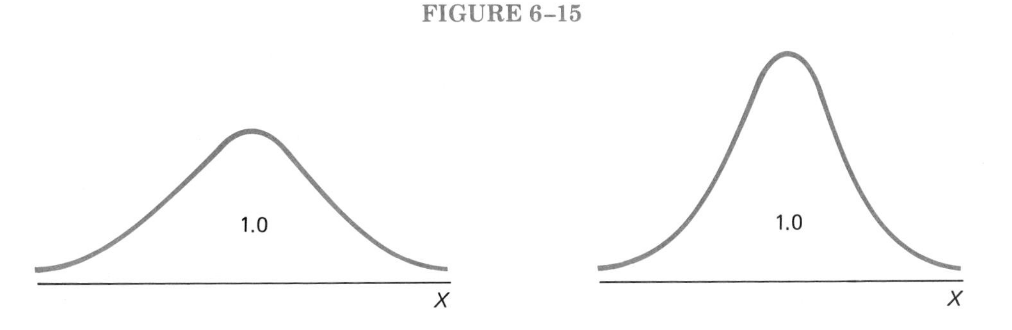

$+ \ P(X_1 < X < X_2)$. When X is continuous, $P(X_1) = 0$ and $P(X_2) = 0$. We can, therefore, write $P(X_1 \leq X \leq X_2)$ as $P(X_1 < X < X_2)$. We will adopt the latter notation. Because the event $3 \leq X \leq 5$ corresponds to the event $3 < X < 5$, a question such as "What is the probability that X is equal to 3 or 5 or something in between?" is identical to "What is the probability that X falls between 3 and 5?"

To judge the adequacy of using the normal distribution as an approximation for a fairly symmetrical and unimodal distribution,

FIGURE 6–16

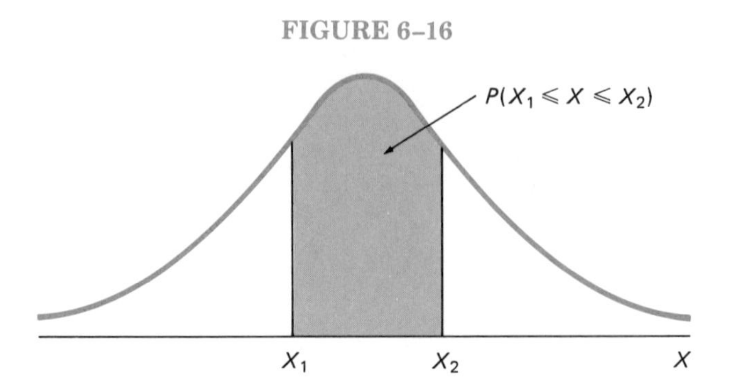

we'll return to the TV-viewing example of Chapter Three. Figure 6–17 is the relative frequency, or probability, histogram of that population.

FIGURE 6–17

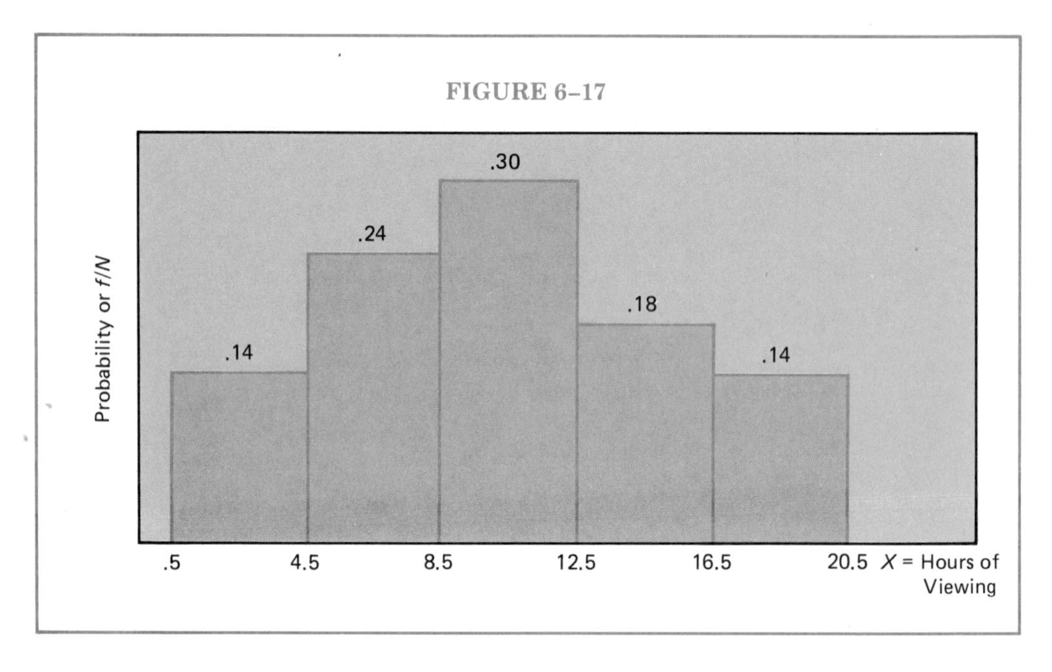

The parameters of the TV-viewing population were $\mu = 10.2$ hours and $\sigma = 4.8$ hours. After these values are entered into Equation 6–1, we can create Figure 6–18 by varying X. (For future reference, the height of the curve at an X one standard deviation greater than μ is the point at which the curve changes from ⌒ to ◡.)

Remember that the probability axis in Figure 6–17 differs from the probability density axis in Figure 6–18. Given the histogram,

FIGURE 6–18

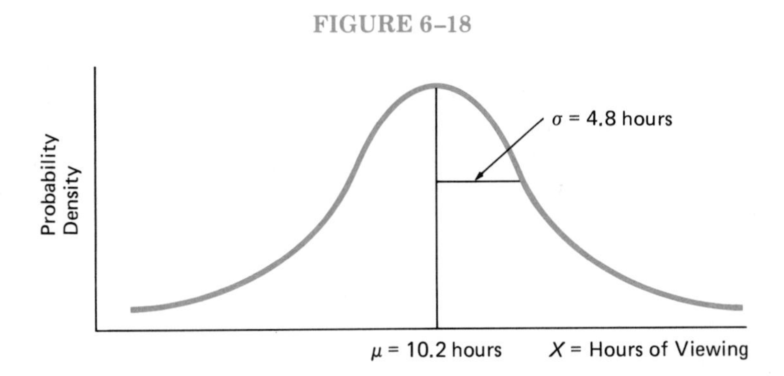

we find that $P(8.5 \leq X \leq 12.5) = .30$. We now want to determine $P(8.5 < X < 12.5)$ using a normal curve. To do so, we must calculate the shaded area in Figure 6–19.

FIGURE 6–19

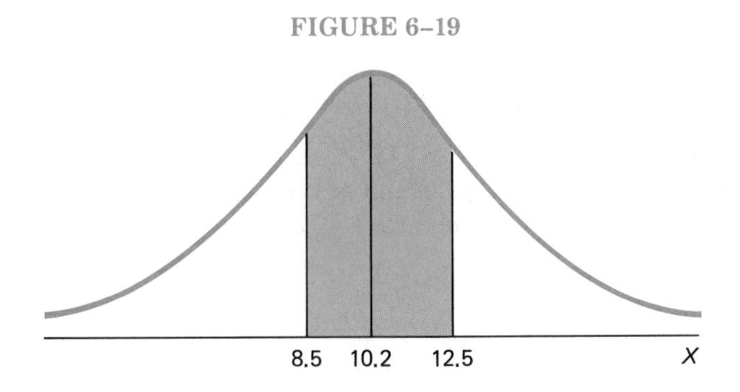

Unlike the rectangular areas of the dial experiment (Figure 6–7), the area between 8.5 and 12.5 (see Figure 6–20) can't be determined according to a simple formula. In fact, no simple formulas

FIGURE 6–20

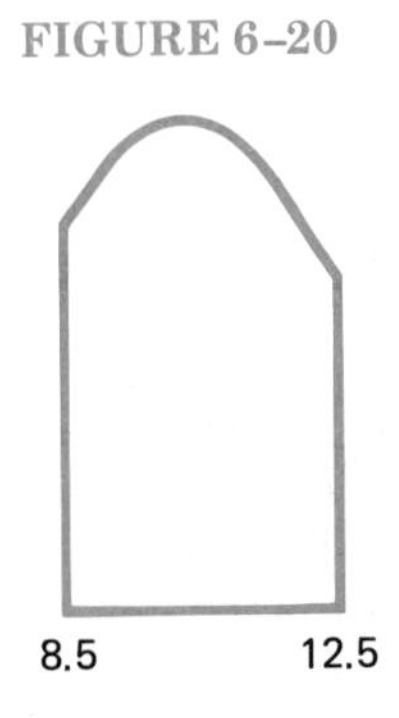

are ever applicable to normal-curve intervals. Although *integral calculus* provides a solution procedure, we'll develop an alternative.

With integral calculus, we confront two problems. Not only is this technique difficult and tedious, but $P(8.5 < X < 12.5)$ when $\mu = 10.2$ and $\sigma = 4.8$ will change when μ and σ are other values. This is apparent in Figure 6–21. The calculus procedure, then, must be re-performed each time there is a new problem.

FIGURE 6–21

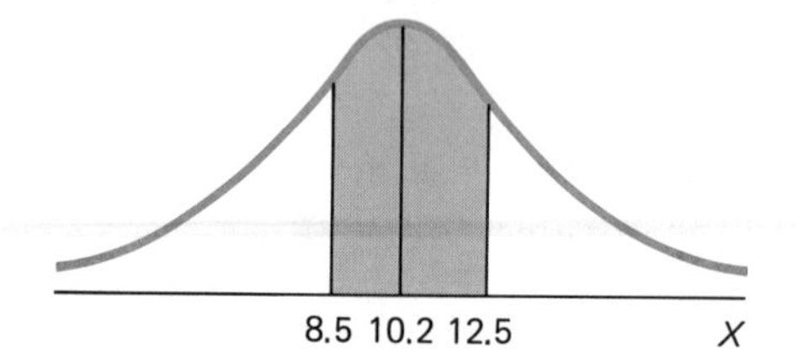

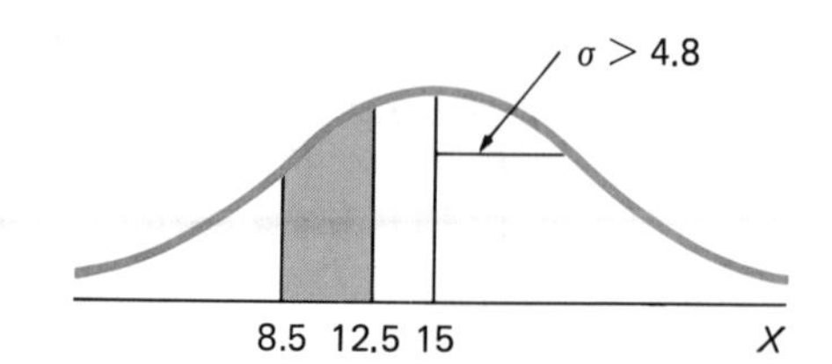

The Standard Normal Curve

We will compute $P(8.5 < X < 12.5)$ or, more generally, $P(X_1 < X < X_2)$ by first converting a normal curve into the *standard normal curve*. To get the standard normal curve, we transform each X into

$$Z = \frac{X - \mu}{\sigma} \qquad (6\text{–}6)$$

The Z value simply translates every X into its *distance from* μ, where distance is measured in standard deviations. Thus, if $X = 8$, $\mu = 6$, and $\sigma = 2$,

$$Z = \frac{8 - 6}{2} = 1$$

This suggests that 8 is *one* standard deviation *larger* than μ. If $X = 4$,

$$Z = \frac{4 - 6}{2} = -1$$

$X = 4$ is, then, *one* standard deviation *smaller* than μ.

When X is normally distributed, the consequence of these conversions is a normal distribution of Zs. Such a distribution has a mean of 0 and a standard deviation of 1.0. Figure 6–22 shows the relationship between the standard normal distribution and two normal distributions of X.

The characteristics of the standard normal curve (e.g., $\sigma_Z = 1$)

FIGURE 6–22

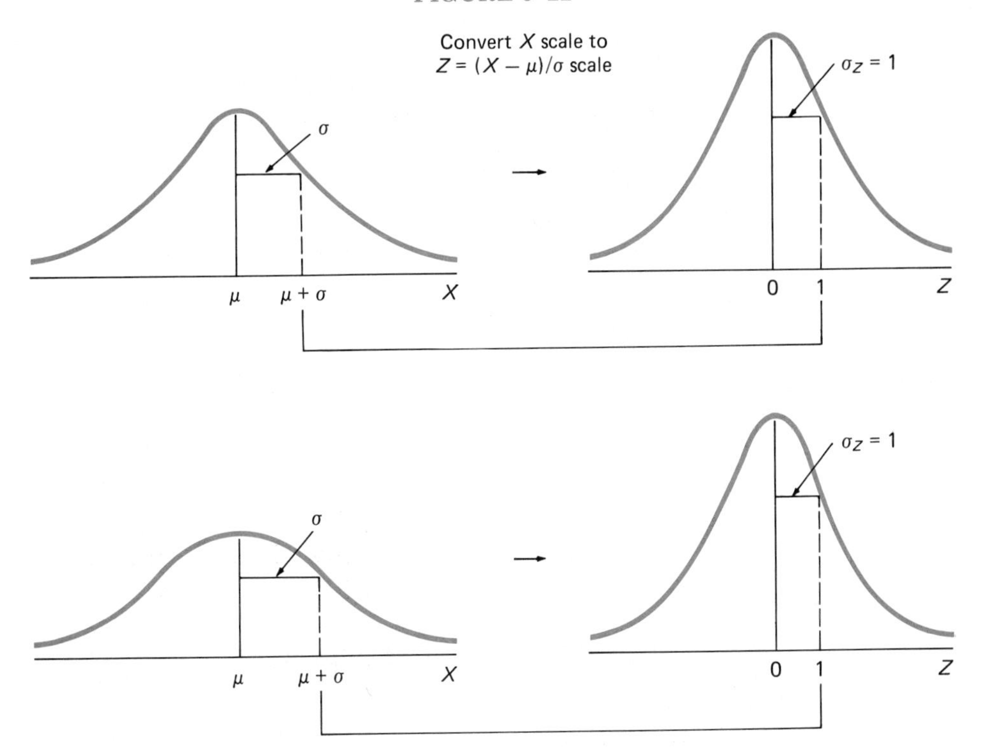

follow from the behavior of random variables. As part of our proof of them, we state the following theorem:

> If X is normally distributed, a *linear function* of X (e.g., $Z = a + bX$) will be normally distributed. Moreover, if X and Y are two independent random variables each of which is normally distributed, a *linear combination* of X and Y will be normally distributed.

Notice that

$$Z = \frac{X - \mu}{\sigma}$$

can be written as

$$Z = \frac{X}{\sigma} - \frac{\mu}{\sigma}$$

$$= \frac{-\mu}{\sigma} + (1/\sigma)X$$

or as

$$Z = a + bX$$

when $a = (-\mu/\sigma)$ and $b = (1/\sigma)$. Since μ and σ are constants, $(-\mu/\sigma)$ and $(1/\sigma)$ are constants as well. Z, therefore, is a linear function of X.

We know that

$$E(Z) = E\left(\frac{X - \mu}{\sigma}\right) = E[(-\mu/\sigma) + (1/\sigma)X]$$

Furthermore,

$$E[(-\mu/\sigma) + (1/\sigma)X] = E(-\mu/\sigma) + E[(1/\sigma)X]$$

By virtue of Equation 5–5

$$= (-\mu/\sigma) + E[(1/\sigma)X]$$

By virtue of Equation 5–4

$$= (-\mu/\sigma) + (1/\sigma)E(X)$$

By virtue of Equation 5–7

However, inasmuch as $E(X) = \mu$,

$$E(Z) = (-\mu/\sigma) + (1/\sigma)(\mu) = 0 \tag{6–7}$$

Finally,

$$V(Z) = V\left(\frac{X - \mu}{\sigma}\right) = V[(-\mu/\sigma) + (1/\sigma)X]$$

Then,

$$V[(-\mu/\sigma) + (1/\sigma)X] = V(-\mu/\sigma) + V[(1/\sigma)X]$$

By virtue of Equation 5–9

$$= 0 + V[(1/\sigma)X]$$

By virtue of Equation 5–8

$$= (1/\sigma)^2 V(X)$$

By virtue of Equation 5–11

Since $V(X) = \sigma^2$,

$$V(Z) = (1/\sigma^2)(\sigma^2) = 1 \tag{6–8}$$

and

$$\sigma_Z = \sqrt{V(Z)} = \sqrt{1} = 1 \tag{6–9}$$

Although the area between X_1 and X_2 will differ from population to population, the area between two particular Zs will not. Once $P(-1.96 < Z < 1.96)$, for example, has been calculated, that probability is still valid the next time we want to derive $P(-1.96 < Z < 1.96)$.

Converting Xs to Zs is justified by the following statement: The area between X_1 and X_2 is *identical* to the area between

$$Z_1 = \frac{X_1 - \mu}{\sigma} \quad \text{and} \quad Z_2 = \frac{X_2 - \mu}{\sigma}$$

We may, then, note that the shaded area in Figure 6–23 is equal to the shaded area in Figure 6–24.

FIGURE 6–23

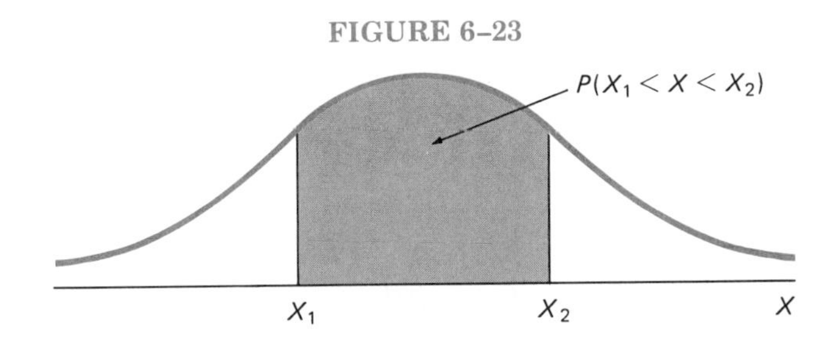

A *table of areas* for the standard normal curve appears in the Appendix. We can quickly determine $P(0 < Z < Z_1)$ from it. Knowledge of this probability makes finding $P(Z_1 < Z < Z_2)$ easy as well. Let's see how to use the table.

The left column in the table lists the units and tenths places for a variety of Zs, while the top row shows the hundredths place. The decimals in the body of the table represent the area between $Z = 0$ and some other Z. To identify $P(0 < Z < 1.63)$, we would move down the left column until we reached 1.6. From there, we would move across the table as far as the .03 heading. Located at the intersection of 1.6 and .03 is .4484. If $Z = 1.63$ corresponds to $X =$

FIGURE 6–24

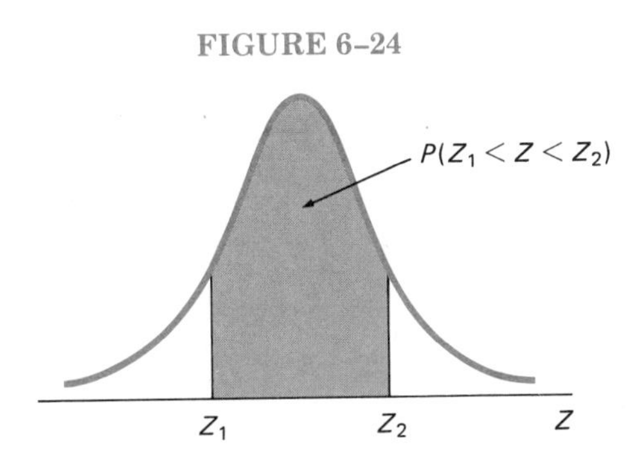

X_1, we can say either $P(0 < Z < 1.63) = .4484$ or $P(\mu < X < X_1) = .4484$ (see Figure 6–25).

FIGURE 6–25

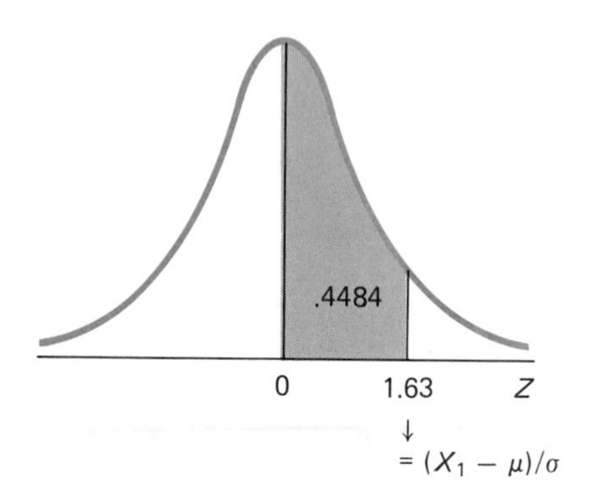

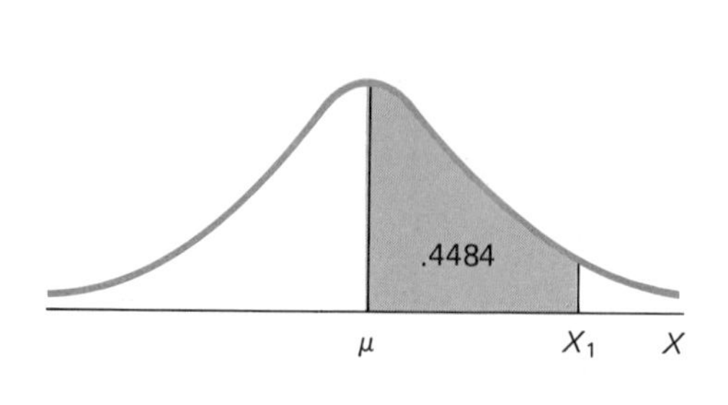

Due to the nature of the table, we have to develop certain strategies whenever we want $P(Z_1 < Z < Z_2)$. As an illustration, consider the shaded area in Figure 6–26. $P(0 < Z < 1.63)$ and $P(0 < Z < 2.04)$ are given explicitly in the table. Such is not the case with $P(1.63 < Z < 2.04)$.

FIGURE 6–26

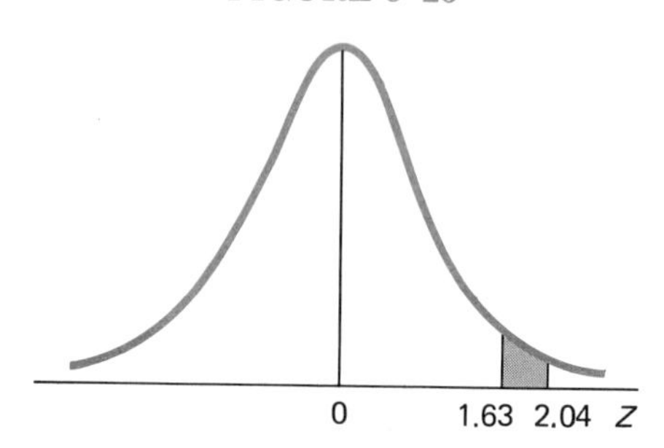

The first two probabilities are indicated in Figure 6–27. We see that $P(1.63 < Z < 2.04)$ is equal to

$$\begin{bmatrix}\text{the area between}\\ Z = 0 \text{ and } Z = 2.04\end{bmatrix} - \begin{bmatrix}\text{the area between}\\ Z = 0 \text{ and } Z = 1.63\end{bmatrix}$$

$$= .4793 - .4484 = .0309$$

Only areas are added or subtracted. We *never* add or subtract the Z values themselves.

We have confined our study to positive Zs. In fact, the left column of the table of areas has no negative Zs. Because the standard normal curve is symmetrical, however, the area between $Z = 0$ and

FIGURE 6–27

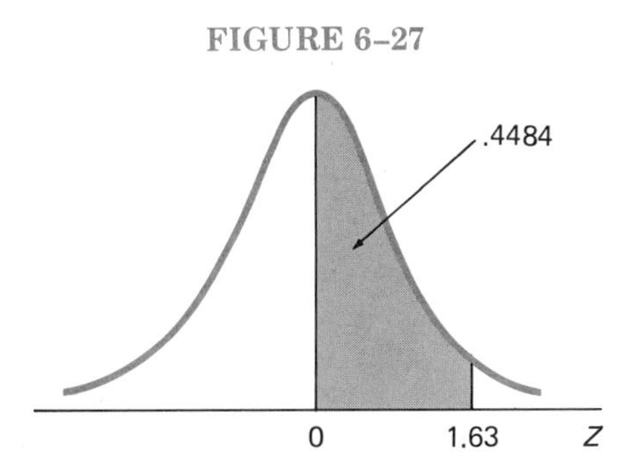

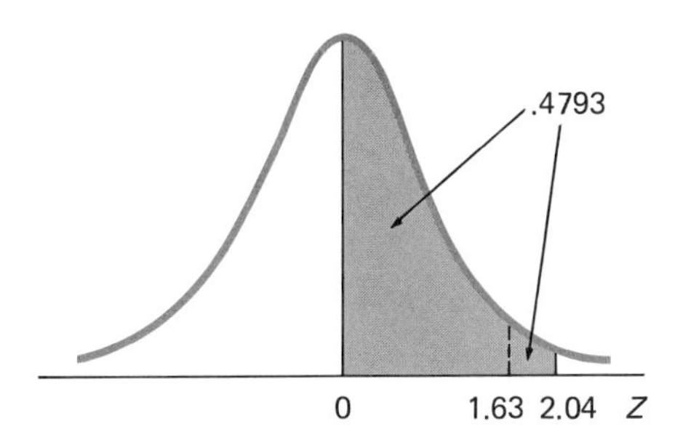

$Z = 1.63$ is the same as the area between $Z = 0$ and $Z = -1.63$ (see Figure 6–28). A negative Z reflects the fact that the area we wish is to the *left* of $Z = 0$. Thus, to derive $P(-Z_1 < Z < 0)$, we look up $P(0 < Z < Z_1)$.

FIGURE 6–28

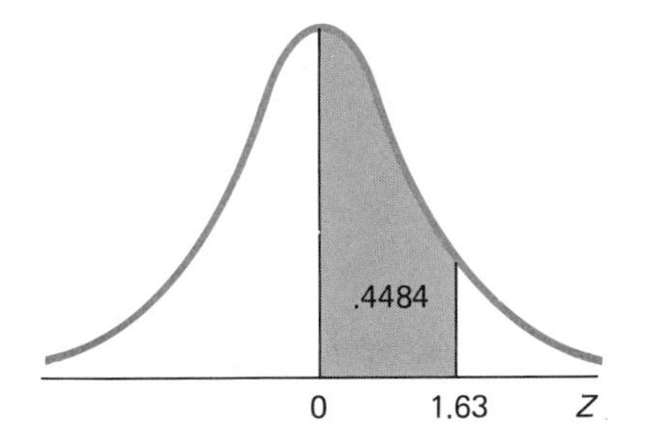

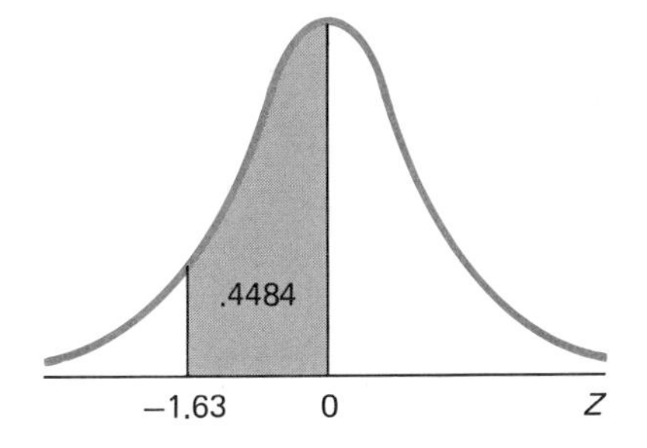

There will always be some area remaining in either *tail,* or half, of the curve beyond Z_1 regardless of the distance between Z_1 and $Z = 0$. Inasmuch as $P(-3 < Z < 3) = .9974$, the probability that Z will lie outside the interval -3 to 3 is near zero. Therefore, we won't need a larger table when we work with $Z = 4, 5, 6$, and so on.

By drawing a diagram of the area between Z_1 and Z_2, we can see how $P(Z_1 < Z < Z_2)$ should be derived. We now want to further our investigation of normal-curve probabilities. To do this, we will return to our TV-viewing example, and we'll compare the probabilities we derive with those associated with the histogram in Figure 6–17.

To compute $P(X_1 < X < X_2)$ when a student is randomly selected from the tenth grade at J. T. High School, let's imagine Figure 6–29. Suppose we want to find $P(8.5 < X < 12.5)$. First, we convert X to Z:

$$Z_1 = \frac{8.5 - 10.2}{4.8} = -.35$$

$$Z_2 = \frac{12.5 - 10.2}{4.8} = .48$$

FIGURE 6–29

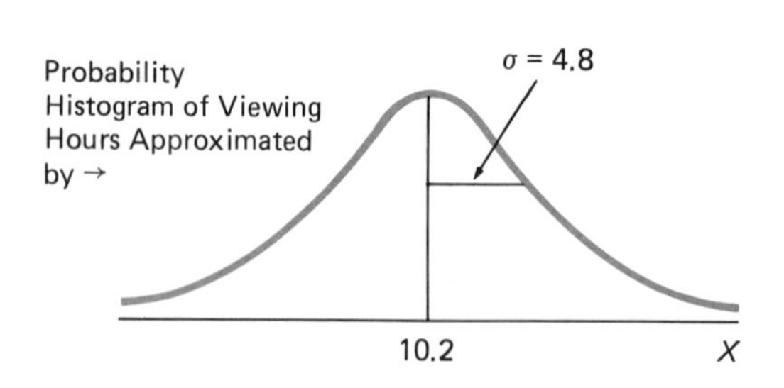

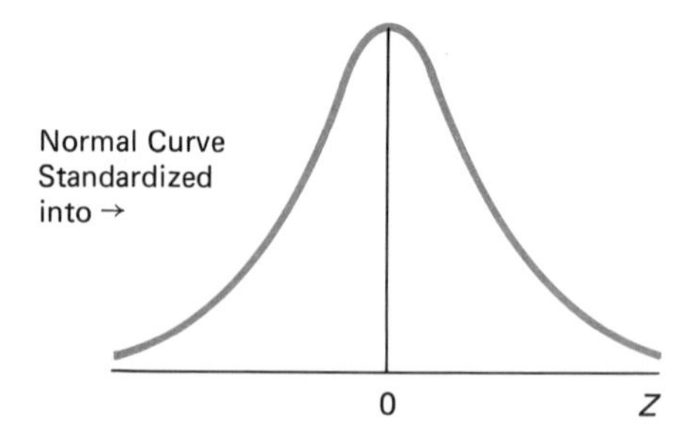

Then we use the table of areas of the standard normal curve to find the areas represented by Z_1 and Z_2. These are .1368 and .1844, respectively. Since $Z_1 = -.35$ and $Z_2 = .48$ lie on opposite sides of $Z = 0$ (see Figure 6–30), we have to *add* the areas. Thus,

$$P(-.35 < Z < .48) = .3212$$

FIGURE 6–30

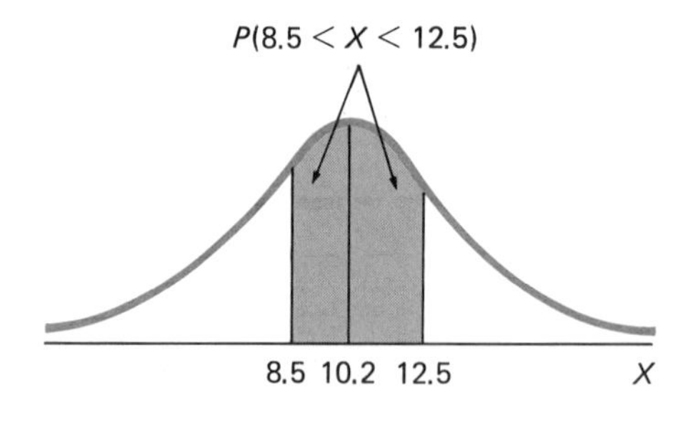

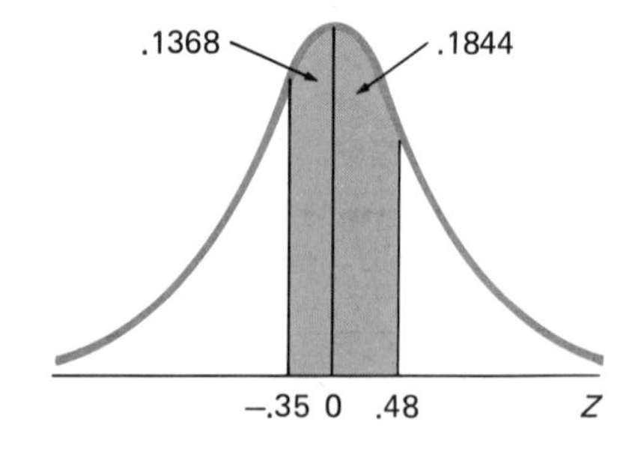

The histogram (Figure 6–17) yields

$$P(8.5 \leq X \leq 12.5) = .30$$

Now let's compute $P(12.5 < X < 16.5)$. Using the same procedure, we find that $Z_1 = .48$ and $Z_2 = 1.31$, and that the areas are .1844 and .4049, respectively. Since $Z_1 = .48$ and $Z_2 = 1.31$ lie on the same side of $Z = 0$ (see Figure 6–31), we have to *subtract* the areas. Thus,

$$P(.48 < Z < 1.31) = .2205$$

FIGURE 6–31

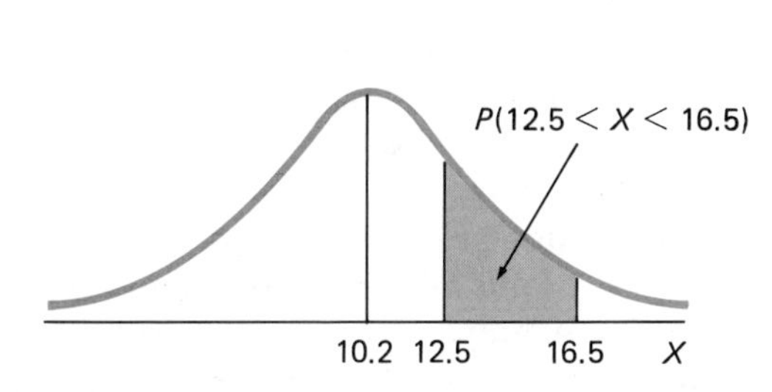

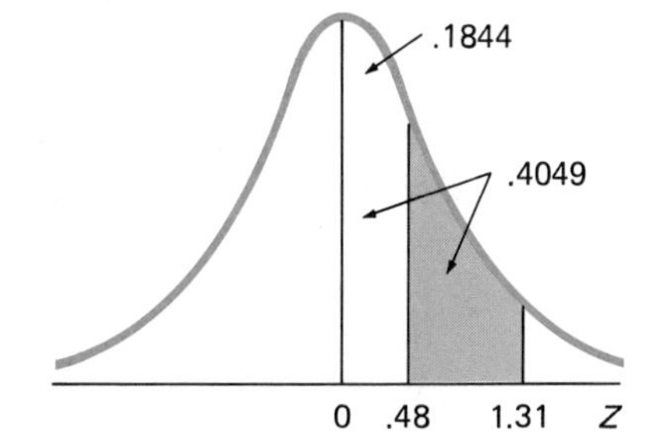

Based on the histogram,

$$P(12.5 \leq X \leq 16.5) = .18$$

Finally, let's find $P(X > 16.5)$. $Z = 1.31$ and the area is .4049. Since each tail of the normal curve represents an area of $1.0/2 = .5000$, we have to *subtract* .4049 from .5000 (see Figure 6–32). Thus,

$$P(Z > 1.31) = .0951$$

Based on the histogram,

$$P(X \geq 16.5) = .14$$

FIGURE 6–32

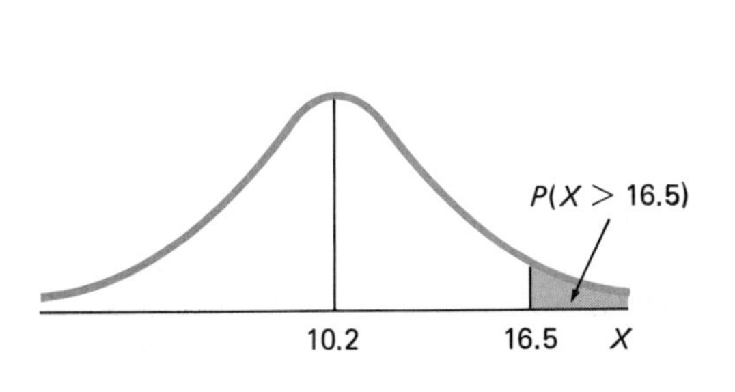

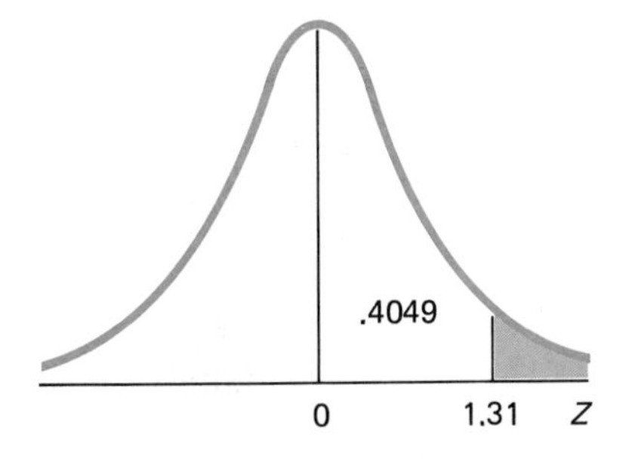

USING THE NORMAL DISTRIBUTION TO APPROXIMATE THE BINOMIAL DISTRIBUTION

The normal distribution is a good approximation for the binomial distribution whenever (1) n is large or (2) π is near .5. Under either of these circumstances, the binomial distribution is fairly symmetrical and unimodal. According to a popular "rule-of-thumb," n is "large" if $n\pi \geq 5$ and $n(1 - \pi) \geq 5$. Consider the following example.

The president of a company wishes to decorate a suite of 14 offices. She has decided on two basic styles. The style selected for any office will be determined by the toss of a coin. If a head appears, the style will be Q. A tail will imply style W.

The probability distribution of style Q offices is binomial because (1) there are only two types of outcomes, (2) $P(\text{style Q}) = .50$ is constant, and (3) the tosses are independent. Note also that $n\pi = (14)(.50) = 7$ and $n(1 - \pi) = (14)(.50) = 7$.

We have seen that the expected value of a binomial probability distribution is $\mu_r = n\pi$ and the standard deviation is $\sigma_r = \sqrt{n\pi(1 - \pi)}$. The parameters in this example are $\mu_r = 7.0$ style Q offices and $\sigma_r = 1.87$ style Q offices. Figure 6–33 is the probability distribution.

We will replace Figure 6–33 with Figure 6–34. In line with the notation used in Chapter Five, r, not X, represents the current random variable. Suppose that we are interested in $P(4 < r < 9)$, illustrated in Figure 6–35.

Inasmuch as

$$Z = \frac{\text{value of the random variable} - \text{mean of the probability distribution}}{\text{standard deviation of the probability distribution}}$$

we can write

$$Z = \frac{r - \mu_r}{\sigma_r} \qquad (6\text{–}10)$$

Therefore, $Z_1 = (4.0 - 7.0)/1.87 = -1.60$, $Z_2 = (9.0 - 7.0)/1.87 = 1.07$, and $P(4 < r < 9) = .4452 + .3577 = .8029$ (see Figure 6–36). The binomial probability is $P(4) + P(5) + P(6) + P(7) + P(8) + P(9) = .88153$.

FIGURE 6–33
Binomial Distribution of the Number of Style Q Offices
where $P(r) = \frac{14!}{r!(14-r)!}(.50)^r(.50)^{14-r}$

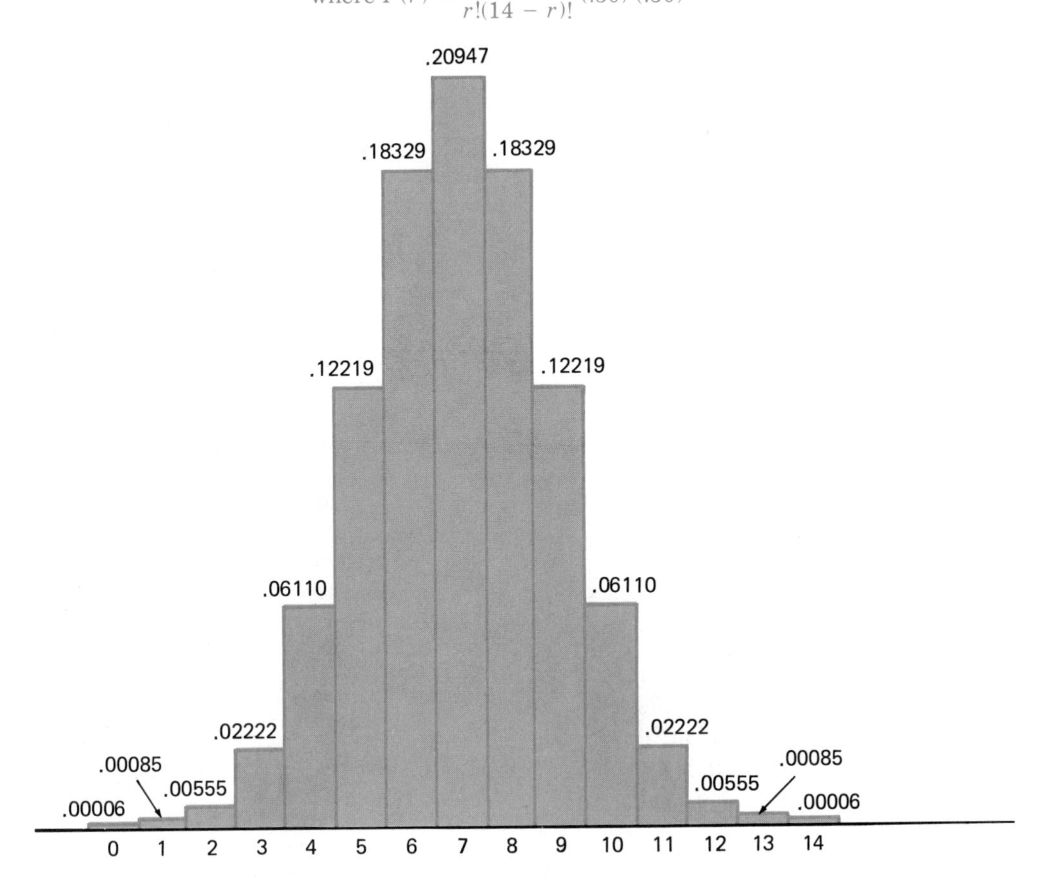

FIGURE 6–34

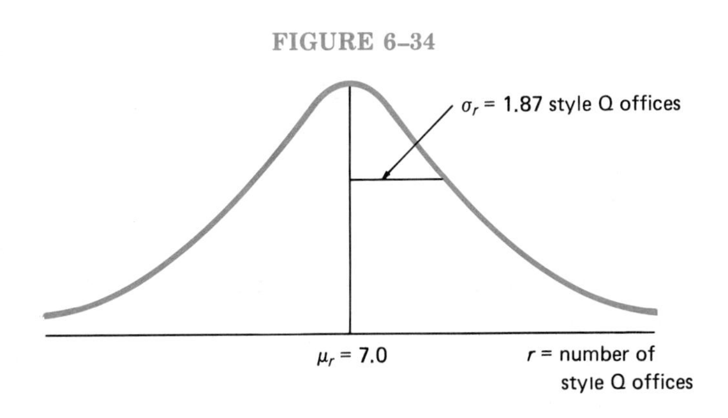

FIGURE 6–35

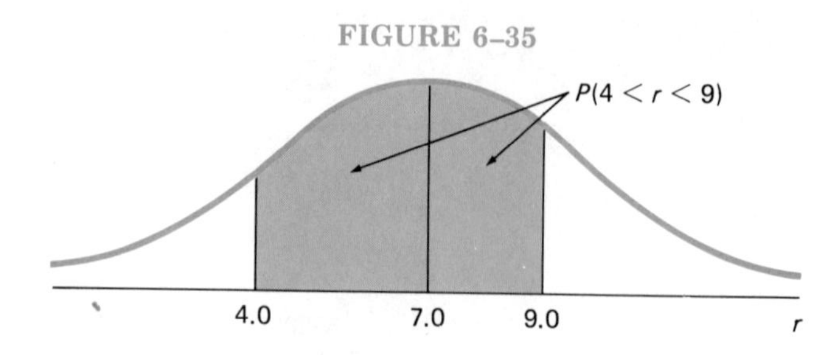

FIGURE 6–36

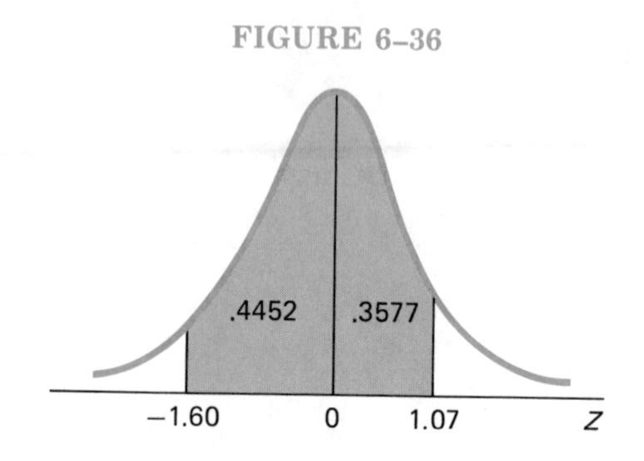

Normal-curve approximations can be improved if a *continuity correction* is introduced. Figure 6–33 indicates that the *discrete* outcomes 4, 5, 6, 7, 8, and 9 form an interval from 3.5 to 9.5. When we calculate $P(3.5 < r < 9.5)$, rather than $P(4 < r < 9)$, we find that the "corrected" probability, that is, $.4693 + .4099 = .8792$ (see Figure 6–37), is much closer to .88153.

FIGURE 6–37

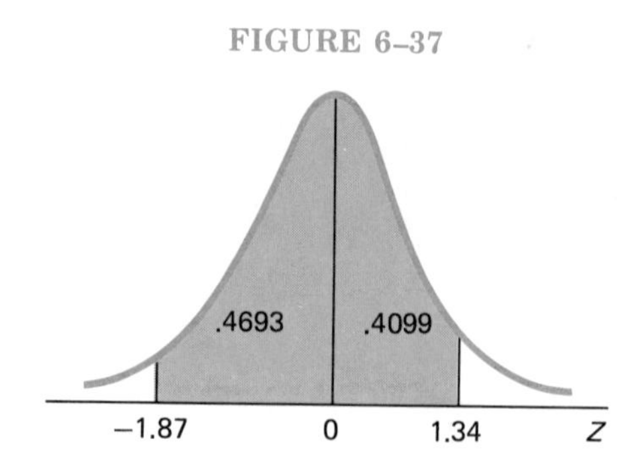

SUMMARY

After clarifying the difference between discrete and continuous random variables, we discussed the normal distribution. Distributions which are roughly symmetrical and unimodal are often approximated by the normal curve. We have seen that normal-curve probabilities can be quite close to the true probabilities.

The preceding chapter examined the following five steps:

1. Create the experiment.
2. Identify the sample space.
3. Define a random variable.
4. Assign numbers to the outcomes in the sample space, where the numbers are values of the random variable.
5. Construct a probability distribution for the random variable.

To this list, we can now add:

6. Approximate the probability distribution with a normal distribution.
7. Convert the normal distribution into the standard normal distribution and determine $P(Z_1 < Z < Z_2) = P(X_1 < X < X_2)$.

We will study this sequence further in Chapter Seven when we perform the experiment "select a sample from a population of N observations."

EXERCISES

1. Suppose a probability density function is $Y = 1/5$, $4 \leq X \leq 9$.

   a. Graph the function.
   b. Show that the "area under the curve" between $X = 4$ and $X = 9$ is 1.0.
   c. Calculate $P(5 \leq X \leq 6)$.
   d. Calculate $P(X \geq 7)$.
   e. Calculate $P(4.5 < X < 8.5)$.

2. The area of a right triangle is equal to $(1/2)bh$, whereas the area of a trapezoid is $(1/2)b(a + c)$. See the illustrations at the top of page 136.

   a. Show that $Y = (1/18)X$, $0 \leq X \leq 6$ is a probability density function.

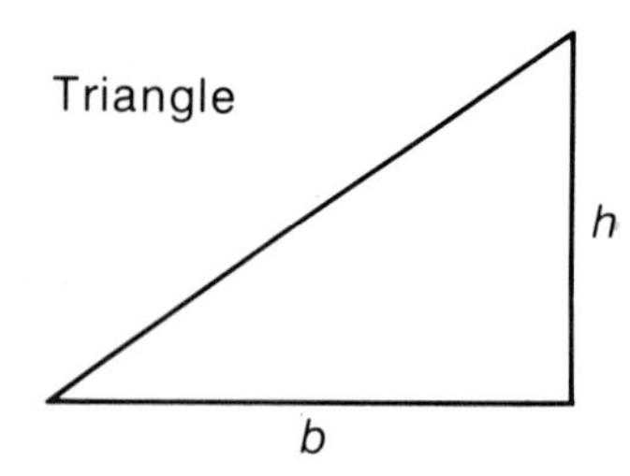

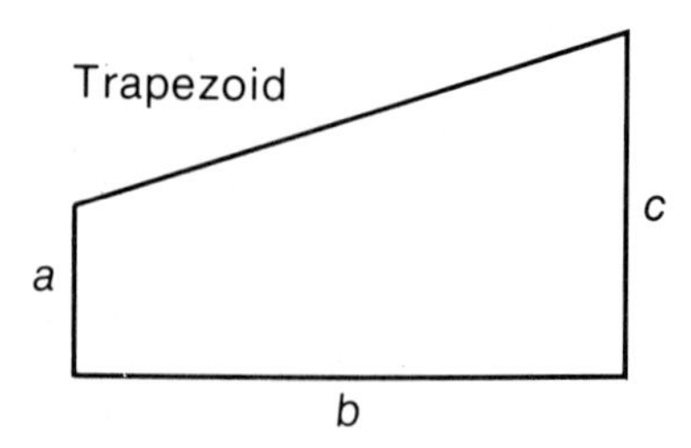

b. Graph the function.
c. What is the probability that the value of X will be greater than 2.5?
d. What is the probability that the value of X will fall between 1.5 and 4.0?
e. Calculate $P(X < 5)$.

3. a. Show that $Y = -.5 + X$, $1 \leq X \leq 2$ is a probability density function.
b. Graph the function.
c. Calculate $P(X > 1.5)$.
d. Calculate $P(1.6 \leq X < 1.8)$.
e. Calculate $P(X < 1.7)$.

4. Which of the following are probability density functions?

a. $Y = (1/8)X$ $\quad 1 \leq X \leq 4$
b. $Y = 1/3$ $\quad 2 \leq X \leq 8$
c. $Y = 1/7$ $\quad 4 \leq X \leq 11$
d. $Y = (2/25)X$ $\quad 0 \leq X \leq 5$

5. Compute the probability density, Y, of the normal curve at $X = \mu$ if (a) $\sigma = 1$, (b) $\sigma = 2$, and (c) $\sigma = 3$.

6. Determine the area between $Z = 0$ and each of the following values of Z:

a. $Z = -2.15$
b. $Z = 4.53$
c. $Z = .16$
d. $Z = -1.47$
e. $Z = 1.08$

7. Calculate the following probabilities:

a. $P(Z > 1.53)$
b. $P(-1.16 \leq Z \leq 1.44)$
c. $P(Z < 2.09)$
d. $P(Z > -2.86)$
e. $P(1.93 < Z < 2.74)$

8. Given normally distributed populations and each of the following sets of parameters, what is the probability that the value of X will be between 3 and 7?

a. $\mu = 4, \sigma = 1$
b. $\mu = 4.5, \sigma = .5$
c. $\mu = 5, \sigma = 1.4$
d. $\mu = 6, \sigma = 1.2$
e. $\mu = 2, \sigma = .4$

9. In a particular normal distribution, $\mu = 60$ and $P(X < 55) = .2389$. What is σ?

10. In a particular normally distributed population, $\sigma = 4$ and $P(X > 17) = .0526$. What is μ?

11. In a particular normally distributed population, $\mu = 12$, $\sigma = 3$, and $P(X > X_1) = .3745$. What is X_1?

12. In a particular normally distributed population, $\mu = 8$, $\sigma = 2$, and $P(X < X_1) = .6554$. What is X_1?

13. The Department of Health, Education, and Welfare announced in 1978 that the average American man had a height of 69 inches and that 15 percent of all men were taller than 72 inches. Suppose that X = height is normally distributed.

a. What is σ?
b. Calculate $P(65 < X < 74)$.
c. Calculate $P(66 < X < 70)$.
d. Calculate $P(X < 66)$.
e. Calculate $P(X > 75)$.

14. Ms. Richards is evaluating three possible investments. The probability distribution of returns for any of them is normally distributed. Consider the following information:

Project A:	μ = \$2 million
	σ = .4 million
Project B:	μ = \$2.1 million
	σ = .3 million
Project C:	μ = \$2.3 million
	σ = .1 million

Richards will invest only if the probability of getting a return greater than \$2.4 million is 15 percent or more. Which project(s) will she choose?

15. A manufacturer produces an electronic component that has an average life of 2,000 hours. Suppose that X = lifetime is normally

distributed with $\sigma = 300$ hours. If the manufacturer guarantees to replace every component that stops working before 1,500 hours, what percentage of his output will ultimately be replaced?

16. While local governments rely on property taxes, most state operations are financed by sales and income taxes. For the 45 states (and the District of Columbia) that used sales taxes in 1975, the rates ranged from 2 to 7 percent. Imagine that the rates are normally distributed with $\mu = 3.85$ percent and $\sigma = .94$ percent. Suppose a state is selected at random.

a. What is the probability that its tax rate will be less than 3.5 percent?
b. What is the probability that its tax rate will be between 2.9 and 4.6 percent?
c. What is the probability that its tax rate won't be greater than 5.2 percent?

17. Part of the earnings of a company are allocated to taxes and dividends. The percentage of after-tax earnings plowed back into the company is referred to as the "retention rate." In a particular industry, the retention rates, X, for 1979 were normally distributed with a mean of 80 percent and a standard deviation of 6 percent. Determine the following probabilities if a company is picked at random:

a. $P(75 < X < 80)$
b. $P(X > 83)$
c. $P(X < 68)$

18. Industry A typifies what economists call the "barometric-firm" model of oligopoly behavior. One company will publicly announce a major price change and, shortly thereafter, other companies in the industry will revise their prices in a similar manner. In the past 40 years, the lag between the initial announcement and the response by a second firm has averaged 18 hours with $\sigma = 3$ hours. Suppose that X = lag time is normally distributed. A researcher is interested in studying one of these pricing episodes. If an episode is selected at random, what is $P(14 < X < 20)$?

19. For each of the following situations, use the normal curve approximation and the continuity correction.

19–1. Many college students qualify for bank loans that are "guaranteed" by the federal government. Under such an arrangement, the government agrees to pay a bank whenever a student defaults on his or her loan. The current rate of default is about 12 percent. If 100 borrowers are interviewed, what is the probability that between 16 and 30 of these people will default?

19–2. Some years ago, college football rules were changed to allow a team to either kick for an extra point or run or pass for two points after getting a touchdown. Suppose that 25 percent of all National Football League fans want the professionals to adopt this option. If 200 fans are chosen,

a. What is the probability that 70 or more of them will be in favor of the change?
b. What is the probability that 60 or fewer of them will be against the change?
c. What is the probability that 50 of them will be in favor of the change?

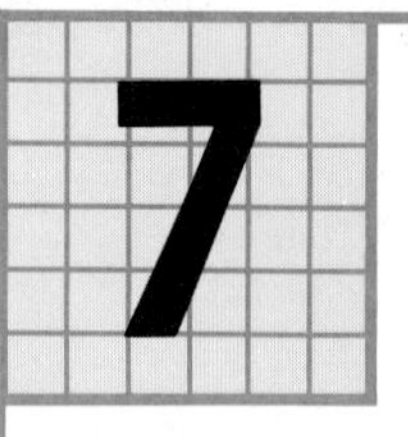

The Sampling Distribution of the Sample Mean

WORKING WITH SAMPLES

With the completion of the section on the normal distribution, we are prepared to study statistical inference. The *sampling distribution* is the major concept introduced in Chapter Seven. This distribution is the probability distribution of a statistic.

As a practical matter, researchers often investigate samples rather than populations. There are several reasons for this:

1. Using samples can *save time*. We noted in Chapter One, for instance, that a manufacturer of jar lids could make decisions about production accuracy in less time if only 50 out of 1,000 lids were measured per hour.
2. Investigating populations can be *impractical*. Suppose the jar-lid manufacturer is concerned about their heat resistance. In tests to determine whether lids maintain their shape at temperatures over 300 degrees Fahrenheit, lids will be destroyed. Destruction is minimized by testing only a sample.
3. Some populations *can't be identified*. For example, a consumer protection agency has pressured Company T to recall its 1977 appliances. The company intends to replace a potentially defective part free of charge. It will mail a notice to every customer who sent in a warranty card following the purchase of an appliance. Since many people don't bother to return warranty cards, the company's list of appliance owners is incomplete.
4. The population may be *infinite*. For instance, if X represents the results (i.e., heads or tails) of flipping a coin, the number of possible observations is infinite because a coin can be flipped repeatedly.

When we must use a statistic to estimate a parameter, the estimate will generally not equal the unknown parameter. The difference between the two is called an *error*. Given all N observations, we make a non-sampling error if we measure the Xs improperly.

Under these circumstances, the computed value of the parameter will not be the "true" value. Any sample chosen from such a population, therefore, would possess measurement errors. Consider the following example.

The S. L. Insurance Co. is revising its filing procedures. Four secretaries were shown the new system, and the company recorded how long they had to practice before they learned the procedure. S. L. was not aware that its stopwatch was 5 seconds fast. The data are contained in Table 7–1.

TABLE 7–1

Employee	*Time on Stopwatch* (μ = *177.5 seconds*)	*Actual Time* (μ = *172.5 seconds*)
A	150	145
B	190	185
C	200	195
D	170	165

According to the stopwatch, $\overline{X}$ is $(150 + 200 + 170)/3 = 173.3$ seconds when a sample of A, C, and D is selected. Using this statistic as an estimate of the "true" population average, we find an error of $173.3 - 172.5 = 0.8$ seconds. Without the non-sampling error, $\overline{X}$ would be $(145 + 195 + 165)/3 = 168.3$ seconds. This estimate deviates from the "true" parameter by $168.3 - 172.5 = -4.2$ seconds. Hence, the total error, 0.8, is composed of a sampling error (i.e., -4.2) and a non-sampling error (i.e., 5.0).

Careful planning is a remedy for non-sampling errors. We will not discuss them further. The probability theory in this chapter applies only to sampling errors. We shall assume that these errors are the sole explanation for any difference between an estimate and a parameter.

SELECTING A RANDOM SAMPLE

We will eventually be able to say such things as "There is a .95 probability that $\overline{X}$ will be no more than 2 units from μ." To make these statements, we have to assume that the researcher chooses the samples *randomly*. Although the dictionary defines "random" as "haphazard" or "purposeless," its technical connotation is quite the opposite.

Recall that we can select a sample with or without replacement. In the latter case, an item is drawn, recorded, and permanently removed from the population. Given a population of N observations

TABLE 7–2
Table of Random Numbers

67245	57739	71894	05092	98422	66427	44532	99528	98140	28542
16668	92606	61965	80165	49762	38869	56878	21188	60837	15300
81072	42106	11961	45102	24938	47764	78635	93276	37506	12058
30978	25139	26356	79764	32142	41757	21431	02019	26488	59223
29627	83125	17542	04131	65456	40501	97604	58716	92269	66697
81962	75304	22151	09897	38030	79085	28701	41588	22546	12761
26296	88598	73403	96617	43268	01470	98074	19969	22792	85476
11146	25544	84381	98928	42862	01967	04583	28670	88746	48857
57117	90192	25254	78992	27324	75203	83820	84260	56712	06536
97513	00339	78752	08299	59886	34316	60136	44376	33010	87203
34249	49500	33957	94626	80843	79329	56928	67173	05498	94094
77756	61009	60548	15162	66132	65045	93348	10605	02498	48439
40571	36272	93886	93664	68719	80015	28345	51392	09187	28382
90087	24569	14500	45689	32876	56768	71861	90872	85153	02809
68470	72812	59247	92965	36492	01564	82282	66677	78747	91349
92314	92521	96195	23104	47846	03038	70660	38955	07479	58041
20675	77855	25127	41707	53922	60349	11610	32152	64094	26517
83013	86452	36206	77551	44833	75023	83774	13586	34596	49473
35944	83776	57641	11694	76808	18707	02818	25940	22639	89168
41641	49817	35066	84171	64106	61938	39751	71367	14302	45560
59131	24022	88481	84407	07186	76409	77997	99118	76609	85909
50483	20272	97072	12145	14267	11918	48839	88105	94849	08017
92044	49651	39029	58146	98605	39318	05544	06006	99686	86441
49084	25574	63204	73486	13897	03045	33080	67900	46838	68163
32447	67437	83344	38746	89235	51922	07933	17686	21388	93225
35656	83624	16225	10824	30288	76696	95626	17603	27278	20472
33939	90576	60557	17891	85294	18528	33618	23047	21159	41620
53132	71864	37661	78843	34824	95848	18205	68886	89177	32559
48656	42723	41890	41573	66283	87294	96486	53435	76962	31992
60131	37548	25942	20221	21199	52813	85833	97845	39473	52592
87753	86939	91368	37994	01473	52708	51653	53636	03576	35186
31519	49224	42553	29513	14715	06673	92863	48713	87600	03697
67784	32191	40336	15042	22340	27932	44842	29116	84322	81967
38817	71055	76042	45593	13220	72254	30991	61345	04309	67486
59398	59634	13215	57218	26355	48081	77237	18034	76210	61453
55215	44403	59066	79667	83179	79595	98577	03862	55429	13817
04460	95197	25214	51106	20173	17018	08238	14692	99356	68749
14252	62973	60027	08104	56222	82763	36385	20833	29628	10087
40240	83556	74334	06092	58657	95385	22749	03571	96578	99525
98743	01514	03616	56372	78053	87064	70998	97591	16926	65779

(continued)

Table 7–2 continued

85240	83785	31102	57306	36277	01340	81035	42910	09632	17791
45574	46659	27270	53948	93560	58240	32977	03306	70135	34785
42465	36649	33992	31040	79312	59165	81152	36392	48492	19199
73627	35535	79488	79938	07219	57037	02070	22286	75668	54172
34886	86421	01357	67274	27030	71650	65300	23664	01896	69378
52998	63612	19651	16074	04575	70509	95420	51569	87284	41693
60937	24831	20441	98220	39065	95945	82663	49286	50481	19663
97357	55869	29861	50831	03127	19918	21486	42788	38729	63074
28874	21369	20578	06009	21097	94368	51062	91612	20575	29354
23613	09659	83689	72036	41942	57834	29378	03434	43779	69085

Source: The RAND Corporation, *A Million Random Digits with 100,000 Normal Deviates* (Glencoe, Ill.: The Free Press, 1955), p. 337.

and a sample of size n, the number of different samples that can be chosen without replacement is the same as the number of combinations of N things taken n at a time, or

$$\frac{N!}{n!(N - n)!}$$

Let M represent this number. We will define *simple random sampling without replacement* as a process which guarantees that P(choose a particular sample) is equal to $1/M$.

We could create randomness in the following way:

1. Write the observations on slips of paper.
2. Put the slips in a hat and mix them.
3. Draw a slip and reshuffle the remaining slips.
4. Repeat step 3 until n slips have been selected.

When N and n are large, this method can be time consuming. Moreover, because slips often stick together, thorough mixing is difficult. Using a *random number table,* such as Table 7–2, is a more efficient approach.

To justify such a table, suppose we construct a wheel containing all ten digits with a peg at each digit (see Figure 7–1). Because of the pegs, the dial will stop only at a digit. Hence, P(dial stops at a particular digit) = 1/10. We should expect each digit to appear 10 percent of the time in the "long run." Since successive spins are independent, the probability of getting a specific pair of digits on two spins would be (1/10)(1/10) = .01. Table 7–2 was developed electronically. As in the wheel experiment, every digit is independent. Thus, P(a specific 3-digit number) = (1/10)(1/10)(1/10) = .001.

Before working with Table 7–2, we have to assign a number to each member of the population. We will then (1) *arbitrarily* pick a

FIGURE 7–1

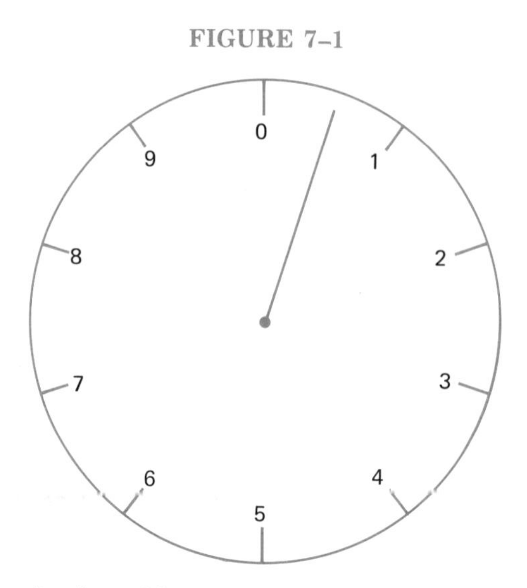

starting point in the table and (2) decide how we want to read the digits (e.g., from left to right or from right to left). Let's look at an example.

A sociologist is studying the personalities of women who support or disapprove of the Equal Rights Amendment. He is especially interested in the attitudes of female executives. He intends to randomly select 40 of the 800 women currently in management positions at certain corporations.

We will give the 800 women numbers from 001 to 800. If we begin at the third row in Table 7–2 and read the first three digits of each 5-digit block from left to right, we see 810, 421, 119, 451, 249, 477, and so on. Since none of the women is #810, we won't write that number down. We will also skip a number when it appears for the second time. After we record 40 numbers, we match them against the numbers assigned to the population. For instance, if 451 identifies a vice-president of IBM, she will be part of the ERA sample.

As the following pages will illustrate, knowing the probabilities of the possible samples is essential for (1) deriving sampling distributions and (2) making objective statements about potential estimation errors.

In contrast to random sampling, *convenience sampling* is a nonprobability technique for choosing n observations. The following example is a case in point.

Ever since the 1973 oil embargo, many Americans have looked upon gasoline shortages as the consequence of a conspiracy by the oil companies rather than the result of a natural depletion of oil reserves. Suspicions are heightened by announcements of substantial oil profits. A newsman wishes to conduct interviews about the "crisis." He will speak to the next 20 people leaving Department Store H. Although the "next 20 people" are determined by chance, "chance" can't be computed in this instance.

CONSTRUCTING THE SAMPLING DISTRIBUTION OF THE SAMPLE MEAN

Each possible random sample has a mean, median, variance, and standard deviation. We now want to study the probability distribution, or *sampling distribution,* of $\overline{X}$. Given its characteristics, we can make a variety of probability statements about potential sampling errors. We'll base our discussion on the following example.

John Warren has come to New York City for several job interviews with B. D. International, Inc. Two of the six department heads are randomly chosen to talk with promising applicants. Although the job pays well, Warren is worried that B. D. may transfer its executives frequently. He has three young children and fears that repeated moving will harm them psychologically.

The six department heads have been with the company for a long time. We'll view these people as a population and assume that Warren can only estimate μ = the average number of transfers in the past 15 years by calculating $\overline{X}$ = the average number of transfers in the past 15 years for two interviewers. The candidate's fears will be allayed if μ is 3.5 transfers or less. At the moment, estimating μ with $\overline{X}$ is sensible on intuitive grounds. The theoretical rationale for $\overline{X}$ will be detailed in Chapter Eight.

Table 7–3 contains the population of transfers. Because $\mu = 4.67$, Warren should decline the job. However, any conclusion that he reaches (i.e., "μ is too high" or "μ isn't too high") will be formed from only two of the observations in Table 7–3. Under these circumstances, he can't be *certain* that his inference is going to be correct.

TABLE 7–3

Employee	*Number of Transfers*
A	1
B	3
C	4
D	5
E	6
F	9
	28

When two executives are selected at random, the number of different possible sets of interviewers is

$$\frac{6!}{2!(6-2)!} = 15 = M$$

The 15 pairs are reproduced in Table 7–4, which also identifies the possible $\overline{X}$s.

TABLE 7–4

Employees in Sample	*Transfers in Sample*	*Sample Average Number of Transfers (estimate of μ)*
A,B	1,3	2.00
A,C	1,4	2.50
A,D	1,5	3.00
A,E	1,6	3.50
A,F	1,9	5.00
B,C	3,4	3.50
B,D	3,5	4.00
B,E	3,6	4.50
B,F	3,9	6.00
C,D	4,5	4.50
C,E	4,6	5.00
C,F	4,9	6.50
D,E	5,6	5.50
D,F	5,9	7.00
E,F	6,9	7.50

Regardless of who interviews Warren, $\overline{X}$ will become his estimate of μ. The sample average is a *random variable* whose value is determined by the experiment "pick two executives at random." Since each sample is equally likely, the probability of getting a particular $\overline{X}$ is the same as the ratio

$$\frac{\text{number of Table 7–4 outcomes corresponding to the particular } \overline{X}}{M}$$

The different values of $\overline{X}$ along with their probabilities are presented in Table 7–5. This probability distribution is the sampling distribution of $\overline{X}$. In general, if c is the value of some statistic (e.g., $\overline{X}$), the sampling distribution of that statistic indicates the probability of c when a sample is chosen at random. Figure 7–2 is a histogram of the sampling distribution in Table 7–5.

Table 7–6 indicates the relation between $\overline{X}$ and the conclusions

TABLE 7–5

$\overline{X}$	$P(\overline{X})$
2.00	1/15
2.50	1/15
3.00	1/15
3.50	2/15
4.00	1/15
4.50	2/15
5.00	2/15
5.50	1/15
6.00	1/15
6.50	1/15
7.00	1/15
7.50	1/15

FIGURE 7–2

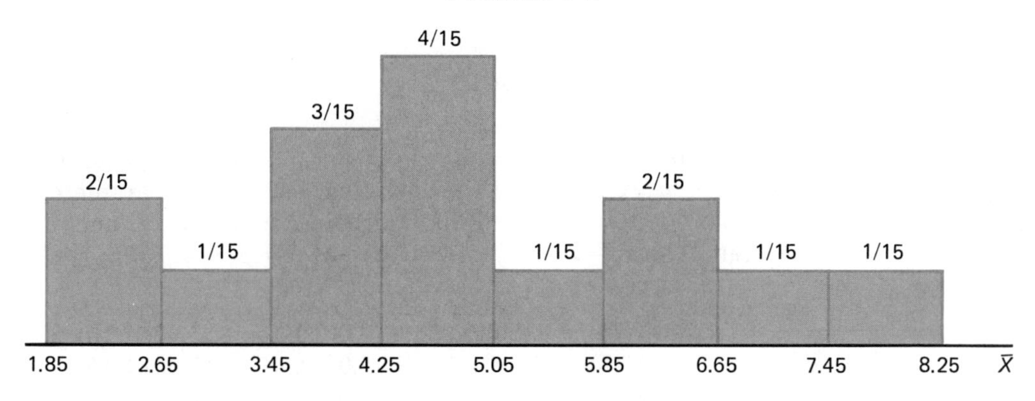

TABLE 7–6

$\overline{X}$	$P(\overline{X})$	*Conclusion*
2.00	1/15	μ isn't high
2.50	1/15	μ isn't high
3.00	1/15	μ isn't high
3.50	2/15	μ isn't high
4.00	1/15	μ is high
4.50	2/15	μ is high
5.00	2/15	μ is high
5.50	1/15	μ is high
6.00	1/15	μ is high
6.50	1/15	μ is high
7.00	1/15	μ is high
7.50	1/15	μ is high

derived from it as long as 3.5 transfers is the cutoff value. Both $P(\overline{X} \leq 3.5)$ and P(incorrect conclusion) are 5/15.

Characteristics

Sampling distributions, like all distributions, possess means and variances. The parameters of any random variable (r.v.) would be

$$E(\text{r.v.}) = \mu_{\text{r.v.}} = \sum (\text{r.v.})P(\text{r.v.}) \qquad (7\text{–}1)$$

$$V(\text{r.v.}) = \sigma^2_{\text{r.v.}} = \sum (\text{r.v.} - \mu_{\text{r.v.}})^2 P(\text{r.v.}) \qquad (7\text{–}2)$$

Equations 7–1 and 7–2 turn into

$$E(X) = \mu_x = \sum XP(X)$$

$$V(X) = \sigma^2_x = \sum (X - \mu_x)^2 P(X)$$

when the random variable is X and into

$$E(\overline{X}) = \mu_{\overline{x}} = \sum \overline{X}P(\overline{X}) \qquad (7\text{–}3)$$

$$V(\overline{X}) = \sigma^2_{\overline{x}} = \sum (\overline{X} - \mu_{\overline{x}})^2 P(\overline{X}) \qquad (7\text{–}4)$$

when the random variable is $\overline{X}$.

We intend to compare two sets of parameters: the population parameters, μ_x, σ^2_x, σ_x, and the sampling distribution parameters, $\mu_{\overline{x}}$, $\sigma^2_{\overline{x}}$, $\sigma_{\overline{x}}$. The former set will be written as simply μ, σ^2, and σ.

Given Tables 7–3 and 7–5, we can compute $\mu = 4.67$, $\sigma^2 = 6.2$, and $\sigma = 2.49$, as well as

$$\mu_{\overline{x}} = (2.00)(1/15) + (2.50)(1/15) + (3.00)(1/15) + (3.50)(2/15) + (4.00)(1/15) + (4.50)(2/15) + (5.00)(2/15) + (5.50)(1/15) + (6.00)(1/15) + (6.50)(1/15) + (7.00)(1/15) + (7.50)(1/15) = 4.67$$

$$\sigma^2_{\overline{x}} = (2.00 - 4.67)^2(1/15) + (2.50 - 4.67)^2(1/15) + (3.00 - 4.67)^2(1/15) + (3.50 - 4.67)^2(2/15) + (4.00 - 4.67)^2(1/15) + (4.50 - 4.67)^2(2/15) + (5.00 - 4.67)^2(2/15) + (5.50 - 4.67)^2(1/15) + (6.00 - 4.67)^2(1/15) + (6.50 - 4.67)^2(1/15) + (7.00 - 4.67)^2(1/15) + (7.50 - 4.67)^2(1/15) = 2.48$$

$$\sigma_{\overline{x}} = \sqrt{\sigma^2_{\overline{x}}} = \sqrt{2.48} = 1.57$$

The parameters are summarized in Table 7–7.

We see that the mean of the sampling distribution is the same as the mean of the population. On the other hand, the standard deviation of the sampling distribution is less than σ. This last condition is the result of averaging. "Extreme" outcomes, such as nine transfers, can't occur in the sampling distribution.

A sampling distribution of $\overline{X}$ exists for each sample size. Let's

TABLE 7–7

Population	*Sampling Distribution when n = 2*	*Comparison of Parameters*
$\mu = 4.67$	$\mu_{\bar{X}} = 4.67$	$\mu_{\bar{X}} = \mu$
$\sigma^2 = 6.20$	$\sigma^2_{\bar{X}} = 2.48$	$\sigma^2_{\bar{X}} < \sigma^2$
$\sigma = 2.49$	$\sigma_{\bar{X}} = 1.57$	$\sigma_{\bar{X}} < \sigma$

assume that B. D. International chooses three interviewers instead of two. The number of different possible samples is

$$\frac{6!}{3!(6-3)!} = 20 = M$$

Table 7–8 lists the M samples and the corresponding $\bar{X}$s.

TABLE 7–8

Employees in Sample	*Transfers in Sample*	*Sample Average Number of Transfers* ($\bar{X}$)
A,B,C	1,3,4	2.67
A,B,D	1,3,5	3.00
A,B,E	1,3,6	3.33
A,B,F	1,3,9	4.33
A,C,D	1,4,5	3.33
A,C,E	1,4,6	3.67
A,C,F	1,4,9	4.67
A,D,E	1,5,6	4.00
A,D,F	1,5,9	5.00
A,E,F	1,6,9	5.33
B,C,D	3,4,5	4.00
B,C,E	3,4,6	4.33
B,C,F	3,4,9	5.33
B,D,E	3,5,6	4.67
B,D,F	3,5,9	5.67
B,E,F	3,6,9	6.00
C,D,E	4,5,6	5.00
C,D,F	4,5,9	6.00
C,E,F	4,6,9	6.33
D,E,F	5,6,9	6.67

The sampling distribution of $\bar{X}$ appears in Table 7–9. Its parameters are $\mu_{\bar{X}} = 4.67$, $\sigma^2_{\bar{X}} = 1.24$, and $\sigma_{\bar{X}} = 1.11$. Table 7–10 combines the new parameters with those of Table 7–7. The means of the three distributions are identical. Additionally, $\sigma_{\bar{X}}$ falls farther below σ as n increases. Whenever more and more observations are averaged, the influence of an "extreme" X is progressively reduced. The $\bar{X}$s, then, will get closer and closer to each other.

TABLE 7–9

$\overline{X}$	$P(\overline{X})$
2.67	1/20
3.00	1/20
3.33	2/20
3.67	1/20
4.00	2/20
4.33	2/20
4.67	2/20
5.00	2/20
5.33	2/20
5.67	1/20
6.00	2/20
6.33	1/20
6.67	1/20

TABLE 7–10

Population	*Sampling Distribution when n = 2*	*Sampling Distribution when n = 3*
$\mu = 4.67$	$\mu_{\bar{x}} = 4.67$	$\mu_{\bar{x}} = 4.67$
$\sigma^2 = 6.20$	$\sigma^2_{\bar{x}} = 2.48$	$\sigma^2_{\bar{x}} = 1.24$
$\sigma = 2.49$	$\sigma_{\bar{x}} = 1.57$	$\sigma_{\bar{x}} = 1.11$

We will formally state three properties of the sampling distribution of $\overline{X}$:

1. Regardless of the size of n, $\mu_{\bar{x}}$ will always equal μ.
2. Regardless of the size of n, $\sigma_{\bar{x}}$ will always be less than σ.
3. As the sample size increases, $\sigma_{\bar{x}}$ will decrease.

The second and third properties are suggested by the following relationship:

$$\sigma^2_{\bar{x}} = \frac{\sigma^2}{n}\left[\frac{N - n}{N - 1}\right] \tag{7–5}$$

In contrast to Equation 7–4, Equation 7–5 relates $\sigma^2_{\bar{x}}$ directly to the population variance. Since

$$\frac{1}{n}\left[\frac{N - n}{N - 1}\right] = \frac{N - n}{nN - n} < 1,$$

the right side of Equation 7–5 must be smaller than σ^2. Moreover,

$$\frac{N - n}{nN - n}$$

will decline if n becomes larger. Under these circumstances, $\sigma^2_{\bar{x}}$ will decline as well. To confirm that Equations 7–4 and 7–5 yield the same values for $\sigma^2_{\bar{x}}$, recall that $\sigma^2_{\bar{x}} = 1.24$ when $n = 3$. With Equation 7–5, we find

$$\sigma^2_{\bar{x}} = \frac{6.2}{3}\left[\frac{6-3}{6-1}\right] = 1.24$$

The bracketed part of Equation 7–5 is called the *finite population correction factor*. If (1) n is a small proportion of the population (say, $< .05$) and (2) N is large, $(N - n) \approx (N - 1)$, $(N - n)/(N - 1) \approx 1$, and $\sigma^2_{\bar{x}} \approx \sigma^2/n$. In the future, we will act as though these conditions were applicable. We can, then, replace Equation 7–5 with the following:

$$\sigma^2_{\bar{x}} = \frac{\sigma^2}{n} \tag{7–6}$$

The other two parameters of the sampling distribution of $\bar{X}$ are

$$\mu_{\bar{x}} = \mu \tag{7–7}$$

$$\sigma_{\bar{x}} = \frac{\sigma}{\sqrt{n}} \tag{7–8}$$

We also want to discuss the shape of the sampling distribution of $\bar{X}$. To do so, we introduce a statistical principle referred to as the *central limit theorem*:

> As n increases, the sampling distribution of $\bar{X}$ approaches a normal distribution.

This theorem is important because of the absence of an assumption about the population. Even if the population looks like Figure 7–3, the sampling distribution will look like Figure 7–4 when n is "large."

FIGURE 7–3

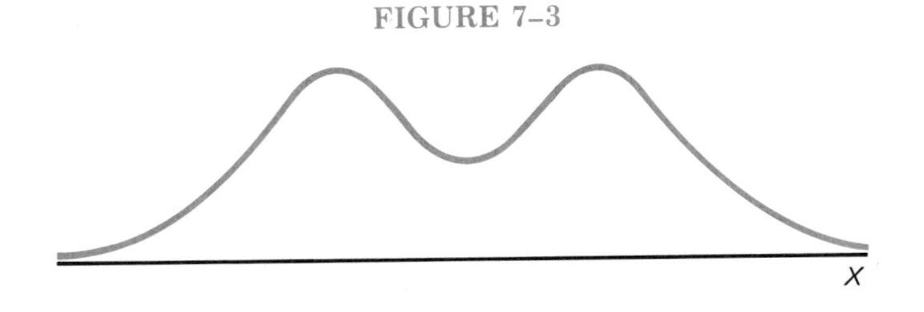

The curve in Figure 7–4 is defined by the density function

$$Y = \frac{1}{\sigma_{\bar{x}}\sqrt{2\pi}} e^{-(1/2)[(\bar{X} - \mu_{\bar{x}})/\sigma_{\bar{x}}]^2} \qquad (7\text{–}9)$$

FIGURE 7–4

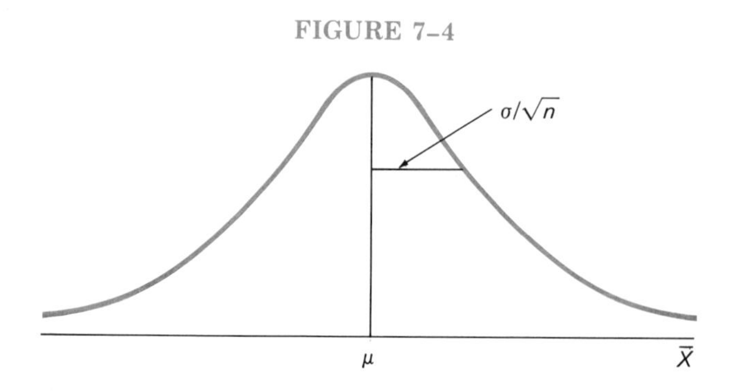

Notice that Equation 7–9 differs from Equation 6–1 since the relevant terms are no longer μ, σ, and X, but $\mu_{\bar{x}}$, $\sigma_{\bar{x}}$, and $\bar{X}$.

If the population is fairly symmetrical and unimodal, a "small" sample is sufficient to produce a normal sampling distribution. In fact, the sampling distribution will be normal for any n (e.g., $n = 2$) provided the population is normally distributed. Unless a population is very skewed, we can be confident that an n of 31 will make the sampling distribution normal. We will say that a sample is "small" when $n \leq 30$ and "large" when $n > 30$.

The relation between a normal population and a normal sampling distribution is shown in Figure 7–5. The sampling distribution is more "pinched in" because $\sigma_{\bar{x}} < \sigma$.

FIGURE 7–5

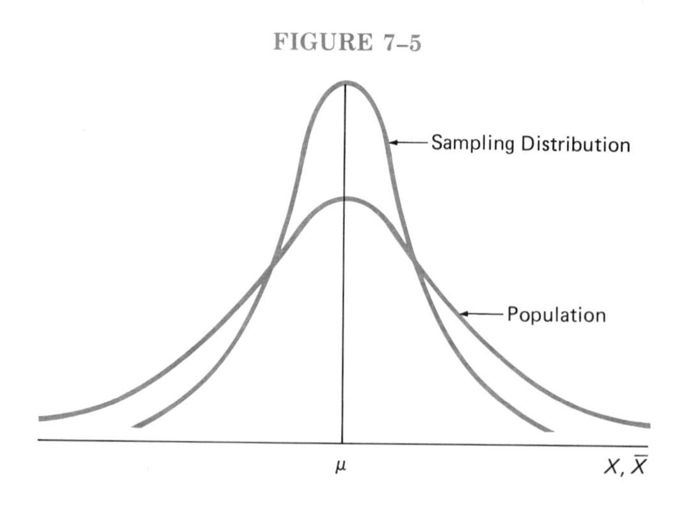

Table 7–11 indicates the distinction between certain parameters and statistics.

TABLE 7–11
Summary of the Relationship Between a Population, a Sample, and the Sampling Distribution of the Sample Mean

Distribution	*Number of Items*	*Mean*	*Standard Deviation*
Population	N	$\mu = \Sigma X/N$	$\sigma = \sqrt{\Sigma(X - \mu)^2/N}$
Sample	n	$\overline{X} = \Sigma X/n$	$s = \sqrt{\Sigma(X - \overline{X})^2/(n - 1)}$
Sampling distribution of the sample mean	M	$\mu_{\overline{x}} = \Sigma \overline{X} P(\overline{X}) = \mu$	$\sigma_{\overline{x}} = \sqrt{\Sigma(\overline{X} - \mu_{\overline{x}})^2 P(\overline{X})}$ $\sigma_{\overline{x}} = (\sigma/\sqrt{n})\sqrt{(N - n)/(N - 1)}$ (if n/N is large and N is small) $\sigma_{\overline{x}} = \sigma/\sqrt{n}$ (if n/N is small and N is large)

Theoretical Support

The theory of statistical inference depends on the properties of random variables. Based on Chapters Five and Six, we can prove (1) $\mu_{\overline{x}} = \mu$ and (2) $\sigma^2_{\overline{x}} = \sigma^2/n$.

Suppose that each observation in the population of transfers (1, 3, 4, 5, 6, 9) is treated as an outcome of the experiment "select *one* employee at random." Table 7–12 would, then, be either the relative frequency distribution of the population or the probability distribution of X = the number of transfers. We can compute $E(X) = \mu = 4.67$ and $V(X) = \sigma^2 = 6.2$.

When we choose $n = 2$ observations *with* replacement, our sample will consist of an X_1 and an X_2. The process of picking X_1, however, is the *same* as selecting one employee at random. Before X_1 is determined, it is a random variable with the probability distri-

TABLE 7–12

X *Number of Transfers*	$P(X)$ *(or f/N)*
1	1/6
3	1/6
4	1/6
5	1/6
6	1/6
9	1/6

bution in Table 7–13. Since Tables 7–12 and 7–13 are not different, $E(X_1) = \mu$ and $V(X_1) = \sigma^2$.

TABLE 7–13

X_1 *Number of Transfers on the First Draw*	$P(X_1)$
1	1/6
3	1/6
4	1/6
5	1/6
6	1/6
9	1/6

After replacing X_1, we can draw X_2. Notice that X_2 will also be the result of selecting one employee at random. Table 7–14 is the probability distribution of X_2. We see that $E(X_2) = \mu$ and $V(X_2) = \sigma^2$. For a "great many" repetitions of "select *two* employees with replacement," the average value on the first draw will be μ, and the average value on the second draw will be μ.

TABLE 7–14

X_2 *Number of Transfers on the Second Draw*	$P(X_2)$
1	1/6
3	1/6
4	1/6
5	1/6
6	1/6
9	1/6

Let's emphasize that X_1 and X_2 represent random variables *prior* to sampling but actual values once a sample has been picked. Inasmuch as $X_1 + X_2$ is the sum of two random variables, we can write

$$\begin{aligned} E(X_1 + X_2) &= E(X_1) + E(X_2) \qquad \text{By virtue of Equation 5–12} \\ &= \mu + \mu \end{aligned}$$

We know that

$$\overline{X} = \frac{\sum^{2} X}{n} = (1/n)(X_1 + X_2)$$

Thus

$$
\begin{aligned}
E(\overline{X}) &= E[(1/n)(X_1 + X_2)] \\
&= (1/n)E(X_1 + X_2) \qquad \text{By virtue of Equation 5–14} \\
&= (1/n)(\mu + \mu) = (1/n)(n\mu) = \mu
\end{aligned}
$$

Provided that sampling is with replacement, X_1 and X_2 are *independent* random variables. The variance of their sum is equal to the sum of their individual variances; that is,

$$
\begin{aligned}
V(X_1 + X_2) &= V(X_1) + V(X_2) \qquad \text{By virtue of Equation 5–18} \\
&= \sigma^2 + \sigma^2
\end{aligned}
$$

TABLE 7–15

Combination	*Order of Selection* (X_1, X_2)
#1	1, 3
	3, 1
#2	1, 4
	4, 1
#3	1, 5
	5, 1
#4	1, 6
	6, 1
#5	1, 9
	9, 1
#6	3, 4
	4, 3
#7	3, 5
	5, 3
#8	3, 6
	6, 3
#9	3, 9
	9, 3
#10	4, 5
	5, 4
#11	4, 6
	6, 4
#12	4, 9
	9, 4
#13	5, 6
	6, 5
#14	5, 9
	9, 5
#15	6, 9
	9, 6

We are, therefore, able to derive

$$\begin{aligned} V(\overline{X}) &= V[(1/n)(X_1 + X_2)] \\ &= (1/n)^2 V(X_1 + X_2) \qquad \text{By virtue of Equation 5–17} \\ &= (1/n^2)(\sigma^2 + \sigma^2) = (1/n^2)(n\sigma^2) = \sigma^2/n \end{aligned}$$

Given any n and replacement,

$$E(\overline{X}) = (1/n)E(X_1 + X_2 + \ldots + X_n) = \mu$$

and

$$V(\overline{X}) = (1/n)^2 V(X_1 + X_2 + \ldots + X_n) = \sigma^2/n$$

If $n = 2$ and sampling is *without* replacement, 15 sample combinations are possible. We can't establish the probability distributions of X_1 and X_2 until we indicate how these combinations occur (see Table 7–15).

Since the selections are random, the probability of any combination is 1/15 and the probability of any permutation is 1/30. The permutations in the right column of Table 7–15 are *joint events* where X_1 = the number of transfers on the first draw and X_2 = the number of transfers on the second draw. Table 7–16 is the joint probability distribution of X_1 and X_2. The main diagonal is blank because an observation chosen on the first draw can't be selected a second time.

The probability distributions of X_1 and X_2 in Table 7–17 are constructed from the marginal probabilities. These distributions and the probability distribution of X (see Table 7–12) are the same.

TABLE 7–16

		X_2 = *Number of Transfers on the Second Draw*						*Marginal Probability of* X_1
		1	3	4	5	6	9	
X_1 = *Number of Transfers on the First Draw*	1	—	1/30	1/30	1/30	1/30	1/30	5/30 = 1/6
	3	1/30	—	1/30	1/30	1/30	1/30	5/30 = 1/6
	4	1/30	1/30	—	1/30	1/30	1/30	5/30 = 1/6
	5	1/30	1/30	1/30	—	1/30	1/30	5/30 = 1/6
	6	1/30	1/30	1/30	1/30	—	1/30	5/30 = 1/6
	9	1/30	1/30	1/30	1/30	1/30	—	5/30 = 1/6
Marginal Probability of X_2		1/6	1/6	1/6	1/6	1/6	1/6	

TABLE 7–17

X_1	$P(X_1)$	X_2	$P(X_2)$
1	1/6	1	1/6
3	1/6	3	1/6
4	1/6	4	1/6
5	1/6	5	1/6
6	1/6	6	1/6
9	1/6	9	1/6

Even though we sampled without replacement, $E(X_1)$, $E(X_2)$, and $E(X)$ still equal μ, and $V(X_1)$, $V(X_2)$, and $V(X)$ still equal σ^2. Therefore,

$$E(\overline{X}) = (1/n)E(X_1 + X_2) = (1/n)(\mu + \mu)$$
$$= (1/n)(n\mu) = \mu$$

In contrast to the replacement situation, however, X_1 and X_2 are no longer independent. The probability of each joint event isn't equal to the product of the marginal probabilities $[1/30 \neq (1/6)(1/6)]$.

When X_1 and X_2 are *dependent,* the variance of $X_1 + X_2$ is

$$V(X_1 + X_2) = V(X_1) + 2\text{Cov}(X_1,X_2) + V(X_2)$$
$$= 6.2 + 2\text{Cov}(X_1,X_2) + 6.2$$

We find that $\text{Cov}(X_1,X_2) = -1.244$:

$$\sum\sum (X_1 - \mu)(X_2 - \mu)P(X_1,X_2) = (3 - 4.67)(1 - 4.67)(1/30) +$$
$$(4 - 4.67)(1 - 4.67)(1/30) +$$
$$(5 - 4.67)(1 - 4.67)(1/30) +$$
$$\vdots$$
$$= -1.244$$

Thus,

$$V(X_1 + X_2) = 6.2 + 2(-1.244) + 6.2 = 9.91$$

and

$$V(\overline{X}) = V[(1/n)(X_1 + X_2)] = (1/n)^2V(X_1 + X_2)$$
$$= (1/4)(9.91) = 2.48$$

The "replacement" $V(\overline{X})$ of $(1/4)(6.2 + 6.2) = 3.10$ is greater than the "without replacement" $V(\overline{X})$ of 2.48 because "extreme" pairings like 9,9 and "extreme" $\overline{X}$s like 9.0 are possible if an observation can be selected twice. The ratio of the "without replacement" to the "with replacement" $V(\overline{X})$ is 2.48/3.10 = .80. Since

$$\frac{N-n}{N-1}=\frac{6-2}{6-1}$$

is also .80, we may conclude that

$$V(\overline{X}) \text{ without replacement} = \left[V(\overline{X}) \text{ with replacement}\right]\left[\frac{N-n}{N-1}\right]$$
$$= \frac{\sigma^2}{n}\left[\frac{N-n}{N-1}\right]$$

In large populations, one draw will not have much of an effect on the probability of a future outcome. We can act as though the selections were independent. Then, σ^2/n will approximate $V(\overline{X})$ without replacement.

Because of its complexity, we won't show the proof of the central limit theorem. When the population is normal, however, the normality of the sampling distribution of $\overline{X}$ is a consequence of the fact that

$$\overline{X} = (1/n)X_1 + (1/n)X_2 + \ldots + (1/n)X_n$$

will be a linear combination of normal random variables.

Normal-Curve Probabilities

We have seen that the sampling distribution of $\overline{X}$ will be normal when (1) a large sample is selected from a nonnormal population or (2) a large or small sample is selected from a normal population. Let's compute $P(a < \overline{X} < b)$, the probability that the random variable $\overline{X}$ will have a value between a and b, for a normal sampling distribution.

In Chapter Six, we selected one X from a distribution such as Figure 7–6. At this stage, we wish to select one $\overline{X}$ from Figure 7–7.

FIGURE 7–6

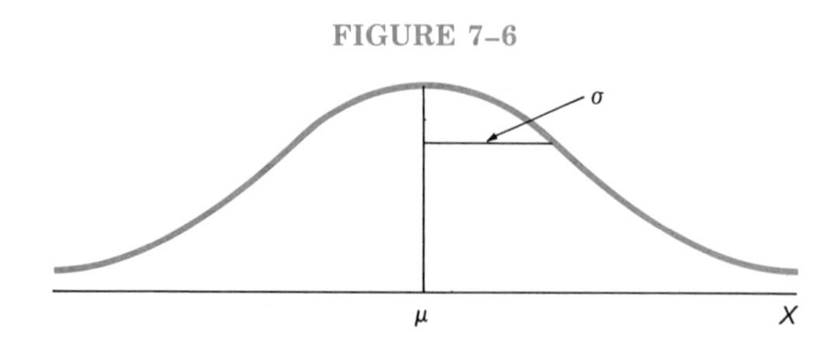

FIGURE 7–7

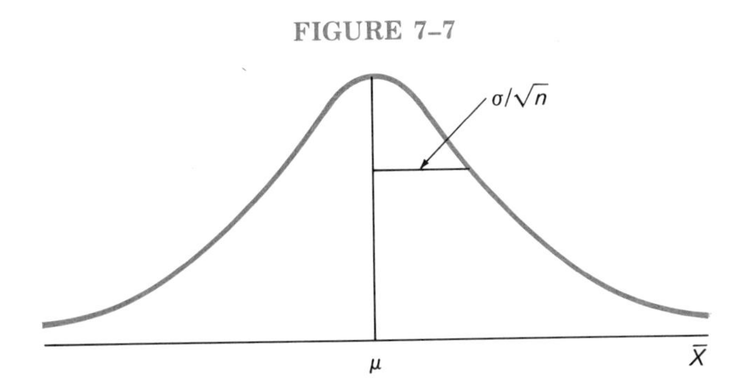

Since

$$Z = \frac{\begin{array}{c}\text{value of the} \\ \text{random variable}\end{array} - \begin{array}{c}\text{mean of the} \\ \text{probability} \\ \text{distribution}\end{array}}{\begin{array}{c}\text{standard deviation} \\ \text{of the probability} \\ \text{distribution}\end{array}},$$

we can change $\overline{X}$ to Z or Figure 7–7 to the standard normal curve by finding

$$Z = \frac{\overline{X} - \mu_{\bar{x}}}{\sigma_{\bar{x}}} = \frac{\overline{X} - \mu}{\dfrac{\sigma}{\sqrt{n}}} \qquad (7\text{–}10)$$

The Z in Equation 7–10 expresses the distance between $\overline{X}$ and μ in terms of $\sigma_{\bar{x}}$s. We'll use the following example to show how $P(a < \overline{X} < b)$ is derived.

E & S Products, Inc., employs a large sales force. Prior to 1979, the sales personnel were paid a straight salary. Management created a "salary + commission" plan in 1979. Sales personnel received a basic monthly salary of $1,000 and a 5 percent commission on monthly sales in excess of $30,000. The distribution of payments for that year had a mean of $16,000 and a standard deviation of $3,000. Management would like to discuss the new plan with a random sample of 36 salespeople.

Inasmuch as n is "large," we know that the sampling distribution of $\overline{X}$ will be normal. Additionally, $\mu_{\bar{x}}$ will equal $16,000 and $\sigma_{\bar{x}}$ will equal $3,000/\sqrt{36}$ = $500 (see Figure 7–8).

Suppose that management is interested in the probability that

FIGURE 7–8

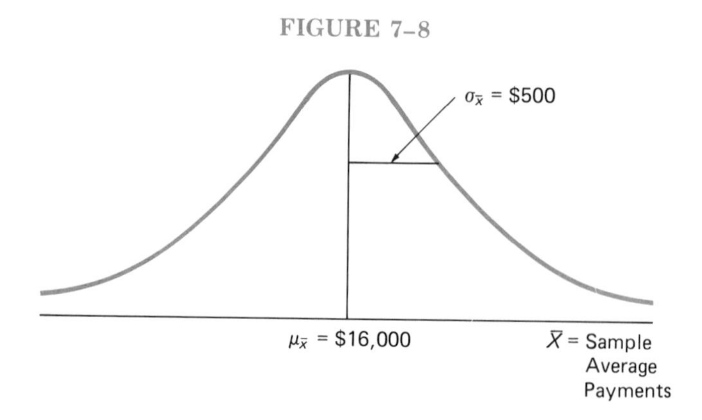

its sample will have average payments between $15,200 and $17,400. This probability corresponds to the shaded area in Figure 7–9.

FIGURE 7–9

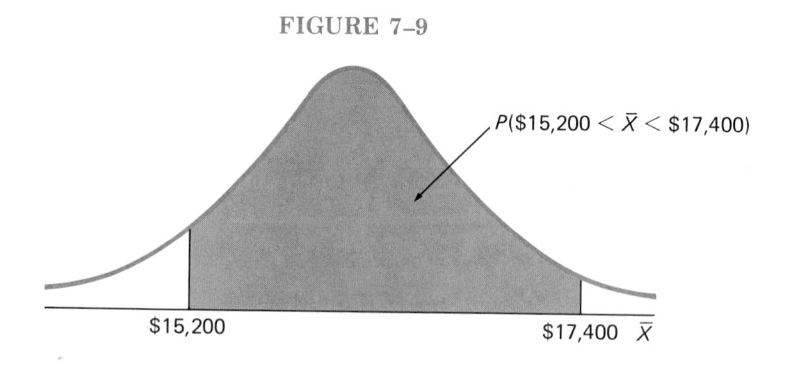

The relevant standard normal-curve area (Figure 7–10) lies between

$$Z_1 = \frac{\$15{,}200 - \$16{,}000}{\$500} = -1.60$$

and

$$Z_2 = \frac{\$17{,}400 - \$16{,}000}{\$500} = 2.80$$

We see, then, that $P(\$15{,}200 < \overline{X} < \$17{,}400) = .4452 + .4974 = .9426$.

Given the population of 1979 payments, 94 percent of all the M different samples of 36 salespeople would have average payments between $15,200 and $17,400. Stated another way, if samples of 36

FIGURE 7–10

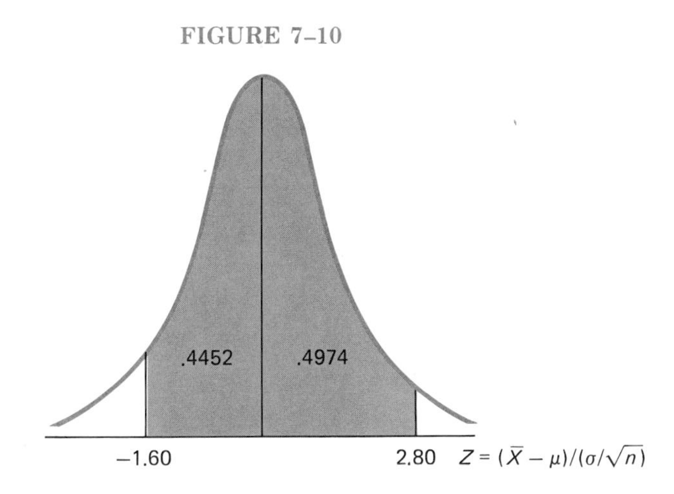

salespeople were repeatedly chosen from the population, 94 percent of the sample averages would fall in the interval \$15,200 $< \bar{X} <$ \$17,400.

SUMMARY

By selecting samples randomly, we were able to develop the sampling distribution of $\bar{X}$. The parameters of this probability distribution are related to the population parameters. The central limit theorem and the equalities $\mu_{\bar{x}} = \mu$ and $\sigma_{\bar{x}} = \sigma/\sqrt{n}$ are the foundation for the techniques of estimation and hypothesis testing that we will study in Chapters Eight and Nine. If observations are not chosen randomly, the sampling distribution characteristics which we have discussed are no longer valid.

EXERCISES

1. The President's Commission on Pornography published a report in the 1970s. As part of its research, suppose the commission selected a random sample of 36 corporate presidents and asked them, "How many obscene movies did you see last year?" What are some of the non-sampling errors associated with this experiment?

2. What factors force the federal government to use a sample to determine the national unemployment rate?

3. A researcher gives each of 5,000 people a number from 0001 to 5,000. Locate the sixth row in Table 7–2. Starting with 2761 and 2546, identify the first twelve people in this sample.

4. If five observations are chosen without replacement from a population of 100 observations, how many different samples are possible?

5. A particular population is large relative to the sample size. If $\sigma = 30$ and $n = 64$, what is the variance of the sampling distribution of $\overline{X}$?

6. A particular population is large relative to the sample size. If $\sigma_{\bar{x}} = 5$ and $n = 36$, what is the standard deviation of the population?

7. Table 7–3 indicates the population of transfers for B. D. International. Suppose that four employees are chosen at random.

   a. Construct the sampling distribution of $\overline{X}$.
   b. Compute $\mu_{\bar{x}}$ and $\sigma_{\bar{x}}$.
   c. Show that the answers in (b) are equivalent to the values derived from Equations 7–7 and 7–5.

8. If B. D. International picks five employees at random, what is the probability that $\overline{X}$ = the sample average number of transfers will be 4.8 or more?

9. We have noted that any statistic can be computed once the M different samples of size n have been determined. Therefore, we can construct the sampling distribution of the sample median, the sampling distribution of the sample variance, and so on. Consider the data in Tables 7–3 and 7–8.

   a. Establish the sampling distribution of the sample median for $n = 3$ transfers.
   b. Calculate the expected value and variance of this distribution.
   c. What is the probability that the sample median will be 4 or under?
   d. What is the relationship between the population median and the expected value of the sample median?

10. Suppose 1, 3, 5, 7 represents a population.

   a. Construct the joint probability distribution of X_1 = the outcome on the first draw and X_2 = the outcome on the second draw when two observations are randomly selected with replacement.
   b. Construct the joint probability distribution under the assumption that sampling is done without replacement.

c. For both cases, show that $(1/n)[E(X_1) + E(X_2)] = \mu$.
d. For (a), show that $(1/n)^2[V(X_1) + V(X_2)] = \sigma^2/n$.

11. In a particular normally distributed population, $\mu = 40$ and $\sigma = 12$. If an observation is selected at random, what are the following probabilities?

a. $P(39.6 < X < 42.3)$
b. $P(X > 40.5)$
c. $P(X < 39.4)$
d. $P(X > 44)$
e. $P(40.7 < X < 42)$

12. Suppose that a random sample of 36 observations is chosen from the population in question 11. Compute the following probabilities:

a. $P(39.6 < \overline{X} < 42.3)$
b. $P(\overline{X} > 40.5)$
c. $P(\overline{X} < 39.4)$
d. $P(\overline{X} > 44)$
e. $P(40.7 < \overline{X} < 42)$

13. Calculate the probabilities in question 12 under the assumption that a random sample of 100 observations is selected.

14. A particular sampling distribution of $\overline{X}$ is derived from a population in which $\mu = 34$ and $\sigma = 5$. Imagine that $n = 49$.

a. Determine $\mu_{\overline{X}}$ and $\sigma_{\overline{X}}$.
b. What is the probability that the value of $\overline{X}$ will be between 32.7 and 33.6?
c. What is the probability that the value of $\overline{X}$ will be greater than 34.9?

15. The earnings of the 2,000 employees at Company Q have a mean of $17,000 and a standard deviation of $2,500. Suppose a random sample of forty-nine employees is chosen.

a. What is the probability that the average earnings of these employees will be less than $16,200?
b. What is the probability that the average earnings of these employees will be between $16,500 and $17,500?

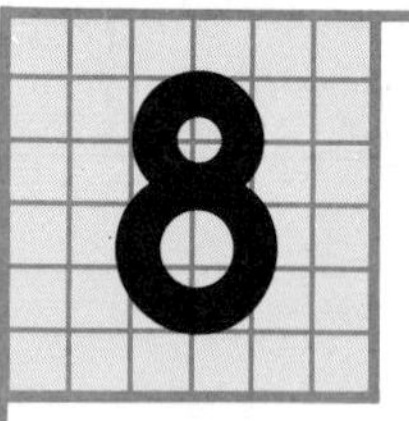

Statistical Inference: *Estimation with One Sample Mean—Large Samples*

PROPERTIES OF ESTIMATORS

Chapter Seven detailed the characteristics of the sampling distribution of $\overline{X}$. Starting with a known population, we proceeded to develop these features according to Figure 8–1.

FIGURE 8–1

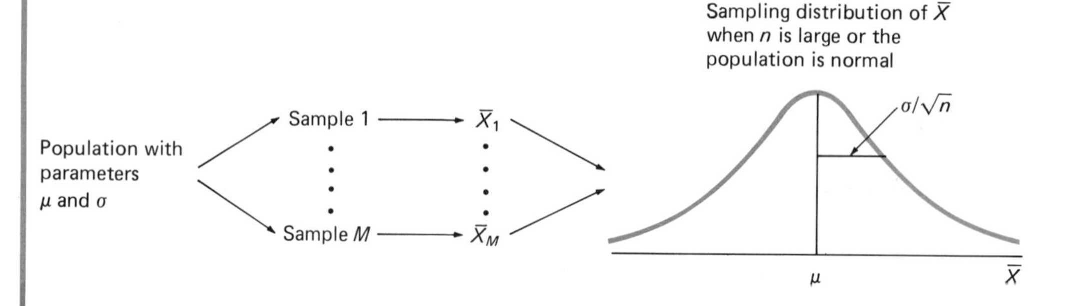

Given the sampling distribution, we can determine the probability that the estimation error will be small. This task is the substance of Chapter Eight. Before computing

$$P(|\overline{X} - \mu| \leq e),$$

that is, the probability that the distance between $\overline{X}$ and μ is no greater than some value called e, we will explore the theoretical justification for using $\overline{X}$ to estimate μ. Throughout this chapter, we will work with the sequence outlined in Table 8–1.

For generality, suppose that the Greek letter *theta*, θ, symbolizes a parameter and that θ^* is the statistic used to estimate θ. We will

TABLE 8–1

Population with an unknown μ	→	Randomly select one sample.	→	Compute $\overline{X}$.	→	Estimate μ.	→	Make an estimation error, $\overline{X} - \mu$.

refer to θ^* as an *estimator* of θ. A value of θ^* is said to be an *estimate*. We recently introduced $\overline{X}$ as an estimator of μ. With $\overline{X} = 2.0$ transfers, for example, the estimate of μ is 2.0 transfers. Based on how estimates behave in repeated random sampling, statisticians have developed some criteria for identifying "good" estimators. The criteria include *unbiasedness, consistency,* and *efficiency*. We shall discuss each of these in turn.

Unbiasedness

Remember that none of the sample averages in the transfers problem in Chapter Seven had the same value as μ. Estimates will often differ from θ. Because of this, we will be interested in the "long-run" (i.e., repeated sampling) relation between estimated values of θ and the actual value of that parameter. Within this long-run context, an estimator is evaluated favorably if the "typical" estimate is the same as θ. When the sampling distribution of θ^* is such that

$$E(\theta^*) = \theta, \qquad (8\text{–}1)$$

θ^* is an *unbiased* estimator of θ.

The sample average is an unbiased estimator of μ since $E(\overline{X}) = \mu$. We can also illustrate that the sample median is an unbiased estimator of the population median. Given the transfers population (1, 3, 4, 5, 6, 9), the median is 4.5. Table 8–2 is the sampling distribution of the sample median when $n = 3$. From the table, we can compute

$$\begin{aligned} E(\text{sample median}) &= \Sigma(\text{sample median})P(\text{sample median}) \\ &= (3)(4/20) + (4)(6/20) + (5)(6/20) + (6)(4/20) \\ &= 4.5 \end{aligned}$$

TABLE 8–2

Sample Median Transfers	*P(Sample Median)*
3	4/20
4	6/20
5	6/20
6	4/20

We can also investigate unbiasedness in connection with the sample variance, s^2. We have calculated s^2 as

$$s^2 = \frac{\Sigma(X - \overline{X})^2}{n - 1}$$

in order to make the sample variance an unbiased estimator of the population variance. If we estimated σ^2 by $\Sigma(X - \overline{X})^2/n$, the "typical" estimate would be smaller than the value of σ^2. Placing $n - 1$, rather than n, in the denominator raises each s^2 enough that $E(s^2) = \sigma^2$.

To prove that s^2 is unbiased, let's consider $E[\Sigma(X - \overline{X})^2/n]$. Since $\Sigma(X - \overline{X})^2/n = (1/n)\Sigma(X - \overline{X})^2$, this expectation is the same as either $(1/n)E[\Sigma(X - \overline{X})^2]$ or $(1/n)E[\Sigma(X - \overline{X} + 0)^2]$. Let's substitute $\mu - \mu$ for 0:

$$(1/n)E[\Sigma(X - \overline{X} + \mu - \mu)^2] = (1/n)E\{\Sigma[(X - \mu) - (\overline{X} - \mu)]^2\}$$

Expanding the square, we get

$$(1/n)E\{\Sigma[(X - \mu)^2 - 2(X - \mu)(\overline{X} - \mu) + (\overline{X} - \mu)^2]\}$$
$$= (1/n)E[\Sigma(X - \mu)^2 - 2\Sigma(X - \mu)(\overline{X} - \mu) + \Sigma(\overline{X} - \mu)^2]$$

Due to the fact that both $\overline{X}$ and μ are constants for a single sample, the right side of this equation is identical to

$$(1/n)E[\Sigma(X - \mu)^2 - 2(\overline{X} - \mu)\Sigma(X - \mu) + n(\overline{X} - \mu)^2]$$

Additionally,

$$\sum^{n}(X - \mu) = \sum^{n}X - \sum^{n}\mu = n\overline{X} - n\mu = n(\overline{X} - \mu)$$

We may now write

$$(1/n)E[\Sigma(X - \mu)^2 - 2n(\overline{X} - \mu)^2 + n(\overline{X} - \mu)^2]$$
$$= (1/n)E[\Sigma(X - \mu)^2 - n(\overline{X} - \mu)^2]$$

Because the expectation of a sum is equal to the sum of the expectations, we are left with

$$(1/n)[\Sigma E(X - \mu)^2 - nE(\overline{X} - \mu)^2]$$

Notice that $E(X - \mu)^2$ is the variance of X and $E(\overline{X} - \mu)^2$ is the variance of $\overline{X}$. As a consequence,

$$(1/n)[\Sigma E(X - \mu)^2 - nE(\overline{X} - \mu)^2] = (1/n)[n\sigma^2 - n(\sigma^2/n)]$$
$$= \sigma^2 - \sigma^2/n$$

With s^2 defined as $\Sigma(X - \overline{X})^2/n$, the "typical" value of s^2 underestimates σ^2 by σ^2/n. When $n - 1$ is used, however,

$$E(s^2) = [1/(n - 1)][n\sigma^2 - n(\sigma^2/n)] = [1/(n - 1)][(n - 1)\sigma^2] = \sigma^2$$

Besides $\overline{X}$, we can construct *many* unbiased estimators of μ. One of these θ^*s is

$$\theta^* = (1/5)X_1 + (4/5)X_2$$

where the first two observations selected are X_1 and X_2. The expected value of θ^* is $E[(1/5)X_1 + (4/5)X_2]$. Inasmuch as $E(aX + bY) = aE(X) + bE(Y)$,

$$E(\theta^*) = (1/5)E(X_1) + (4/5)E(X_2)$$

When sampling is random, $E(X_1) = E(X_2) = \ldots = E(X_n) = \mu$, and, therefore,

$$E(\theta^*) = (1/5)\mu + (4/5)\mu = \mu$$

If the population is symmetrical, the sample median will also be an unbiased estimator of μ. We can show this by introducing the population 1, 4, 6, 9. Table 8–3 is the sampling distribution of the sample median for $n = 3$. Notice that

$$E(\text{sample median}) = (4)(2/4) + (6)(2/4) = 5 = \mu$$

Since $\overline{X}$ is not the only unbiased estimator of μ, any preference for $\overline{X}$ must rest on other considerations.

TABLE 8–3

Sample Median	*P(Sample Median)*
4	2/4
6	2/4

Before moving on, however, let's explore another estimator. If $\theta^* = (1/5)X_1 + (6/5)X_2$, we would say that θ^* was a *biased* estimator of μ because

$$E(\theta^*) = (1/5)\mu + (6/5)\mu = 1.4\mu$$

We can visualize the "bias" if we assume that X_1 and X_2 come from a normal population. The new estimator will, then, be a linear combination of normal random variables, and its sampling distribution will be normal as well (see Figure 8–2). The difference between $E(\theta^*)$ and θ (i.e., $1.4\mu - \mu = 0.4\mu$) is the "bias" of $(1/5)X_1 + (6/5)X_2$. When we repeatedly estimate μ by this estimator, the "typical" estimate will be 0.4μ larger than μ.

Consistency

We next want to investigate consistency. Inasmuch as the size of a sample influences sampling costs, researchers are also concerned about whether a large n "pays off." We can regard the "payoff" as an increase in the probability that θ^* will be near the unknown

FIGURE 8–2

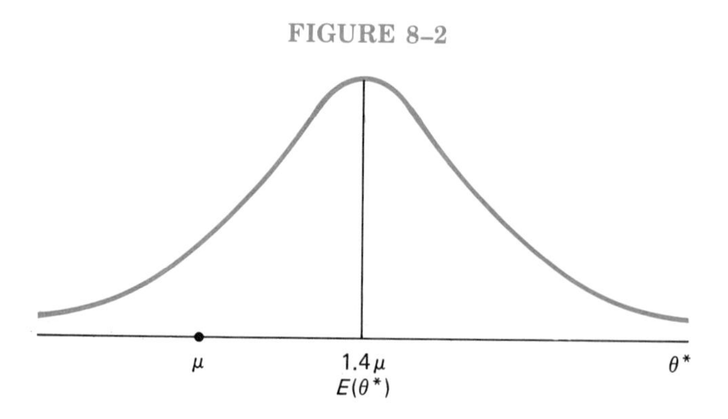

parameter. An estimator is called *consistent* if, for some positive value e,

$$P(|\theta^* - \theta| < e) \text{ approaches 1 as } n \text{ approaches } \infty \qquad (8\text{–}2)$$

Condition 8–2 implies that the chance of incurring a big estimation error is reduced when n increases.

To clarify this idea, consider the normal sampling distribution in Figure 8–3. Every sample average that has a value between $\overline{X}_1$ and $\overline{X}_2$ is no more than a distance of e from μ. Hence, $P(\overline{X}_1 < \overline{X} < \overline{X}_2) = P(|\overline{X} - \mu| < e)$.

FIGURE 8–3

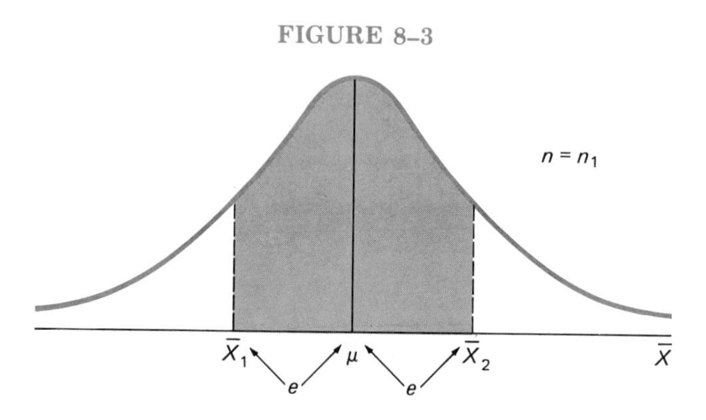

For $\overline{X}$ to be a consistent estimator, $P(\overline{X}_1 < \overline{X} < \overline{X}_2)$ or $P(|\overline{X} - \mu| < e)$ must increase as n does. Since $\sigma_{\bar{x}} = \sigma/\sqrt{n}$, the sampling distribution drawn for $n_2 > n_1$ will be more pinched-in than that of Figure 8–3, resembling Figure 8–4 instead. $P(|\overline{X} - \mu| < e)$ rises because a greater area now lies in the interval $\overline{X}_1$ to $\overline{X}_2$. When we select a sample of n_2 observations compared to n_1, we can be more confident that the corresponding estimation error will be less than e.

FIGURE 8–4

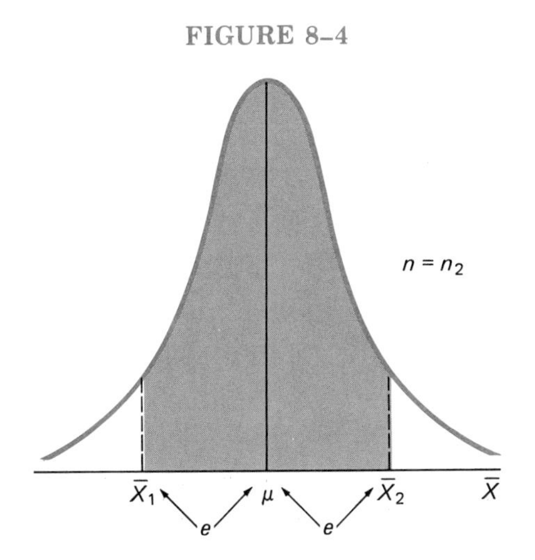

To expand our discussion of consistency, we'll state the following theorem:

> If a population is normal, the sampling distribution of the *sample median* will be normal with a mean of μ and a standard deviation approximately equal to $1.253(\sigma/\sqrt{n})$.

Since the normal distribution is symmetrical, our previous statement that E(sample median) $= \mu$ whenever a population is symmetrical is reinforced by the theorem.

A normal sampling distribution is presented in Figure 8–5. Increases in n cause $1.253(\sigma/\sqrt{n})$ to decline. The sampling distribution, therefore, closes in on μ, as in Figure 8–6. Thus we can conclude that when the population is normal, the sample median is a consistent estimator of μ.

Finally, let's study the unbiased estimator $(1/5)X_1 + (4/5)X_2$. Its variance is

$$V[(1/5)X_1 + (4/5)X_2] = (1/5)^2V(X_1) + (4/5)^2V(X_2)$$

FIGURE 8–5

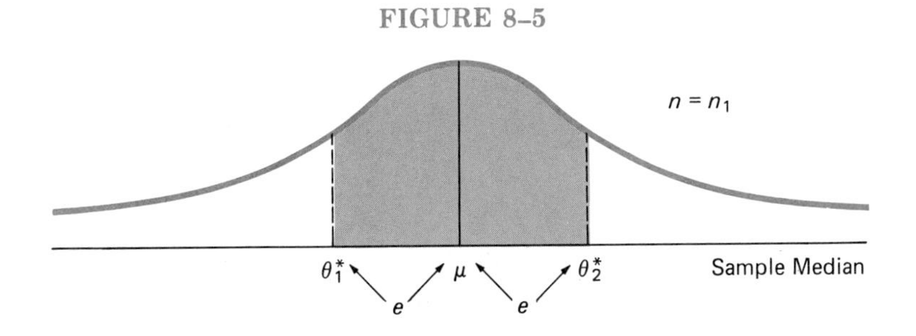

FIGURE 8–6

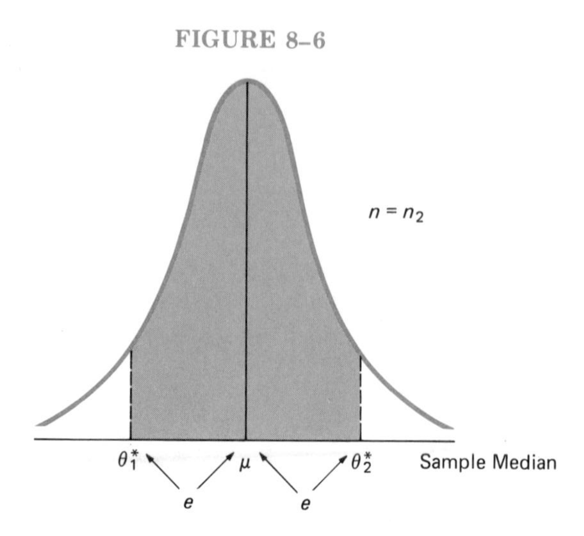

Given random sampling, $V(X_1) = V(X_2) = \ldots = V(X_n) = \sigma^2$. Thus,

$$V[(1/5)X_1 + (4/5)X_2] = (1/5)^2\sigma^2 + (4/5)^2\sigma^2 = (17/25)\sigma^2 = .68\sigma^2$$

Assuming a normal population, the sampling distribution of this last estimator would look like Figure 8–7.

FIGURE 8–7

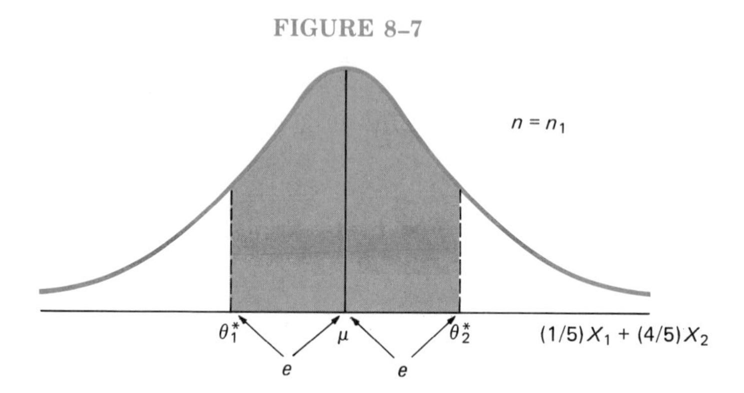

For illustrative purposes only, imagine that a researcher is willing to select larger and larger samples even though the estimator will remain as $(1/5)X_1 + (4/5)X_2$. Since such an estimator uses only two observations, the standard deviation of the sampling distribution (i.e., $.82\sigma$) will be the same for $n = 3$, $n = 30$, $n = 300$, and so on. Hence, $P[|(1/5)X_1 + (4/5)X_2 - \mu| < e]$ is constant as well. Because the probability doesn't approach 1 as n increases, $(1/5)X_1 + (4/5)X_2$ is an *inconsistent* estimator of μ.

Efficiency

Although $\overline{X}$ and the sample median are both unbiased and consistent, the sample mean has an advantage. In Figure 8–8, we have superimposed Figure 8–5 on Figure 8–3. Each distribution is constructed from samples of $n = n_1$ observations.

FIGURE 8–8

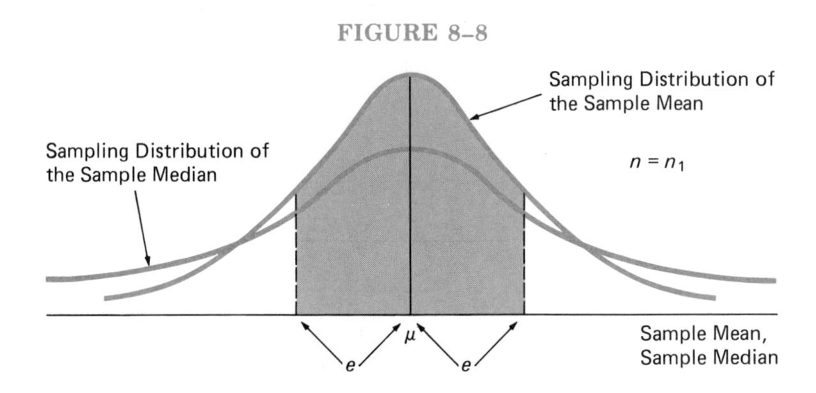

Since the sampling distribution of $\overline{X}$ is more concentrated around μ, $P(|\overline{X} - \mu| < e)$ is greater than $P(|\text{sample median} - \mu| < e)$. If we intend to choose a sample and derive estimates of μ based on (1) the sample mean and (2) the sample median, $\overline{X}$ will have a higher probability of being close to μ. We express this fact by saying that $\overline{X}$ is a *relatively more efficient* estimator than the sample median.

Where θ^* and θ^{**} are two unbiased estimators of θ, the efficiency of θ^* relative to θ^{**} is given by the ratio

$$\text{Relative efficiency of } \theta^* = V(\theta^{**})/V(\theta^*) \qquad (8\text{–}3)$$

The estimator θ^* is relatively more efficient than θ^{**} when this value exceeds 1.0. In particular, $\overline{X}$ is

$$\frac{V(\text{sample median})}{V(\overline{X})} = \frac{(1.253\sigma/\sqrt{n})^2}{(\sigma/\sqrt{n})^2} = 1.57$$

times more efficient than the sample median.

Due to the fact that the variance of the sample median is 57 percent larger than the variance of $\overline{X}$, $P(|\text{sample median} - \mu| < e)$ would equal $P(|\overline{X} - \mu| < e)$ only if $\overline{X}$ were to come from a sample of n and the sample median were to come from a sample of $1.57n$ (illustrated in Figure 8–9). By using the sample mean rather than the sample median, a given $P(|\theta^* - \mu| < e)$ can be achieved with less sampling expense. In fact, among all unbiased estimators of μ, $\overline{X}$ has the smallest, or *minimum*, variance.

FIGURE 8–9

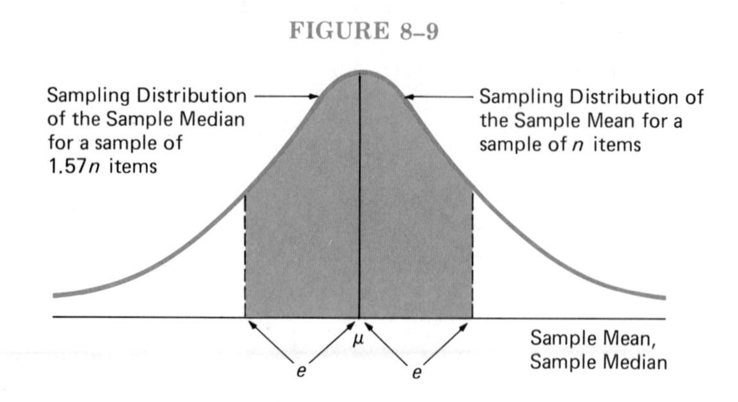

Based on our examination of estimators, we can note that (1) $\overline{X}$ is unbiased; (2) $\overline{X}$ has a higher probability of being close to μ as n increases; and (3) among unbiased estimators of μ derived from samples of the same size, $\overline{X}$ has the greatest probability of lying near μ. Table 8–4 summarizes why $\overline{X}$ is the "best" estimator of μ.

TABLE 8–4

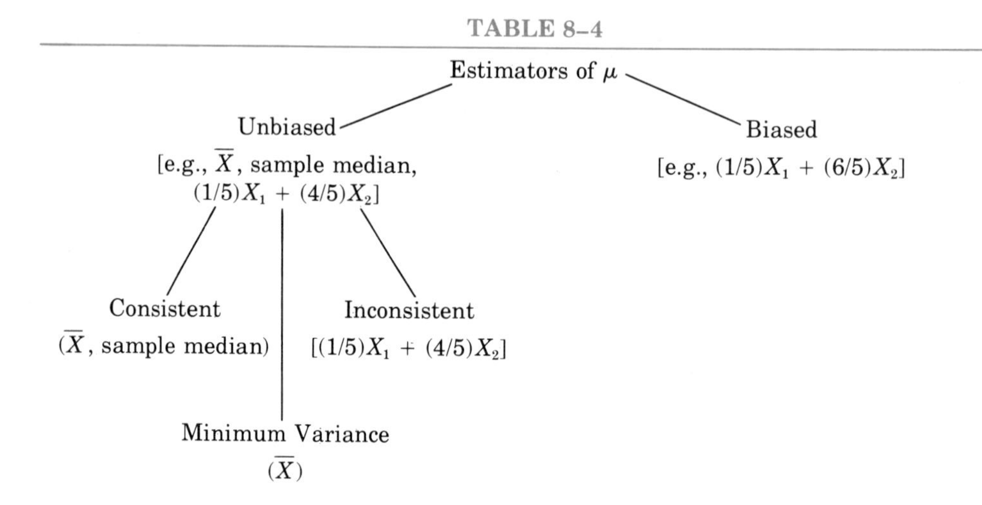

In studying estimation, remember that statistical inference is rarely an end in and of itself. Certain actions will evolve from estimates. When inferences are in error, these actions could be extremely inappropriate. A scenario involving the consequences of an incorrect inference appears in Table 8–5.

TABLE 8–5

A sample of American households is selected by the federal government.	→	The sample proportion (a statistic) of unemployed is calculated.	→	The sample proportion is used to estimate the national rate of unemployment (a parameter).	→	Various monetary and fiscal strategies are implemented when the estimate indicates substantial unemployment. If the actual rate of unemployment is .05 (i.e., 5%), but the estimate is .12 (i.e., 12%), the government will believe that antirecession measures (e.g., tax reductions and increased government spending) are required. Since 5% is near "full employment," the government's actions would be quite inflationary.

DETERMINING THE PROBABILITY OF AN ERROR

If we're going to act on the basis of an estimate, we should naturally be concerned about the distance between θ^* and θ. We have defined the estimation error as $\theta^* - \theta$. In the case of $\overline{X}$ and μ, this error is $\overline{X} - \mu$. Since inferences are developed within the context of uncertainty, we'll never know what error is *actually* associated with an observed $\overline{X}$. We wish to show that researchers can, however, derive the probability that a potential error will be small. To detail the approach, we will use the following example.

Aware of the growing number of job-related injuries, Congress authorized the creation of the Occupational Safety and Health Administration (OSHA) in 1970. This agency regulates working conditions. Because of budgetary and personnel restrictions, OSHA must establish priorities. Suppose that it will postpone revising safety standards for industry H unless μ = H's average number of accidents per plant is excessive.

A researcher intends to estimate this average by selecting a random sample of 36 industry H plants. From 1978 data, the re-

FIGURE 8–10

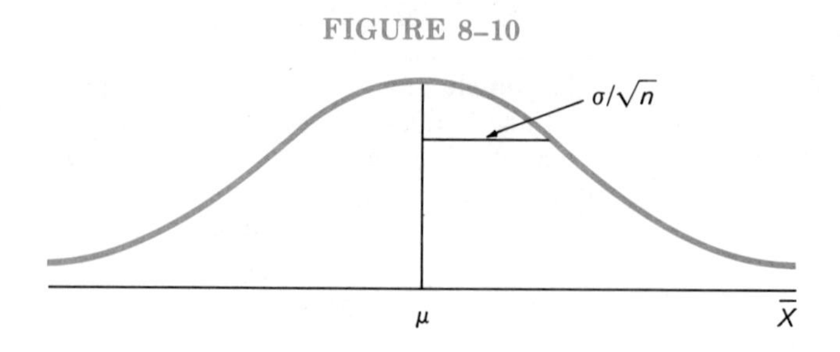

searcher will compute $\overline{X}$ = sample average number of accidents per plant. He assumes that $\sigma = 2.0$ accidents.

Since $n = 36$ is "large," the central limit theorem makes either Figure 8–10 or Figure 8–11 applicable for inferences. If the researcher wishes to (1) choose a random sample, (2) calculate $\overline{X}$, and (3) find Z, $P(-1.96 < Z < 1.96)$ will be .95 (see Figure 8–12).

FIGURE 8–11

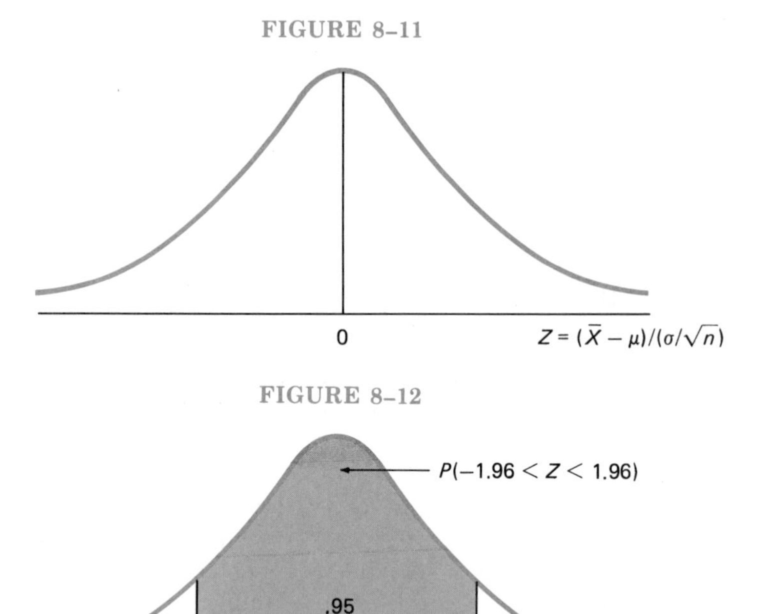

FIGURE 8–12

We shall let the Greek letter *alpha*, α, represent the probability that Z will *not* lie in the interval $-Z^*$ to Z^*. We can, then, state

$$P(-Z^* < Z < Z^*) = 1 - \alpha \qquad (8\text{–}4)$$

Figure 8–13 identifies the area between $-Z^*$ and Z^*. We want to examine $-Z^* < Z < Z^*$ further.

FIGURE 8–13

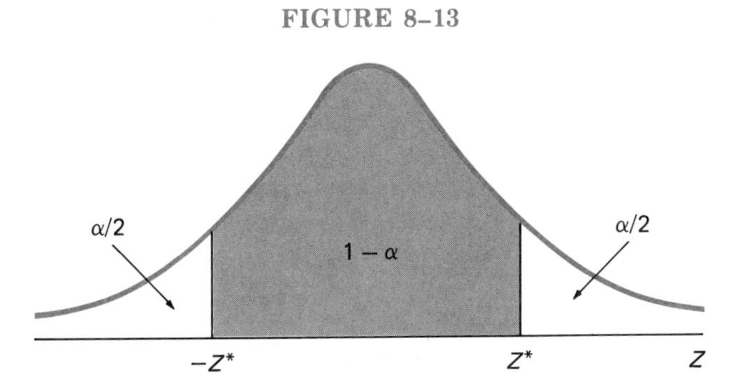

Any discussion of (1) $-Z^* < Z < Z^*$ is just as relevant to

$$(2)\ -Z^* < \frac{\overline{X} - \mu}{\sigma/\sqrt{n}} < Z^*$$

because $Z = (\overline{X} - \mu)/(\sigma/\sqrt{n})$. Multiplying each part of (2) by $\sigma/\sqrt{n}$ yields

$$(3)\ -Z^*(\sigma/\sqrt{n}) < (\overline{X} - \mu) < Z^*(\sigma/\sqrt{n})$$

Inasmuch as (3) is a rearrangement of (1), we can also say

$$P[-Z^*(\sigma/\sqrt{n}) < (\overline{X} - \mu) < Z^*(\sigma/\sqrt{n})] = 1 - \alpha \qquad (8\text{–}5)$$

Notice, though, that $(\overline{X} - \mu)$ is the estimation error. Equation 8–5 indicates the following: Before sampling, there is a $1 - \alpha$ probability that $\overline{X}$ will neither underestimate nor overestimate μ by more than $Z^*(\sigma/\sqrt{n})$; that is,

$$P[\textit{maximum error} \text{ in either direction is } Z^*(\sigma/\sqrt{n})] = 1 - \alpha$$

Figure 8–14 illustrates this probability.

FIGURE 8–14

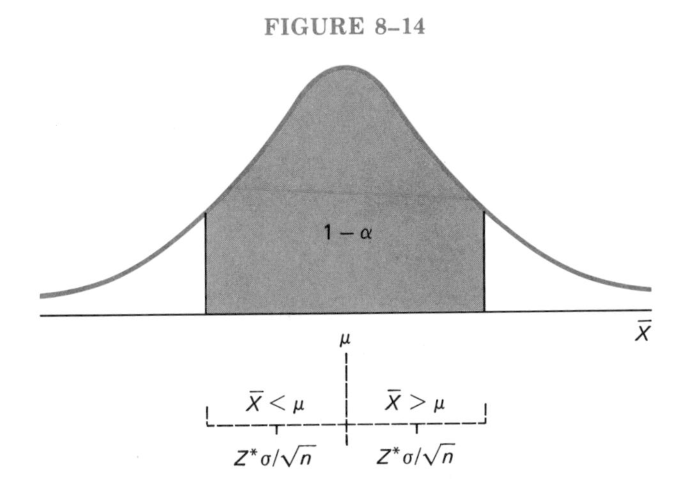

To appreciate Equation 8–5 conceptually, imagine that we identify the different samples that make up the sample space for "select 36 plants at random." Suppose we then list the samples in the order of their $\overline{X}$s (i.e., the sample with the smallest average is first, etc.). Of all the Zs that correspond to these averages, $100(1 - \alpha)$ percent are between $-Z^*$ and Z^*. The percentage of errors between $-Z^*(\sigma/\sqrt{n})$ and $Z^*(\sigma/\sqrt{n})$ is $100(1 - \alpha)$ percent as well. Figure 8–15 outlines the implications of our discussion.

From $P(-1.96 < Z < 1.96) = .95$, we can derive

$$P[-1.96(\sigma/\sqrt{n}) < (\overline{X} - \mu) < 1.96(\sigma/\sqrt{n})] = .95$$

FIGURE 8–15
A Frame of Reference for Estimation

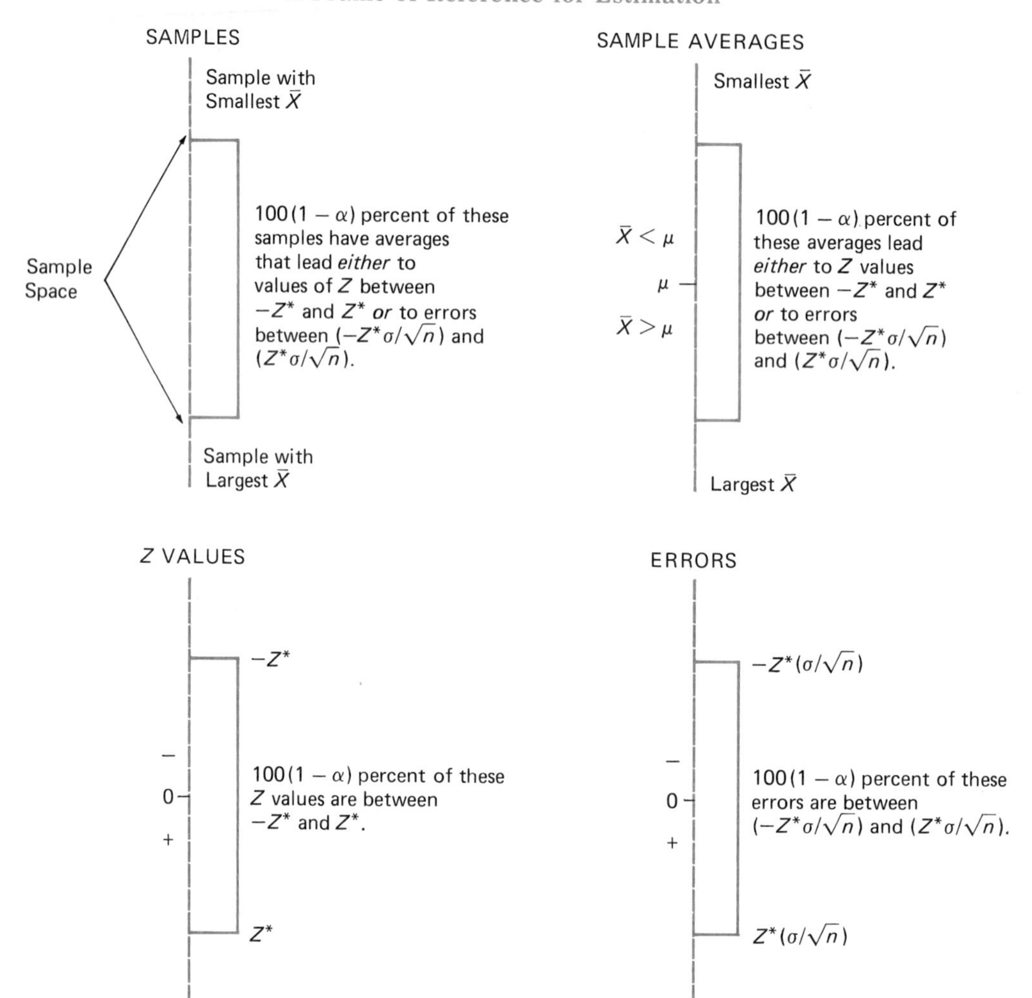

In the accident scenario, $n = 36$ plants and $\sigma = 2$ accidents. Therefore, given random sampling,

$$P[\text{maximum error is } Z^*(\sigma/\sqrt{n})]$$
$$= P[\text{maximum error is } (1.96)(2/\sqrt{36})]$$
$$= P(\text{maximum error is .65 accidents})$$
$$= .95$$

Given many repetitions of "select 36 plants at random," we would expect that the distance between μ and 95 percent of the estimates of μ would not exceed .65 accidents (see Figure 8–16).

FIGURE 8–16

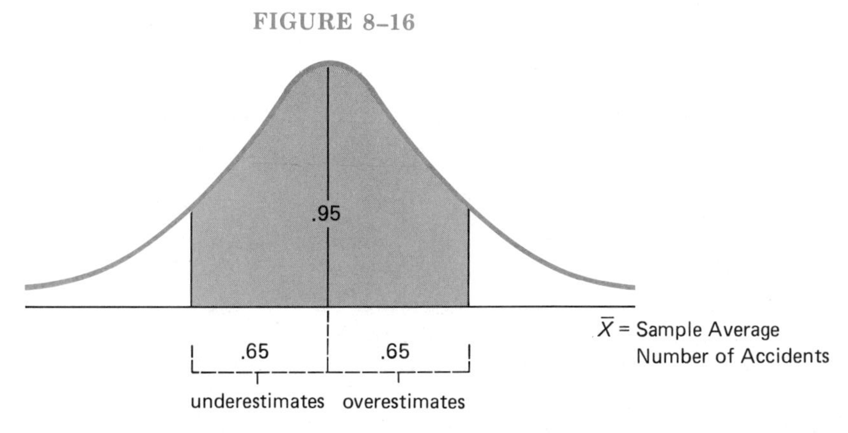

Additionally, because $P(-1.64 < Z < 1.64) = .90$, we can compute

$$P[-1.64(\sigma/\sqrt{n}) < (\overline{X} - \mu) < 1.64(\sigma/\sqrt{n})] = .90$$

The probability is .90 that a random sample of 36 plants will have an average which is less than or equal to $(1.64)(2/\sqrt{36}) = .55$ accidents from μ (see Figure 8–17). The bracketed sections of Figure 8–15 shrink if we shift from $P(-Z^* < Z < Z^*) = .95$ to $P(-Z^* < Z < Z^*) = .90$.

A one-value estimate of θ is called a *point estimate.* Prior to selecting a sample, we can view both $\overline{X}$ and $(\overline{X} - \mu)$ as random variables. We can, then, calculate the probability that $|\overline{X} - \mu|$ will not have a value larger than $Z^*(\sigma/\sqrt{n})$. After we have chosen the sample, however, |point estimate $- \theta|$ or $|\overline{X} - \mu|$ is the *actual* sampling error. This error either is or isn't greater than $Z^*(\sigma/\sqrt{n})$. We can't determine the true situation because μ is unknown.

The quantity $|\overline{X} - \mu|$ is now fixed. On the other hand,

$$P[|\overline{X} - \mu| \leq Z^*(\sigma/\sqrt{n})] = 1 - \alpha$$

is valid only when $|\overline{X} - \mu|$ is random. Since any statement about

FIGURE 8–17

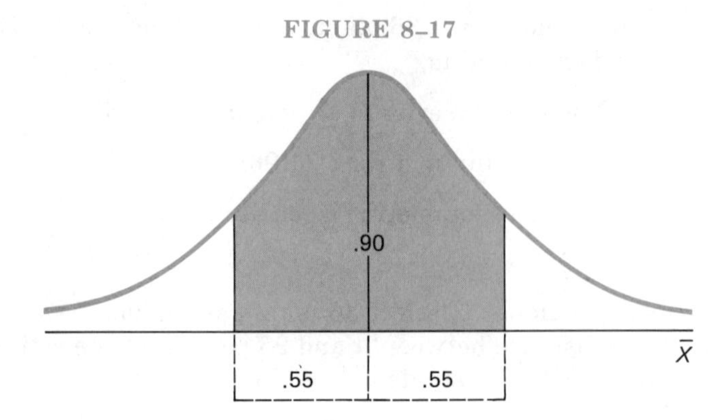

a $1 - \alpha$ probability is inappropriate once $\bar{X}$ is computed, we will discuss an observed $\bar{X}$ in a different context.

We know that, over the "long run," random sampling produces the event

$$|\bar{X} - \mu| \leq Z^*(\sigma/\sqrt{n})$$

in $100(1 - \alpha)$ percent of the trials. Therefore, if we obtain a sample and state that $|\bar{X} - \mu| \leq Z^*(\sigma/\sqrt{n})$, we will be correct $100(1 - \alpha)$ percent of the time. The "track record" of such statements is the *measure of confidence* that we have in them.

When 36 plants have been chosen, we can be 95 percent "confident" that $|\bar{X} - \mu| \leq .65$ accidents or 90 percent "confident" that $|\bar{X} - \mu| \leq .55$ accidents. Since the researcher can set α at any value, he is free to make a variety of "confidence" statements about $|\bar{X} - \mu|$. The *actual* error itself, however, *cannot* change.

Confidence statements include an indication of both the *reliability* [i.e., $100(1 - \alpha)$ percent] and the *precision* [i.e., $Z^*(\sigma/\sqrt{n})$] of the estimate. We hope to design our research so that $100(1 - \alpha)$ percent is high and $Z^*(\sigma/\sqrt{n})$ is small (see Table 8–6).

"High," "low," "large," and "small" are clearly subjective. Every researcher will evaluate these concepts differently. Given that a trade-off exists between the size of $100(1 - \alpha)$ percent and the size

TABLE 8–6

		$Z^*(\sigma/\sqrt{n})$	
		Large	Small
$100(1 - \alpha)\%$	High	—	Ideal situation
	Low	—	—

of $Z^*(\sigma/\sqrt{n})$, the question arises as to how the "high-small" case in Table 8–6 is achieved. While 95 percent is higher than 90 percent, .65 accidents is larger than .55 accidents. A researcher who believes that 95 percent is "high" may not feel that .65 accidents is "small." We now want to examine the way in which Z^* and n affect reliability and precision.

With $\sigma = 2$ accidents and $n = 36$, Equation 8–5 can be written as

$$P[|\overline{X} - \mu| \leq Z^*(2/\sqrt{36})] = 1 - \alpha$$

Recall that the area between $-Z^*$ and Z^* is $1 - \alpha$. When $Z^* = 2.33$, $1 - \alpha = .98$, and $Z^*(2/\sqrt{36}) = (2.33)(2/\sqrt{36}) = .78$ accidents. Table 8–7 shows the impact of varying Z^* if σ and n are held constant. Reading down the table, we see that reliability increases but precision decreases (i.e., the maximum error increases) as Z^* increases.

TABLE 8–7

Z^*	*Probability* $1 - \alpha$	*Reliability or Confidence* $100(1 - \alpha)\%$	*Precision or Maximum Error* $Z^*(2/\sqrt{36})$
1.00	.68	68%	.33 accidents
1.64	.90	90%	.55 accidents
1.96	.95	95%	.65 accidents
2.33	.98	98%	.78 accidents
2.58	.99	99%	.86 accidents

In Table 8–8, Z^* is held at 1.96 as n changes. Under these circumstances, precision increases when n increases. If the researcher makes $100(1 - \alpha)$ percent "high," Tables 8–7 and 8–8 suggest that he can make $Z^*(\sigma/\sqrt{n})$ "small" only by selecting a large sample.

The OSHA investigator thought that $100(1 - \alpha)\% = 95\%$ was "high." Moreover, he chose a sample of 36 plants because $(1.96)(2/\sqrt{36}) = .65$ accidents is "small." The random sample he selected

TABLE 8–8

n	*Reliability* $100(1 - \alpha)\%$	*Precision* $1.96(2/\sqrt{n})$
36	95%	.65 accidents
49	95%	.56 accidents
64	95%	.49 accidents
81	95%	.44 accidents

had an average of 7.3 accidents. To him, $\overline{X} = 7.3$ accidents is alarming. Inasmuch as this researcher is confident that 7.3 is near μ, he recommended that OSHA should publish stricter safety standards for industry H.

Since inferences pertain to populations with unknown parameters, we will rarely know σ if μ is unknown. A dilemma arises because a value of σ is used in computing $Z^*(\sigma/\sqrt{n})$. When n is large (i.e., > 30), the sample standard deviation, s, is a good estimator of σ. *Approximately* $100(1 - \alpha)$ percent of the sampling errors will fall between $-Z^*(s/\sqrt{n})$ and $Z^*(s/\sqrt{n})$. In Chapter Ten, we'll explain what happens if s is substituted for σ when a small sample is selected from a normal population.

DETERMINING THE SAMPLE SIZE

Of the three factors that make up e, namely, Z^*, σ, and n, the researcher can control both Z^* and n. Given any Z^*, we have seen that e declines as the sample gets larger. Since sampling costs increase along with n, researchers cannot be indifferent to the sample size.

In the planning stage of a statistical study, an investigator will identify a "high" value for $100(1 - \alpha)$ percent and a "low" value for e. Imagine that the OSHA researcher originally picked $100(1 - \alpha)\% = 95\%$ and $e = .30$ accidents. The sample size for which

$$P(|\overline{X} - \mu| \leq .30) = .95$$

must be derived from

$$\text{Maximum error} = e = .30 = Z^*(\sigma/\sqrt{n}) = 1.96(2/\sqrt{n})$$

Thus,

$$.30(\sqrt{n}) = (1.96)(2) = 3.92$$

$$\sqrt{n} = 3.92/.30 = 13.1$$

$$n = 13.1^2 = 172$$

The logic of choosing a sample of 172 observations arises from the fact that 95 percent of the $\overline{X}$s in the corresponding sampling distribution of $\overline{X}$ are within .30 accidents of μ (see Figure 8–18). Since the OSHA data consisted of only 36 observations, the researcher apparently felt that reducing e to .30 didn't justify the expense and effort involved in looking up 172 records.

FIGURE 8–18

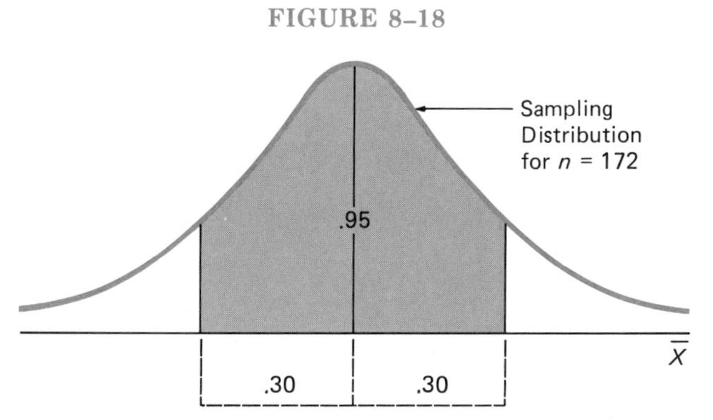

The procedure for finding the required n is generalized below, where $P(0 < Z < Z^*) = (1 - \alpha)/2$:

$$e = Z^*(\sigma/\sqrt{n})$$

$$e(\sqrt{n}) = Z^*(\sigma)$$

$$\sqrt{n} = Z^*(\sigma)/e$$

$$n = [Z^*(\sigma)/e]^2 \qquad (8\text{–}6)$$

In the previous section, we said that s will replace σ when σ is unknown. During the *presampling* period in which Equation 8–6 is relevant, however, we do not even have s. We can approach the problem in several ways:

1. Clues as to a reasonable value for σ may appear in published studies of the population.
2. A preliminary, or *pilot,* project could be conducted with a small sample. Some thoughts about σ could be established in the light of s.
3. Personal experience could be helpful.

If our estimate of σ is bigger than the unknown σ,

$$n = [Z^*(\text{estimate of } \sigma)/e]^2$$

will be greater than the n actually needed to state $P(|\overline{X} - \mu| \le e) = 1 - \alpha$. When the estimate is smaller than σ, $P(|\overline{X} - \mu| \le e) < 1 - \alpha$. By intentionally overestimating σ, the researcher can avoid the possibility of not getting the desired reliability.

CONSTRUCTING A CONFIDENCE INTERVAL

We have already computed (1) a single-value estimate of μ and (2) $P(|\overline{X} - \mu| \le e)$. A *confidence interval* expresses an estimate directly in terms of a *range* of possible values of the parameter. The following statements are consistent with each other:

1. $100(1 - \alpha)$ percent of the time, $\overline{X}$ will not miss μ by more than e; that is,

$$-Z^*(\sigma/\sqrt{n}) < (\overline{X} - \mu) < Z^*(\sigma/\sqrt{n}) \qquad (8\text{–}7)$$

2. $100(1 - \alpha)$ percent of the time, μ will fall between $\overline{X} - e$ and $\overline{X} + e$; that is,

$$\overline{X} - Z^*(\sigma/\sqrt{n}) < \mu < \overline{X} + Z^*(\sigma/\sqrt{n}) \qquad (8\text{–}8)$$

Let's subtract $\overline{X}$ from every part of Relationship 8–7:

$$-Z^*(\sigma/\sqrt{n}) - \overline{X} < (\overline{X} - \mu) - \overline{X} < Z^*(\sigma/\sqrt{n}) - \overline{X}$$

$$-Z^*(\sigma/\sqrt{n}) - \overline{X} < -\mu < Z^*(\sigma/\sqrt{n}) - \overline{X} \qquad (8\text{–}9)$$

Suppose we multiply Relationship 8–9 by -1:

$$(-1)(-Z^*)(\sigma/\sqrt{n}) - (-1)(\overline{X}) > (-1)(-\mu)$$

$$> (-1)(Z^*)(\sigma/\sqrt{n}) - (-1)(\overline{X})$$

$$Z^*(\sigma/\sqrt{n}) + \overline{X} > \mu > -Z^*(\sigma/\sqrt{n}) + \overline{X} \qquad (8\text{–}10)$$

In so doing, we have changed the "$<$" signs to "$>$" signs. Recall that $-4 < -3$, but $(-1)(-4) = 4$ is greater than $(-1)(-3) = 3$. Because $4 > 3$ implies $3 < 4$, we can express Relationship 8–10 as Relationship 8–8.

Statisticians refer to Relationship 8–8 as a $100(1 - \alpha)$ percent *confidence interval estimator* of μ. Since the OSHA researcher calculated $\overline{X} = 7.3$ accidents, he could have reported his estimate as

$$\overline{X} - 1.96(2/\sqrt{36}) < \mu < \overline{X} + 1.96(2/\sqrt{36})$$

$$7.3 - .65 < \mu < 7.3 + .65$$

$$6.65 \text{ accidents} < \mu < 7.95 \text{ accidents}$$

He would, then, be 95 percent confident that μ = industry H's average number of accidents per plant was somewhere between 6.65 and 7.95.

The length of an interval is

$$[\overline{X} + Z^*(\sigma/\sqrt{n})] - [\overline{X} - Z^*(\sigma/\sqrt{n})] = 2Z^*(\sigma/\sqrt{n}) = 2e$$

Due to the fact that $100(1 - \alpha)$ percent and e represent the reliability and precision of interval estimates, anything which affects the maximum error (e.g., an increase in n) influences Relationship 8–8 as well.

The end-points will change from sample to sample because $\overline{X}$ varies. Table 8–9 reinforces the *randomness* of $\overline{X}$ and the interval.

TABLE 8–9

Experiment: Select a random sample	→ Random sample of *n* observations	→ Value for the random variable $\overline{X}$	→ Values for the random end-points $\overline{X} - Z^*(\sigma/\sqrt{n})$ $\overline{X} + Z^*(\sigma/\sqrt{n})$

Given additional samples, the OSHA researcher could construct many different confidence intervals. We have reproduced a few of these in Table 8–10. If μ is actually 7.1 accidents, two of the three intervals in Table 8–10 would include μ, as Figure 8–19 illustrates.

TABLE 8–10

Sample Average	*95% Confidence Interval Estimate of* μ
6.8 accidents	$6.15 < \mu < 7.45$
7.6 accidents	$6.95 < \mu < 8.25$
8.3 accidents	$7.65 < \mu < 8.95$

The researcher knows that estimating μ according to

$$\overline{X} - 1.96(2/\sqrt{36}) < \mu < \overline{X} + 1.96(2/\sqrt{36})$$

will "work" 95 percent of the time and "fail" 5 percent of the time.

FIGURE 8–19

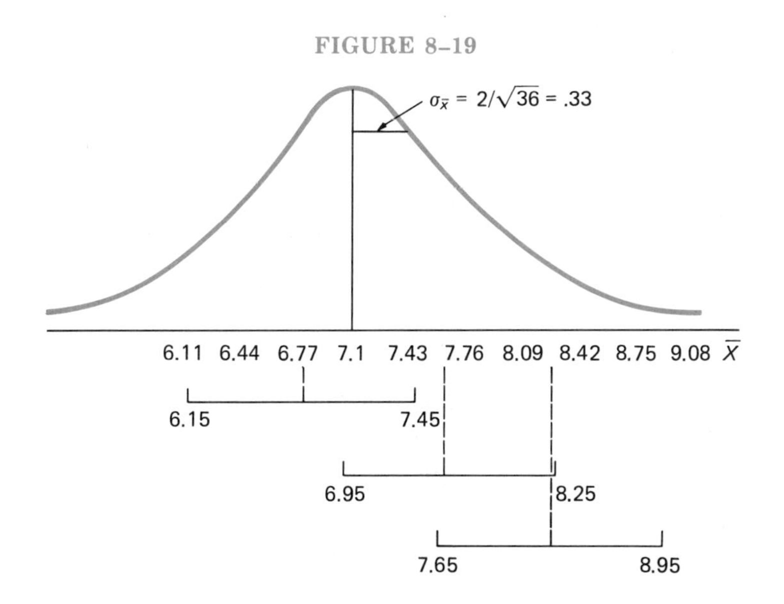

Whenever $100(1 - \alpha)$ percent is "high," he will view his computed interval as one of the "successful" outcomes. The following example reviews the principles of interval estimation.

Company G produces a 39¢ pen called Smooth-Riter. G would like its advertisements to focus on μ = the average life of the pens. The value of μ will determine which ad is chosen. A special machine will guide a random sample of pens over a large roll of paper. When a pen dries up, the line it traced on the paper will be measured. The measurement identifies the "life" of the pen.

G wishes to derive a 99 percent confidence interval estimate of μ where the interval is 60 feet. As the result of a pilot study, 80 feet will be substituted for σ.

Since $100(1 - \alpha)\% = 99\%$, $P(0 < Z < Z^*) = .99/2 = .4950$ and $Z^* = 2.58$. Inasmuch as $60 = 2e$, the maximum error, e, is $60/2 = 30 = 2.58(80/\sqrt{n})$. We can determine the required sample size by solving Equation 8–6:

$$n = [(2.58)(80)/30]^2 = 48$$

After measuring the lives of 48 pens, G calculated $\overline{X} = 5{,}300$. The 99 percent confidence interval is

$$5{,}300 - 30 < \mu < 5{,}300 + 30$$

$$5{,}270 \text{ feet} < \mu < 5{,}330 \text{ feet}$$

Because management believes that this interval contains μ, it must also believe that μ is close to 5,280 feet (i.e., 1 mile). The company will, therefore, publish its 1-mile ad—a cartoon of a pen running around a track and shouting "Smooth-Riter can break the 40¢ mile."

SUMMARY

We have shown that $\overline{X}$ is an unbiased, consistent, and efficient estimator of μ. We can develop either point or interval estimates of parameters. Since sampling errors are unavoidable, researchers have to evaluate the reliability and precision of their estimates. Unless the sample size is increased, raising $100(1 - \alpha)$ percent will diminish the precision of an estimate. In the planning stage of an experiment, researchers will specify preferred values for Z^* and e. Thereafter, they can identify the required n. Chapter Nine will examine another way to make inferences.

EXERCISES

1. Which of the following estimators of μ are unbiased?

   a. $(1/6)X_1 + (4/6)X_2$
   b. $(3/7)X_1 + (4/7)X_2$
   c. $(3/5)X_1 + (3/5)X_2$
   d. $(4/9)X_1 + (5/9)X_2$
   e. $(5/8)X_1 + (3/8)X_2$

2. Determine the "bias" of the biased estimators in question 1.

3. Which unbiased estimator in question 1 has the smallest variance?

4. Using $n = 3$, show that the sample median is a biased estimator of μ if the population is 2, 4, 9, and 10.

5. Using $n = 3$, show that the sample median is an unbiased estimator of μ if the population is 2, 3, 9, and 10.

6. Using $n = 3$, show that the sample median is an unbiased estimator of the population median in questions 4 and 5.

7. Given a normal population, identify the efficiency of $\overline{X}$ relative to the sample median when $n = 100$.

8. In a particular population, $\mu = 20$ and $\sigma = 3$. If a random sample of 36 observations is selected, what are the following probabilities?

   a. $P(|\overline{X} - \mu| < .10)$
   b. $P(|\overline{X} - \mu| < 1.1)$
   c. $P(|\overline{X} - \mu| < 1.4)$
   d. $P(|\overline{X} - \mu| < .15)$
   e. $P(|\overline{X} - \mu| < 1.46)$

9. In a particular population, $\mu = 32$ and $\sigma = 5$. If a random sample of 64 observations is selected, what are the following probabilities?

   a. $P(|\overline{X} - \mu| < .25)$
   b. $P(|\overline{X} - \mu| < .75)$
   c. $P(|\overline{X} - \mu| < 1.4)$
   d. $P(|\overline{X} - \mu| < 1.8)$
   e. $P(|\overline{X} - \mu| < 1.9)$

10. If a random sample of 100 observations is chosen from a particular population, the probability that the value of $\overline{X} - \mu$ will fall between $-.32$ and $.32$ is .5762. What is σ?

11. In a particular population, $\sigma = 11.5$. For what sample size is $P(|\overline{X} - \mu| < .30)$ equal to .8740?

12. A researcher would like $P(|\theta^* - \mu| < 5)$ to equal .90. If θ^* is $\overline{X}$, he will have to select a random sample of 64 observations. Suppose that the population is normal and the researcher makes θ^* the sample median. How large a sample must he select under these circumstances?

13. Given $n = 49$ and $\sigma = 3$, a researcher computes $P(|\overline{X} - \mu| < a) = .95$. What is the value of a?

14. A researcher has picked a random sample of 64 observations and finds $\sigma = 30$. With what level of confidence can she state the following?

a. The sampling error is no more than 1.0.
b. The sampling error is no more than 0.6.
c. The sampling error is less than 0.8.

15. A researcher is using $\overline{X}$ to estimate μ. If he increases the sample size from $n = 64$ to $n = b$, his estimate will be ten times more precise. What is the value of b?

16. The length of a particular 95 percent confidence interval estimate of μ is 6. Determine n if $\sigma = 50$.

17. The length of a particular 99 percent confidence interval estimate of μ is 8. Determine n if $\sigma = 38$.

18. Suppose that the average attendance at National League baseball games in 1978 was 18,000 with $\sigma = 3{,}000$ and that an individual chooses a random sample of 36 games.

a. What is the probability that $\overline{X}$ = the sample average attendance will not be more than 1,000 people from μ?
b. What is the probability that $|\overline{X} - \mu|$ won't be greater than 800?

19. Imagine that the Food and Drug Administration is investigating calcium propionate, a preservative found in bread. As part of its study, the FDA wants to determine μ = the average number of slices of bread consumed by adults per week. The agency believes that five slices is a reasonable value for σ. If a researcher wishes to be 95 percent confident that $\overline{X}$ won't be "off target" by more than 0.7 slices, how many adults should he interview?

20. The medical director of a large corporation is exploring the relationship between exercise and health. She wants to estimate how much time, on the average, employees devote to jogging, tennis, bicycling, or swimming. From a random sample of 64 workers, she computes $\overline{X} = 32$ minutes per day and $s = 10$ minutes per day.

a. Calculate a 90 percent confidence interval estimate of μ.
b. Calculate a 95 percent confidence interval estimate of μ.
c. Calculate a 99 percent confidence interval estimate of μ.

21. With the closing of a local steel mill in 1978, the unemployment rate for City D increased dramatically. To determine the effect of unemployment on retail sales, an investigator randomly asks 49 of D's businesses to compare their 1979 and 1978 sales. He finds that $\overline{X} = -\$34{,}000$ change in sales and $s = \$6{,}200$.

a. Calculate a 90 percent confidence interval estimate of the average change among all businesses.
b. What level of confidence can be associated with the claim that $\overline{X}$ is no more than \$2,000 from μ?

22. Company R keeps screws in metal containers. After moving to its new warehouse, the company decides to store the containers on shelves instead of on the floor. To determine the weight that the shelves must support, management selects 36 drums at random, weighs them, and calculates $\overline{X} = 44$ pounds and $s = 0.6$ pounds.

a. Compute a 95 percent confidence interval estimate of μ.
b. Compute a 99 percent confidence interval estimate of μ.

23. A particular community is subjected to erratic service by the nearest utility company. Residents are frequently without power several times a week. Since refrigerators don't operate during these periods, residents fear that some of their food will spoil. A dairy wants to inform the community about the spoilage time of one of its products. For a random sample of 81 items, the dairy computes $\overline{X} = 15$ hours before spoiling due to lack of refrigeration and $s =$ 3 hours.

a. Calculate a 95 percent confidence interval estimate of μ.
b. Calculate a 99 percent confidence interval estimate of μ.

24. In the mid-1960s the Department of Labor began to identify the "poverty level" by determining the cost of a "subsistence" diet and, thereafter, multiplying such a cost by three. The factor, 3, derived from studies indicating that poor families allocate about one-third of their incomes to food. Imagine that a subsistence diet for a family of four costs \$2,400. The poverty level would be (3)(\$2,400) = \$7,200. This is approximately the 1979 poverty level. A researcher believes that \$7,200 is unrealistically low. He selects a random sample of 49 members of the American Economic Association and asks each of them to complete the following sentence: "I would consider a family of four to be poor if it earned less than . . ." The researcher notes that $\overline{X} = \$8{,}900$ and $s = \$700$. Calculate a 95 percent confidence interval estimate of $\mu =$ the average response of all members of the American Economic Association.

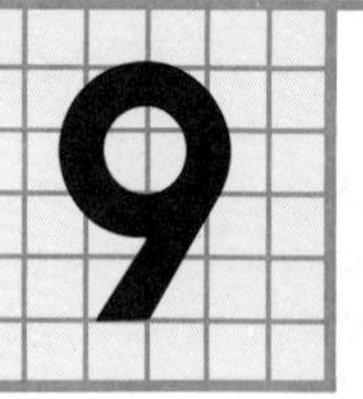

Statistical Inference: *Hypothesis Testing with One Sample Mean— Large Samples*

THE HYPOTHESIS-TESTING PERSPECTIVE

We have just established a confidence interval estimate of the population mean. With repeated random sampling, a $100(1 - \alpha)$ percent confidence interval will include μ $100(1 - \alpha)$ percent of the time. Only one interval, of course, is ever formed. Chapter Nine explores statistical inference from a different point of view. Before sampling, we will assume that μ equals a value called μ_0. Given a random sample, we will subsequently determine whether or not we are willing to believe that $\mu = \mu_0$. The observed $\overline{X}$ will be crucial to this determination. The following example illustrates the fundamentals of *hypothesis testing*.

A certain baseball team ended 1978 in last place because the infielders were poor. The team looked so shabby, in fact, that attendance dropped substantially. To improve the team and increase the number of fans, the owners spent the entire "off-season" searching for a shortstop. A former college player tried out for the position in spring training and was immediately signed by the club. Thereafter, the team officials wasted no time in publicly praising the shortstop's "golden hands."

Impressed by all the publicity, James Meade decided to take his sons to the ballpark soon after the 1979 season got underway. Since the sons were Little Leaguers, Meade felt obligated to show them how a tremendous pro functioned. The father left for the stadium assuming, or hypothesizing, that "the new shortstop for the home team is great."

Twelve balls were hit to the shortstop during the game. He fielded eight cleanly, but the other four slipped through his legs. Because of this performance, Meade rejected the assumption that the shortstop was good and concluded the opposite. The fielder's four errors represented the "evidence" for this conclusion. Meade's experience as a baseball fan convinced him that the event "four errors" was extremely unlikely *given* that a shortstop was good. The father's conclusion might be correct or it might be wrong. Even outstanding shortstops can have "off" nights. At any rate, Meade vowed that he would attend no more of the team's games.

We want to identify four elements of hypothesis testing that are apparent in this scenario:

1. A hypothesis is developed.
2. Information is collected.
3. The validity of the hypothesis is judged in terms of the information.
4. Some action is undertaken as the result of a conclusion.

Table 9–1 shows how these elements relate to the shortstop example.

TABLE 9–1

Elements of Hypothesis Testing	*Shortstop Example*
Original assumption	The shortstop is a fine defensive player.
Alternative assumption which is accepted if the original is rejected by the evidence	The shortstop is not a fine defensive player.
Basis for making a judgment	Fans know that great infielders will not make many errors.
Evidence	The performance of the shortstop during the game.
Conclusion	Someone of star caliber would not generally make four errors in a game. Although a great player might occasionally commit four errors, the claim that "the shortstop is a fine defensive player" is suspicious. Four errors seem more consistent with the assumption that "the shortstop is not a fine defensive player."
Action stemming from conclusion	Stay away from the ballpark.

If the shortstop situation had been introduced in Chapter Eight, it might have appeared as follows. James Meade heard that his local team had acquired a new shortstop. Knowing nothing about this player's talents, Meade took his children to a game. He wanted to see if the shortstop was good. Meade had *no* preconceptions. Based on the shortstop's four errors, Meade "estimated" that the shortstop was incompetent. He declared that he would not return to the stadium.

In general, *estimation* is used for inferences whenever we have no basis for formulating an impression of a parameter prior to sampling. *Hypothesis testing,* on the other hand, is used if we can form an opinion about θ *before* evidence is collected. Under either circumstance, our inferences may be in error. At the end of the next section, we'll indicate the parallels between these two procedures.

TWO-TAILED HYPOTHESIS TESTS

Whenever we estimate θ or test hypotheses about that parameter, we have to know the characteristics of the sampling distribution of θ^*. If the statistic is $\overline{X}$, these characteristics are:

1. The sampling distribution is normal for $n > 30$.
2. The mean of the sampling distribution is equal to the population mean.
3. The standard deviation of the sampling distribution is equal to $\sigma/\sqrt{n}$.

Because of the second characteristic, an assumption about the value of μ, say, $\mu = \mu_0$, is also an assumption about the mean of the sampling distribution of $\overline{X}$ (see Figure 9–1). We will refer to this assumption as the *null hypothesis*. (Note: One of the connotations of "null" is "ineffective." Suppose that a certain kind of rash typically lasts fourteen days if left untreated. A pharmaceutical company has given a new ointment to a random sample of 49 people suffering from this rash. To determine whether the ointment speeds

FIGURE 9–1

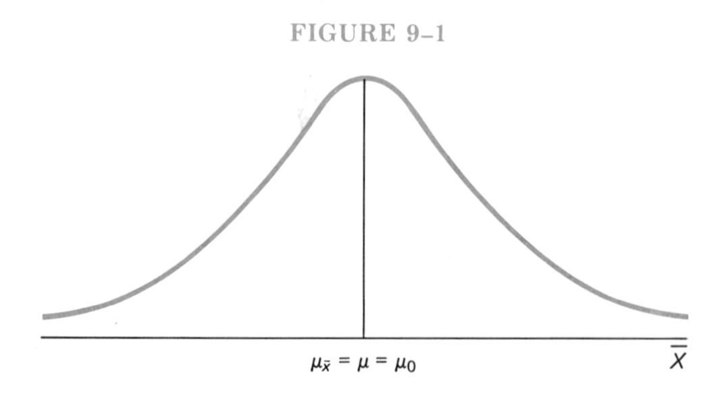

the disappearance of the rash, researchers would first assume that the ointment is ineffective, or "null"; that is, $\mu_0 = 14$ days. Although any initial assumption about μ is called the "null hypothesis," the term derives from tests like the preceding one.) Inasmuch as the horizontal axis extends indefinitely in both directions, *any* observed $\overline{X}$ could belong to Figure 9–1. We want to discriminate between (1) $\overline{X}$s that are likely given $\mu = \mu_0$ and (2) $\overline{X}$s that are unlikely.

The basis for separating likely and unlikely $\overline{X}$s is the *level of significance, α*. Provided that α is "low," $\overline{X}$s in the top and bottom $\alpha/2$ of a sampling distribution are "rare" (see Figure 9–2). As with the reliability of a confidence interval, the level of significance is determined subjectively.

FIGURE 9–2

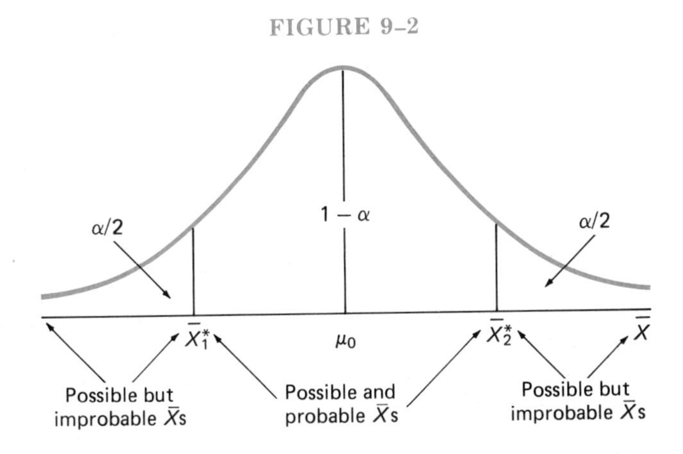

An observed $\overline{X}$ won't equal μ_0 for one of two reasons:

1. *Chance.* A sample was selected from a population having $\mu = \mu_0$, but sampling error exists.
2. *An incorrect null hypothesis.* A sample was selected from a population having $\mu \neq \mu_0$.

We will associate "possible and probable" $\overline{X}$s with the first reason and "possible but improbable" $\overline{X}$s with the second. We say that "possible but improbable" $\overline{X}$s differ *significantly* from μ_0.

Hypothesis testing is a formal procedure for evaluating whether $\mu = \mu_0$ seems sensible. By "rejecting" μ_0, we *haven't proved* that our hypothesis is false. Remember, $\overline{X} < \overline{X}_1^*$ or $\overline{X} > \overline{X}_2^*$ is still possible when $\mu = \mu_0$.

We'll replace Figure 9–2 with Figure 9–3, where

$$Z = \frac{\overline{X} - \mu_0}{\dfrac{\sigma}{\sqrt{n}}} \qquad (9\text{–}1)$$

FIGURE 9–3

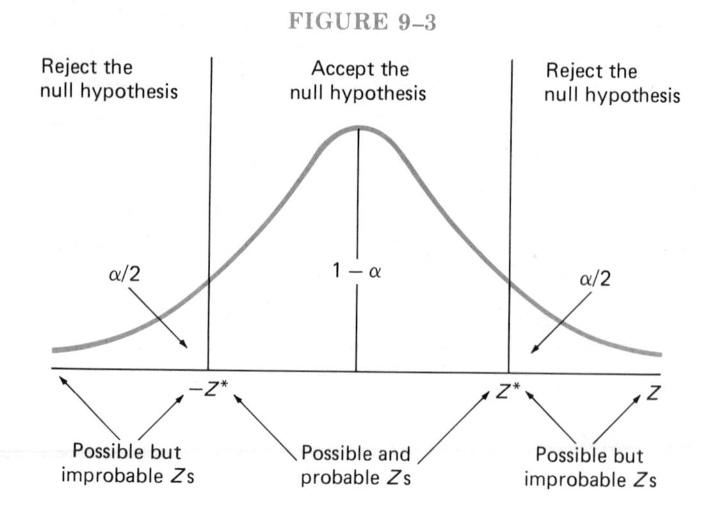

Both $-Z^*$ and Z^* are *critical values*. Table 9–2 compares Meade's reasoning in the shortstop example with the approach we are going to follow.

Two-tailed hypothesis testing proceeds as follows:

1. Form the null hypothesis $\mu = \mu_0$. This hypothesis arises from experience or from familiarity with a situation.
2. State the alternative hypothesis $\mu \neq \mu_0$. The alternative is accepted when the null is rejected.
3. Identify the level of significance. The choice of α is a personal decision. The most often used αs are .01, .05, and .10.
4. Determine the critical values of Z. The critical values are established by α since $P(Z < -Z^*) = \alpha/2$ and $P(Z > Z^*) = \alpha/2$.
5. Select a large random sample. The sample represents the "evidence."
6. Calculate $\overline{X}$.
7. Convert $\overline{X}$ to Z. If σ is unknown, we'll substitute s for σ and compute Z in terms of

$$Z = \frac{\overline{X} - \mu_0}{\frac{s}{\sqrt{n}}} \tag{9–2}$$

8. Compare the computed Z to the critical values. When $-Z^* < Z < Z^*$, accept the null hypothesis; otherwise, reject $\mu = \mu_0$.

The following example illustrates these steps.

Property taxes are the major source of local tax revenue. City J periodically estimates the highest price that each piece of private property would receive if it were offered for sale. This price is mul-

TABLE 9–2

Meade	*A Researcher*
Question: If the shortstop is great, would he make four errors?	**Question:** If $\mu = \mu_0$, is $Z < -Z^*$ or $Z > Z^*$ likely?
Answer: Four errors are "possible but improbable." (Three, perhaps, is a "critical" value.)	**Answer:** Such Zs are "possible but improbable."
Conclusion: Based on the evidence, reject "the shortstop is great."	**Conclusion:** Based on a computed $Z < -Z^*$ or $Z > Z^*$, reject $\mu = \mu_0$.

tiplied by a certain fraction, or "assessment ratio," to obtain "assessed value."

When a home is assessed at $10,000 and the tax rate is 50 mills (i.e., 5%), the owner pays $500 in taxes. Inasmuch as calculating assessments is somewhat subjective, homeowners occasionally complain about the process.

The chairman of J's Property-Tax Appeals Board claims, "Owners spend an average of 35 minutes arguing their case." To test his "off-the-cuff" remark at the .05 level of significance, the chairman picked a random sample of 36 appeals and found $\overline{X} = 31$ minutes and $s = 10$ minutes.

The null and alternative hypotheses can be written as

$$H_0 \text{ (null hypothesis): } \mu = 35 \text{ minutes}$$

$$H_1 \text{ (alternative hypothesis): } \mu \neq 35 \text{ minutes}$$

(Note: Due to the fact that we must convert a sampling distribution centered at a *particular* μ into the Z distribution, the hypotheses can't be

$$H_0\text{: } \mu \neq 35 \text{ minutes}$$

$$H_1\text{: } \mu = 35 \text{ minutes}$$

The "=" sign always appears in H_0.)

Because $P(Z > Z^*) = \alpha/2 = .05/2 = .025 = P(Z > 1.96)$, Figure 9–4 is the "frame of reference" for the test. The calculated value of Z is

$$Z = \frac{31 - 35}{\dfrac{10}{\sqrt{36}}} = -2.40$$

FIGURE 9–4

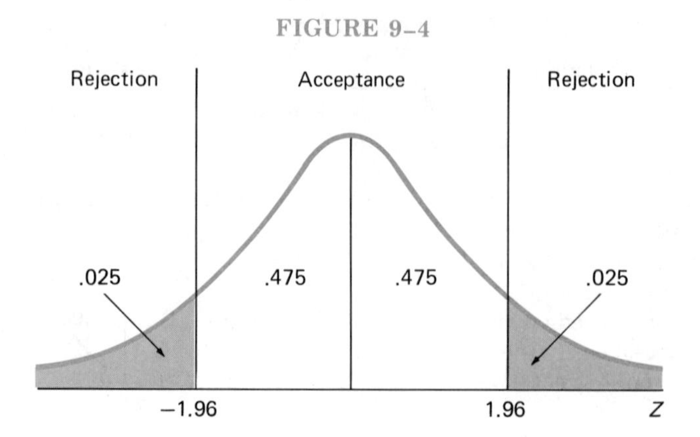

Since $Z = -2.40$ lies in the rejection region, the chairman feels uneasy about H_0. He thus concludes $\mu \neq 35$ minutes. Thirty-one minutes is significantly different from $\mu = 35$.

The chairman's conclusion could be a mistake. This type of mistake, rejecting a true H_0, is called a *Type I error*. Prior to sampling,

$$P(\text{Type I error}) = P(\text{reject } H_0 | H_0 \text{ is true}) = \alpha \qquad (9\text{–}3)$$

The probability of a Type I error is the same as the level of significance. Such a probability will increase as α increases.

Note that α is stated at the *beginning* of the test. If it were set after step #7, a researcher might be tempted to "control" the outcome. To clarify this, we see that $P(Z < -2.40) = .5000 - .4918 = .0082$. Suppose the chairman wished to "rig" his test in favor of accepting H_0. He would choose $\alpha < 2(.0082)$, for example, $\alpha = .01$. Under these circumstances, hypothesis testing would be a "test" in name only. Let's look at another example.

In the draft of an article for the alumni newsletter, the dean of a business school wrote:

> The administration is dedicated to supplying quality education. Furthermore, we try to help students secure lucrative jobs. The average starting salary for our 1979 MBAs was $22,000.

Before submitting the article, the dean intends to test $\mu = \$22{,}000$ at the .01 level of significance. Using a random sample of 36 graduates, he calculated $\overline{X} = \$21{,}400$ and $s = \$2{,}300$.

Figure 9–5 shows the acceptance and rejection regions for

$$H_0: \mu = \$22{,}000$$

$$H_1: \mu \neq \$22{,}000$$

FIGURE 9–5

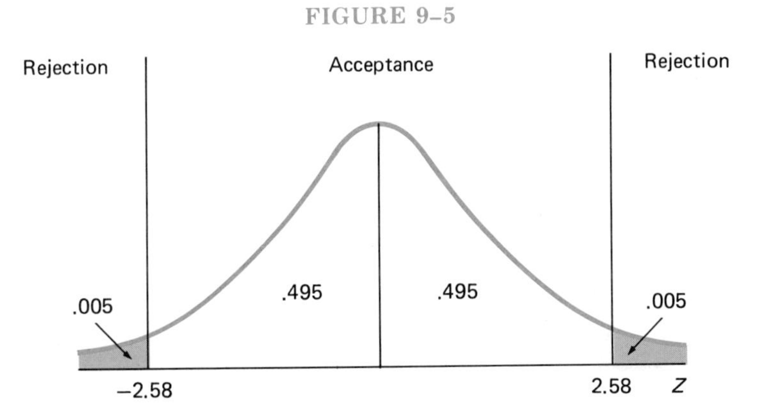

when $\alpha = .01$. Inasmuch as

$$Z = \frac{21{,}400 - 22{,}000}{\frac{2{,}300}{\sqrt{36}}} = -1.57,$$

the dean will accept H_0. He concludes that the difference between \$21,400 and \$22,000 is the result of sampling error. Since μ might *not* equal \$22,000, the dean could be committing a *Type II error,* that is, accepting a false H_0. We will investigate these errors in the following section.

Recall the jar-lid example in Chapter One. When the machines are working properly, μ = the average lid diameter, and the specifications (e.g., 2.5 inches) should be identical. We can suggest two conditions that would yield this equality:

1. If there are no defectives, $X_1 = X_2 = \ldots = X_N = \mu$.
2. If there are some defectives, oversized and undersized lids may "cancel out."

Management could test

$$H_0: \mu = 2.5 \text{ inches}$$

$$H_1: \mu \neq 2.5 \text{ inches}$$

A computed Z value in the acceptance region will convince management that an inspection isn't needed. Zs in the rejection region, though, will force the company to shut down production and inspect the machinery (see Figure 9–6).

We will assume that $\alpha = .05$. Then, $-Z^*$ is -1.96 and Z^* is 1.96.

FIGURE 9–6

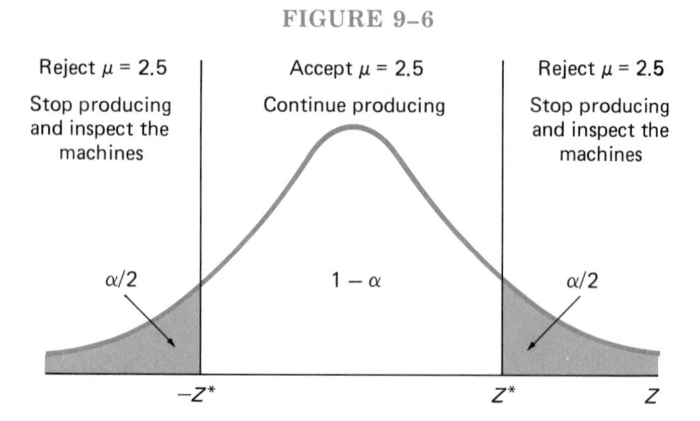

Suppose that $n = 36$, $\overline{X} = 2.43$ inches, and $s = 0.64$ inches. The computed Z, that is,

$$Z = \frac{2.43 - 2.50}{\frac{.64}{\sqrt{36}}} = -0.66,$$

is in the acceptance region. Thus management doesn't feel that $\mu = 2.5$ inches is questionable. Nonetheless, by allowing production to continue without an inspection, the company risks a Type II error. If a Type II error does occur (i.e., if the 36 lids were drawn from a population where $\mu \neq 2.5$), the next run of lids will possess many defectives.

We have already indicated that hypothesis testing and estimation are connected. Let's be more explicit. At the α level of significance, we will accept H_0 as long as

$$-Z^* < \frac{\overline{X} - \mu_0}{\frac{\sigma}{\sqrt{n}}} < Z^* \tag{9–4}$$

After rearranging Relationship 9–4, we get

$$\overline{X} - Z^*(\sigma/\sqrt{n}) < \mu_0 < \overline{X} + Z^*(\sigma/\sqrt{n}) \tag{9–5}$$

Whenever Relationship 9–4 is true, the $100(1 - \alpha)$ percent confidence interval derived from Relationship 9–5 will include μ_0.

Consider the jar-lid decision in terms of estimation:

$$\overline{X} - (1.96)(s/\sqrt{n}) < \mu < \overline{X} + (1.96)(s/\sqrt{n})$$

$$2.43 - (1.96)(.64/\sqrt{36}) < \mu < 2.43 + (1.96)(.64/\sqrt{36})$$

$$2.43 - .209 < \mu < 2.43 + .209$$

$$2.22 \text{ inches} < \mu < 2.64 \text{ inches}$$

If H_0 is correct and we sampled repeatedly, 95 percent of the confidence intervals would contain $\mu = 2.5$. Since 2.5 falls between 2.22 and 2.64, we aren't suspicious of the null hypothesis.

TYPE II ERRORS

Instead of using

$$-Z^* < \frac{\overline{X} - \mu_0}{\frac{\sigma}{\sqrt{n}}} < Z^*$$

as the acceptance region, we can establish critical values in terms of $\overline{X}$ (i.e., $\overline{X}_1^*$ and $\overline{X}_2^*$). By manipulating Relationship 9–4, we find

$$\underbrace{\mu_0 - Z^*(\sigma/\sqrt{n})}_{\overline{X}_1^*} < \overline{X} < \underbrace{\mu_0 + Z^*(\sigma/\sqrt{n})}_{\overline{X}_2^*} \qquad (9\text{–}6)$$

Although an observed $\overline{X}$ may fall between $\overline{X}_1^*$ and $\overline{X}_2^*$, we can't be certain that $\mu = \mu_0$. If $\mu \neq \mu_0$ and we accept H_0, we make a Type II error. The probability of such an error is represented by the Greek letter *beta*, β:

$$P(\text{Type II error}) = P(\text{accept } H_0 | H_0 \text{ is false}) = \beta \qquad (9\text{–}7)$$

Like α, β is a *conditional* probability. Consider the following example.

The lid manufacturer has determined that σ is always .5 inches, regardless of the state of the equipment. To test $\mu = 2.5$ inches, he sets α at .05 and n at 64.

By virtue of Relationships 9–4 and 9–6, the entrepreneur will accept his null hypothesis every time (1) the computed Z is between -1.96 and 1.96 or (2) the computed $\overline{X}$ is between

$$2.5 - (1.96)(.5/\sqrt{64}) < \overline{X} < 2.5 + (1.96)(.5/\sqrt{64})$$

$$2.38 \text{ inches} < \overline{X} < 2.62 \text{ inches}$$

This interval is shown in Figure 9–7.

If, in fact, $\mu = 2.5$ inches, the manufacturer can't commit a Type II error. However, when $\mu \neq 2.5$ but $2.38 < \overline{X} < 2.62$, he will accept a hypothesis that isn't correct. The probability of $2.38 < \overline{X} < 2.62$ given $\mu \neq \mu_0$ varies with μ. We will confirm that β is high when |the *actual* $\mu - \mu_0$| is small and that β is low when |the *actual* $\mu - \mu_0$| is large.

FIGURE 9–7

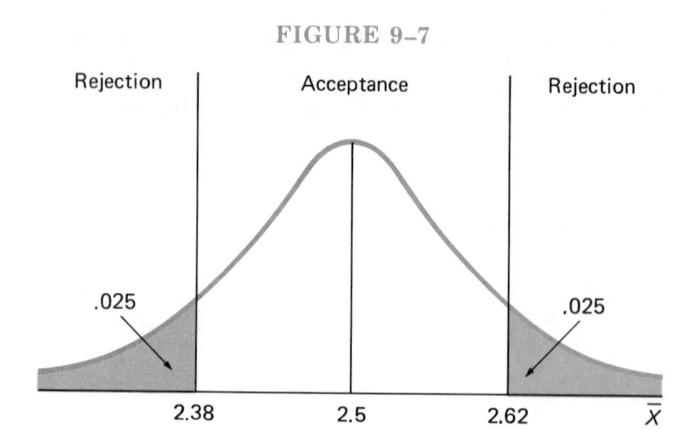

We now want to construct a "blueprint" for hypothesis testing. This blueprint will identify the risks associated with sampling. If the probability of these risks is too great, we can form a new blueprint by adjusting α and n. (Note: Estimation blueprints specify reliability and precision.)

The entrepreneur's decision will place him in one of the four situations shown in Table 9–3. He hopes to avoid actions that generate #2 or #3. Yet, as long as his conclusions stem from samples, he can *never* eliminate the possibility of those outcomes. He is, though, able to influence their probability.

TABLE 9–3

		Decision	
		Continue Producing	Stop Producing
State of the Equipment	Machines Are Working Properly	#1	#2
	Machines Are Working Improperly	#3	#4

Table 9–4 translates Table 9–3 into the language of hypothesis testing and indicates the probability of each situation. Since α and β are conditional probabilities, their sum is not restricted to 1.0. We will demonstrate, however, that a change in α will change β as

TABLE 9–4

		Decision	
		Accept H_0	Reject H_0
State of the Equipment	H_0 is true ($\mu = \mu_0$)	$P(\#1) = 1 - \alpha$	$P(\#2) = P(\text{Type I error}) = \alpha$
	H_0 is false ($\mu \neq \mu_0$)	$P(\#3) = P(\text{Type II error}) = \beta$	$P(\#4) = 1 - \beta$

well. We should stress that a Type I and a Type II error can't be made simultaneously (e.g., if the entrepreneur accepts H_0, he has either acted correctly or he has suffered a Type II error).

We have seen that $2.38 < \overline{X} < 2.62$ is the acceptance region when $\mu_0 = 2.5$, $\sigma = .5$, $\alpha = .05$, and $n = 64$. Suppose that the lids are picked from a population having $\mu = 2.6$ and $\sigma = .5$. Under these circumstances, the probability of falsely concluding $\mu = 2.5$ is $P(2.38 < \overline{X} < 2.62 | \mu = 2.6)$, or the shaded area in Figure 9–8. Due to the fact that

$$Z_1 = \frac{2.38 - 2.6}{\frac{.5}{\sqrt{64}}} = -3.52$$

and

$$Z_2 = \frac{2.62 - 2.6}{\frac{.5}{\sqrt{64}}} = 0.32,$$

FIGURE 9–8

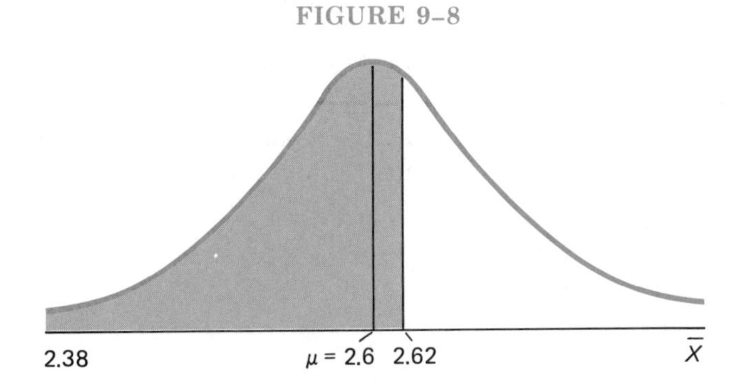

FIGURE 9–9

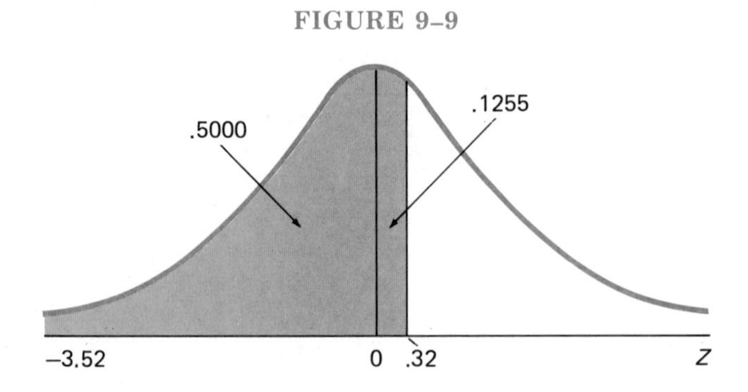

this probability is .5000 + .1255 = β = .6255 (see Figure 9–9). In repeated random sampling from a population with μ = 2.6, 62.55 percent of the $\overline{X}$s would not force the manufacturer to doubt μ = 2.5. One blueprint, then, would look like Table 9–5.

TABLE 9–5

		Decision	
		Accept H_0	Reject H_0
State of the Equipment	H_0 is true (μ = 2.5)	$1 - \alpha = .95$	$\alpha = .05$
	H_0 is false (μ = 2.6)	$\beta = .6255$	$1 - \beta = .3745$

We will present different βs in the context of a graph called an *operating characteristic curve.* Points on the curve pair "the probability of accepting H_0" and μ. Other than at $\mu = \mu_0 = 2.5$, where the probability is $1 - \alpha$, $P(2.38 < \overline{X} < 2.62|\mu)$ equals β.

Figure 9–10 contains three sampling distributions for our jar-lid example (σ = .5, α = .05, and n = 64). They have different means but the same standard deviation, namely, $\sigma_{\overline{x}} = \sigma/\sqrt{n} = .5/\sqrt{64} = .0625$. The figure also indicates the value of β given each distribution (i.e., given each μ). Each pair of μ, β values represents a point on the operating characteristic curve. The three points identified in Figure 9–10 are:

μ = 2.30, β = .1003
μ = 2.55, β = .8653
μ = 2.80, β = .0020

FIGURE 9–10

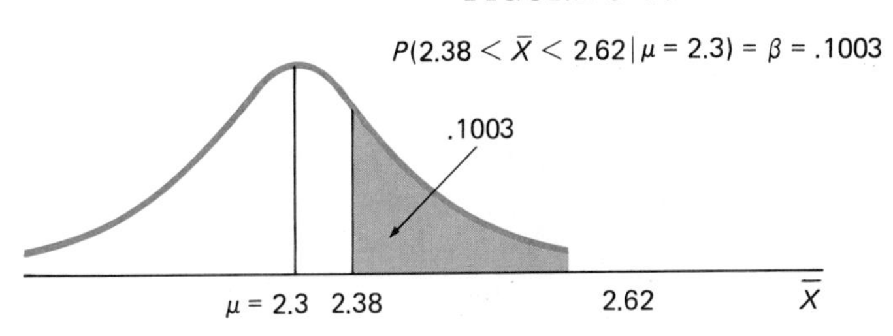

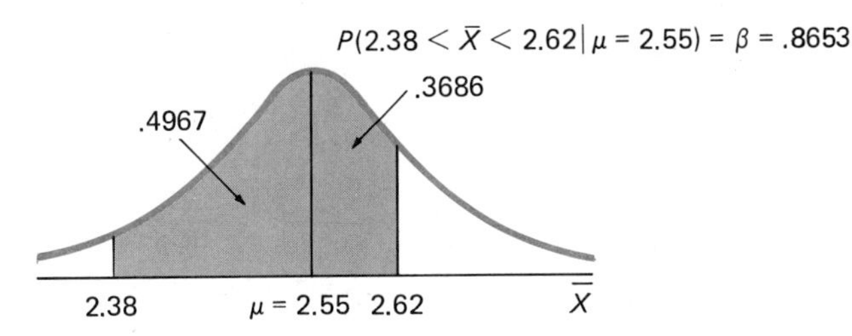

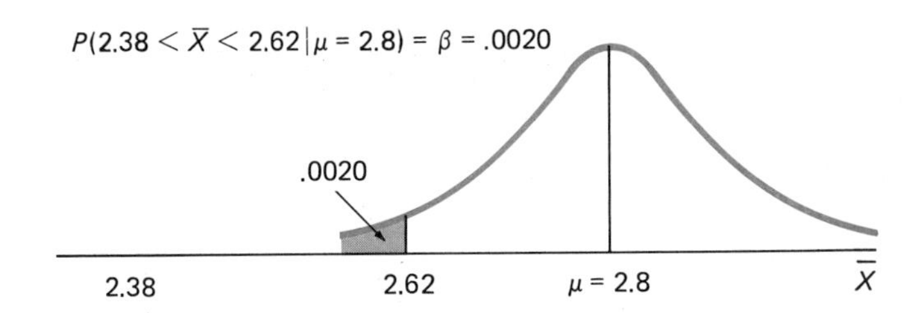

After collecting a host of μ, β pairs, we are able to plot Figure 9–11. P(Type II error) declines as we move farther from μ_0 because the "$2.38 < \bar{X} < 2.62$" event becomes progressively rarer.

Inasmuch as β is the probability that we will accept a false H_0, $1 - \beta$ is the probability that we will reject it. The expression $1 - \beta$ is referred to as the *power* of a hypothesis test. A *power curve* such as Figure 9–12 pairs $P(\text{reject } H_0|\mu)$ and μ. At $\mu = \mu_0$, $P(\text{reject } H_0|\mu)$ is the level of significance, or α.

Assume that the lid manufacturer is especially afraid of making a Type II error when $\mu = 2.55$ inches. If he feels that $\beta = .8653$ is higher, or $1 - \beta$ is lower, than he would like, he will redesign his hypothesis test. Let's explore such a procedure.

The shaded portion of each curve in Figure 9–10 and, hence, each β will decrease provided the acceptance region,

$$\mu_0 - Z^*(\sigma/\sqrt{n}) < \bar{X} < \mu_0 + Z^*(\sigma/\sqrt{n}),$$

FIGURE 9–11
Operating Characteristic Curve
(α = .05, σ = .5, n = 64)

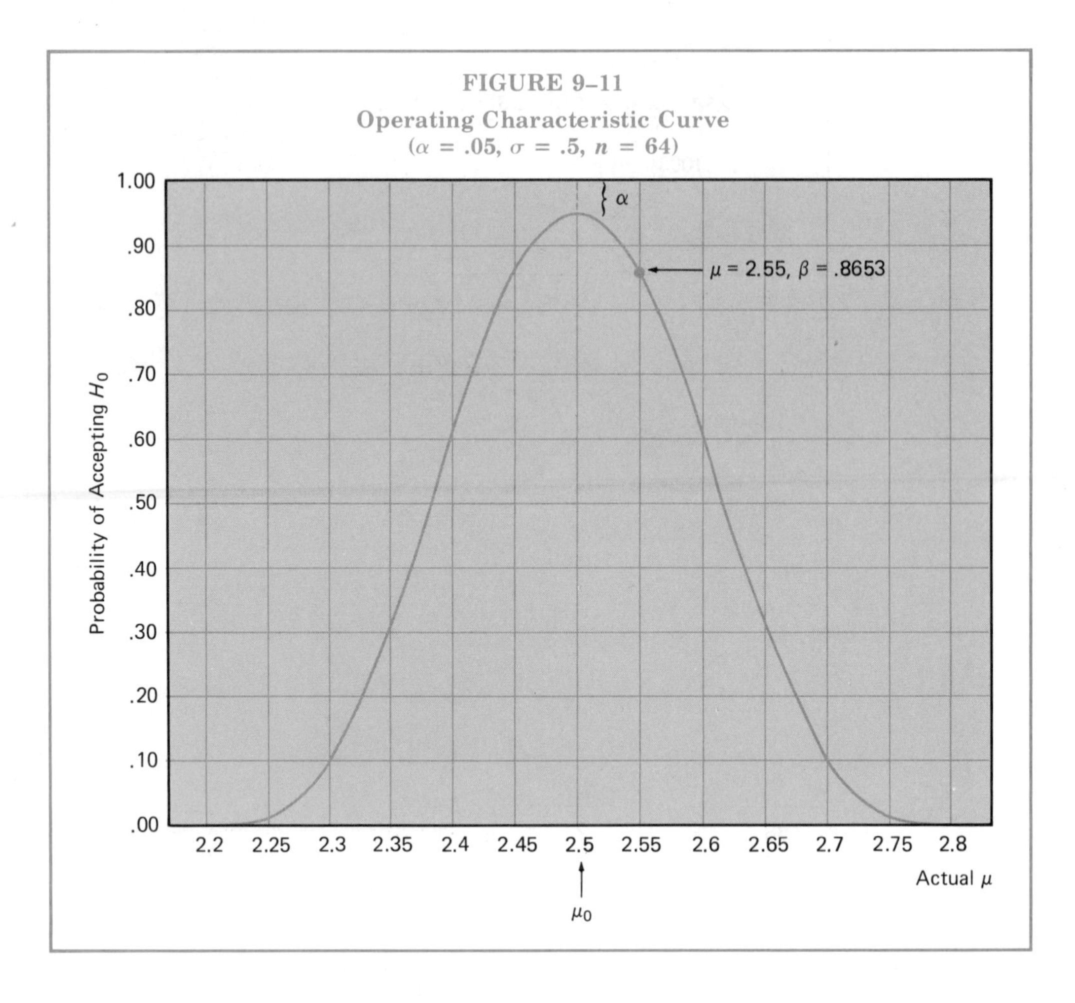

shrinks. With μ_0 and σ fixed, we can accomplish this by reducing Z^* and/or increasing n. We know that Z^* falls when $1 - \alpha$ falls (or α rises). Unfortunately, a larger α increases P(Type I error). Using $\alpha = .10$, the new acceptance region will be

$$2.5 - (1.64)(.5/\sqrt{64}) < \overline{X} < 2.5 + (1.64)(.5/\sqrt{64})$$

$$2.4 < \overline{X} < 2.6$$

We have developed the corresponding operating characteristic curve, Figure 9–13, where $P(2.4 < \overline{X} < 2.6 | \mu \neq \mu_0) = \beta$.

When α stays at .05 but the sample size is increased to 81, the acceptance region will be

$$2.5 - (1.96)(.5/\sqrt{81}) < \overline{X} < 2.5 + (1.96)(.5/\sqrt{81})$$

$$2.39 < \overline{X} < 2.61$$

FIGURE 9–12
Power Curve
(α = .05, σ = .5, n = 64)

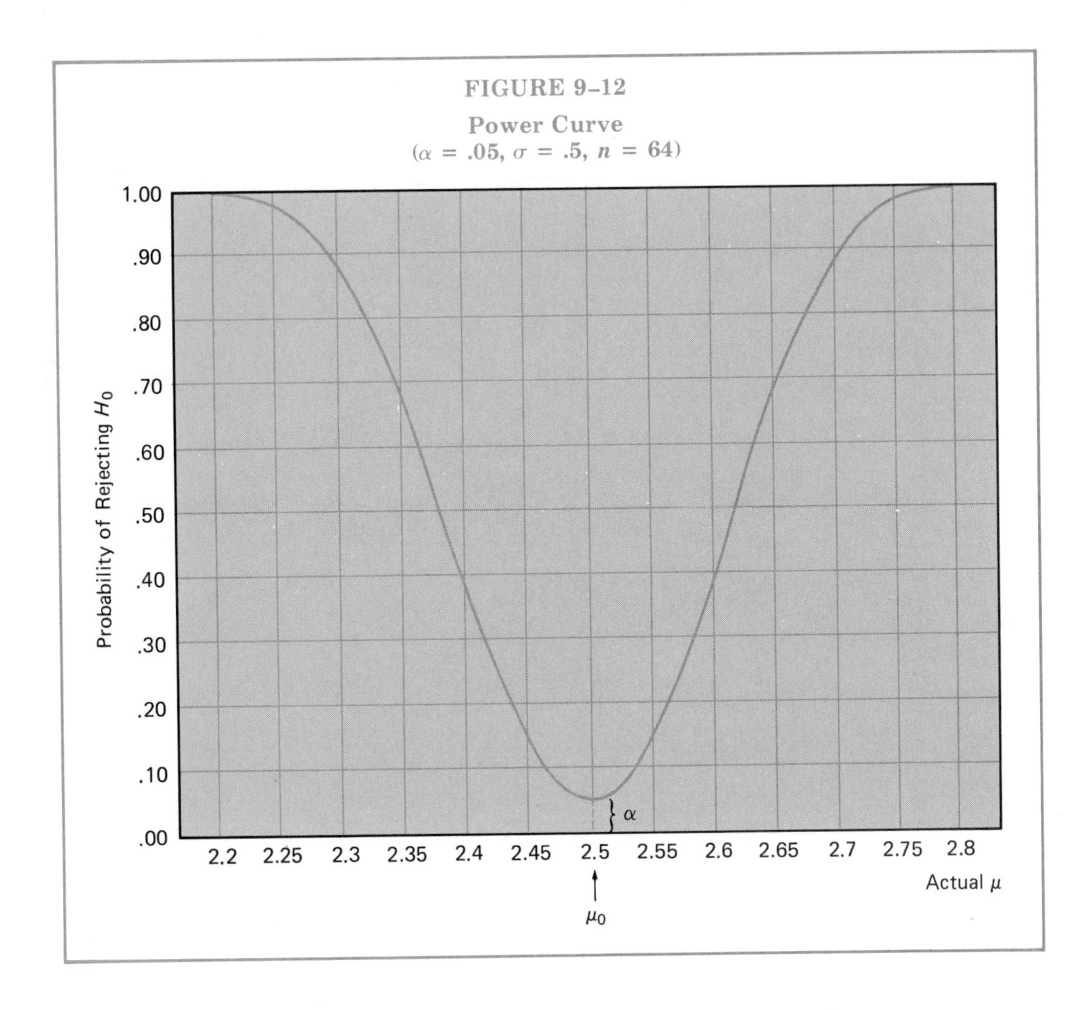

Large values of n, of course, make sampling more expensive. For Figure 9–14, $P(2.39 < \overline{X} < 2.61|\mu \neq 2.5) = \beta$.

As in estimation, the researcher will settle for a blueprint that reflects his concept of a "high" or a "low" probability. Given μ_0 and σ, he must determine an α, n combination that provides "adequate" protection against Type I and Type II errors. We will investigate the financial implications of these errors in Chapter Thirteen.

ONE-TAILED HYPOTHESIS TESTS

In the previous examples, the rejection region included parts of both tails. As such, very small *or* very large $\overline{X}$s would lead us to reject H_0. Researchers, though, may not care whether $\mu > \mu_0$ since the

FIGURE 9–13
Operating Characteristic Curve
($\alpha = .10$, $\sigma = .5$, $n = 64$)

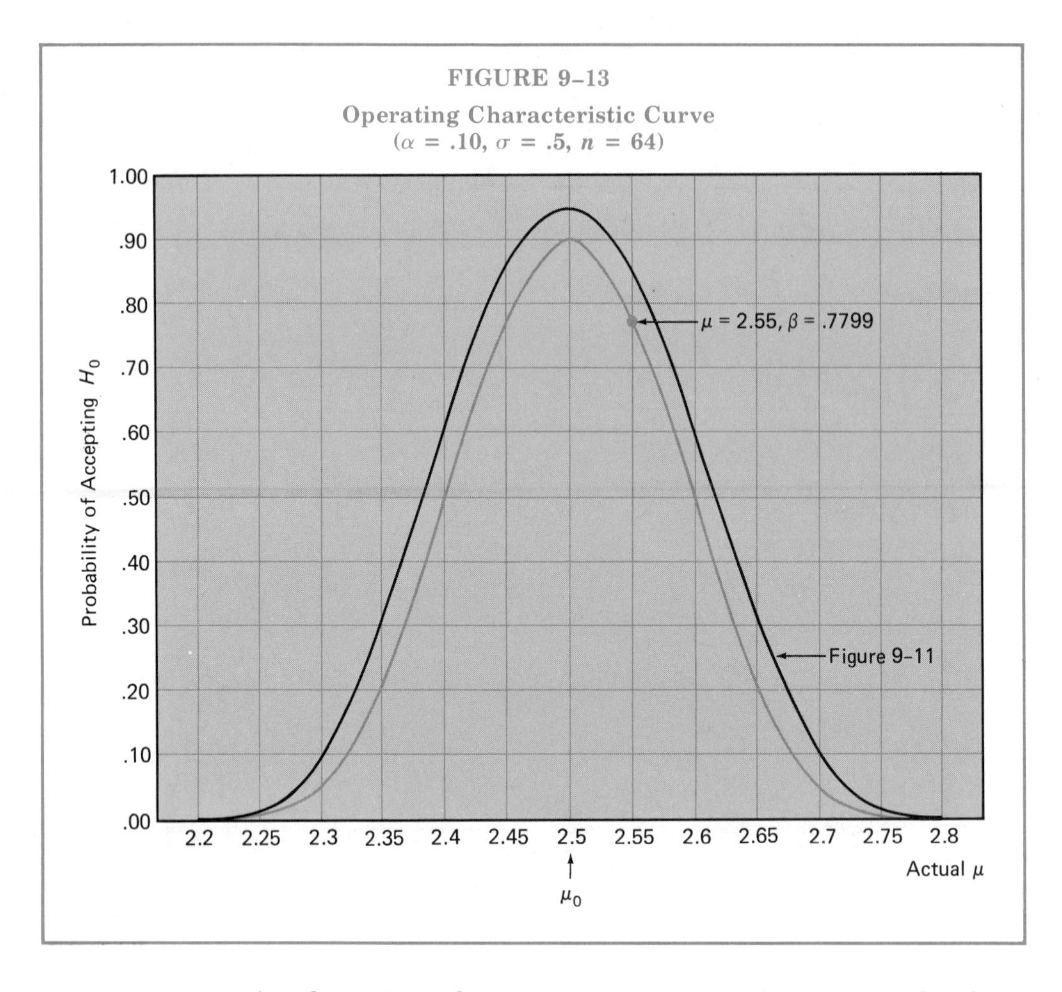

action that arises when $\mu = \mu_0$ or $\mu > \mu_0$ is the same. Under these circumstances, observing large $\overline{X}$s will not be harmful to their hypothesis. To clarify this, let's look at an example.

CG Magazine sponsors an annual conference on the "Future of the American Economy." Over 1,000 economists are invited. Details of the conference are reported by *CG*. One writer, charged with summarizing the speeches dealing with inflation, has tentatively titled the article "The War on Double-Digit Inflation." She intends to discuss the difficulty of maintaining price stability without sacrificing employment. If the economists predict a substantial reduction in inflation, however, she will call her article "Winning the Inflation War."

Prior to the meeting, the writer wants to know whether she should continue with her preliminary work on the "double-digit" theme. She randomly phones 36 of the participants and asks them

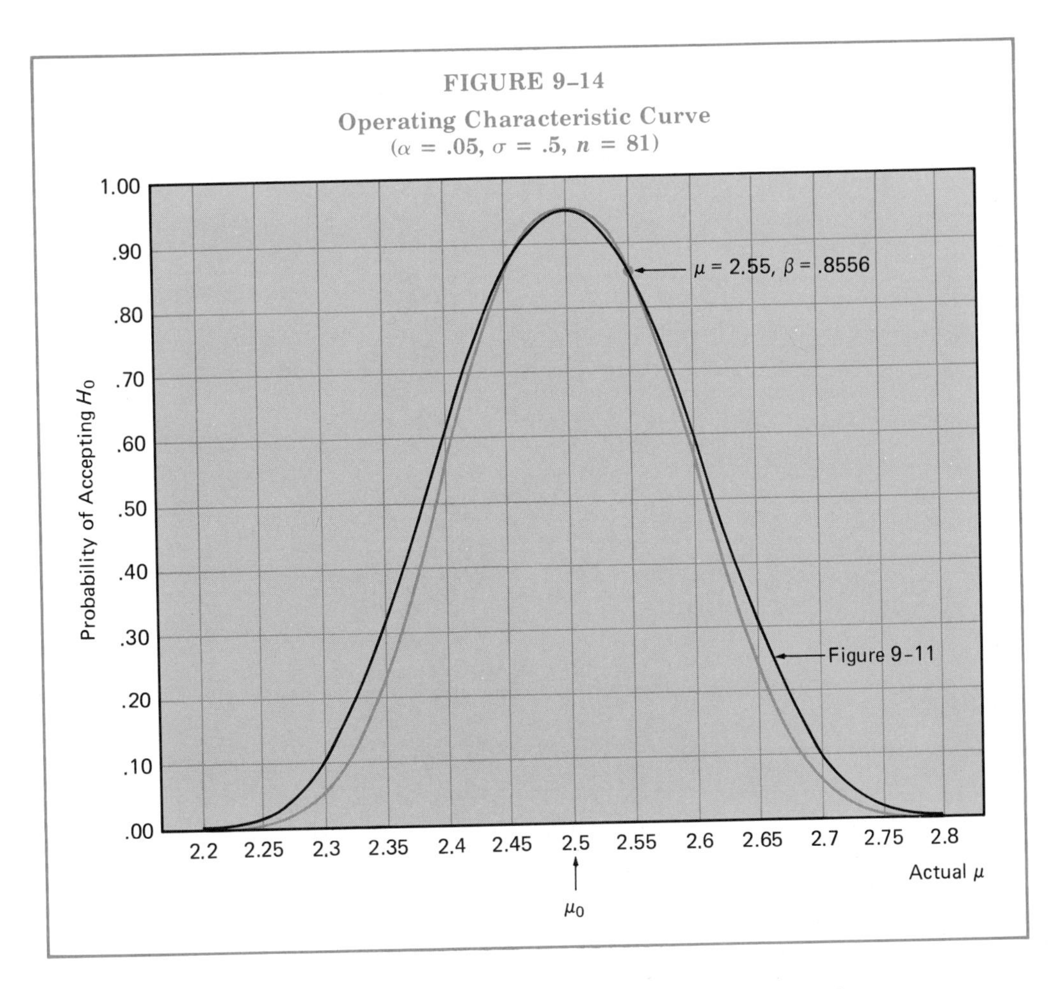

FIGURE 9–14
Operating Characteristic Curve
(α = .05, σ = .5, n = 81)

what inflation rate they predict for the next 12 months. She finds that $\overline{X}$ = 8.9 percent with s = 1.8 percent. Based on this sample, she will evaluate the hypothesis that the average predicted rate of inflation is 10 percent or more.

In contrast to the procedure in two-tailed tests, here the writer will accept or reject a *range* of μs; that is, μ = *any* double-digit number. Let's test

$$H_0: \mu \geq 10\%$$

$$H_1: \mu < 10\%$$

Our frame of reference will be a sampling distribution in which $\mu = \mu_0$ = the *smallest* value in H_0. The rationale for this will be explained soon. Because large $\overline{X}$s "don't matter," we'll place α in the *left* tail. The acceptance and rejection regions appear in Figure 9–15.

FIGURE 9–15

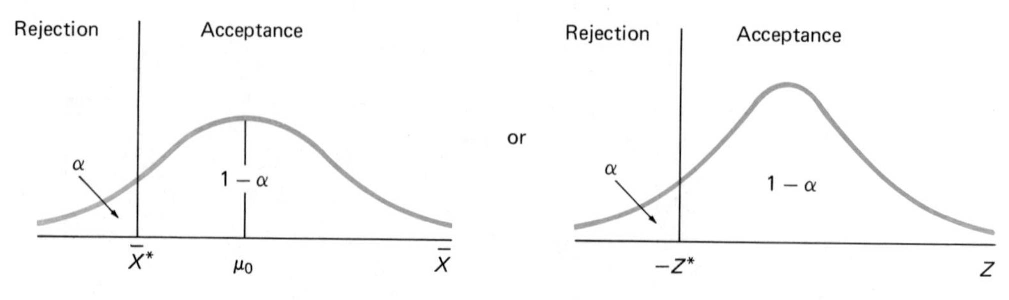

For $\alpha = .05$, we can derive Figure 9–16. The computed Z is

$$Z = \frac{8.9 - 10.0}{\frac{1.8}{\sqrt{36}}} = -3.67$$

FIGURE 9–16

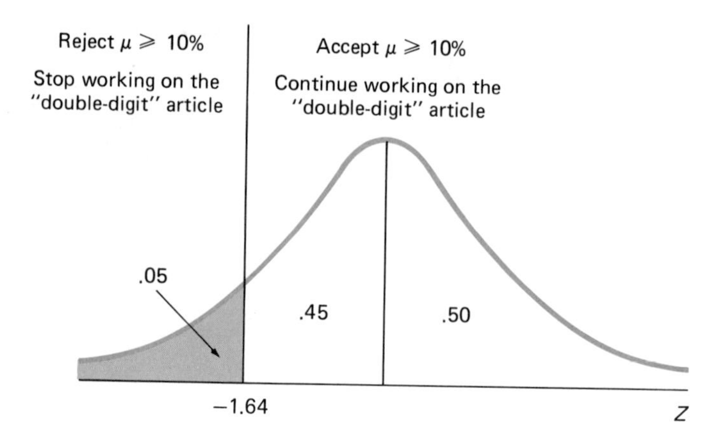

In a sampling distribution having $\mu = 10.0$ and $\sigma \approx 1.8$, $\overline{X} = 8.9$ is an "unlikely" event. Zs lower than -1.64 would occur only 5 percent of the time. The writer will reject H_0 and conclude $\mu < 10.0$. Thus, she will shift from the "double-digit" to the "winning" article.

Notice that $\mu = 10.2$, $\mu = 11.4$, $\mu = 12.0$, . . . are rejected along with $\mu = 10.0$. If we had used one of these other μs in our test, say $\mu = 10.2$, Z would fall farther to the left of $-Z^*$:

$$Z = \frac{8.9 - 10.2}{\frac{1.8}{\sqrt{36}}} = -4.33$$

In the planning stage of her experiment, the writer could specify a value for σ and determine $P(\text{accept } H_0|\mu)$ when $\alpha = .05$, $n = 36$,

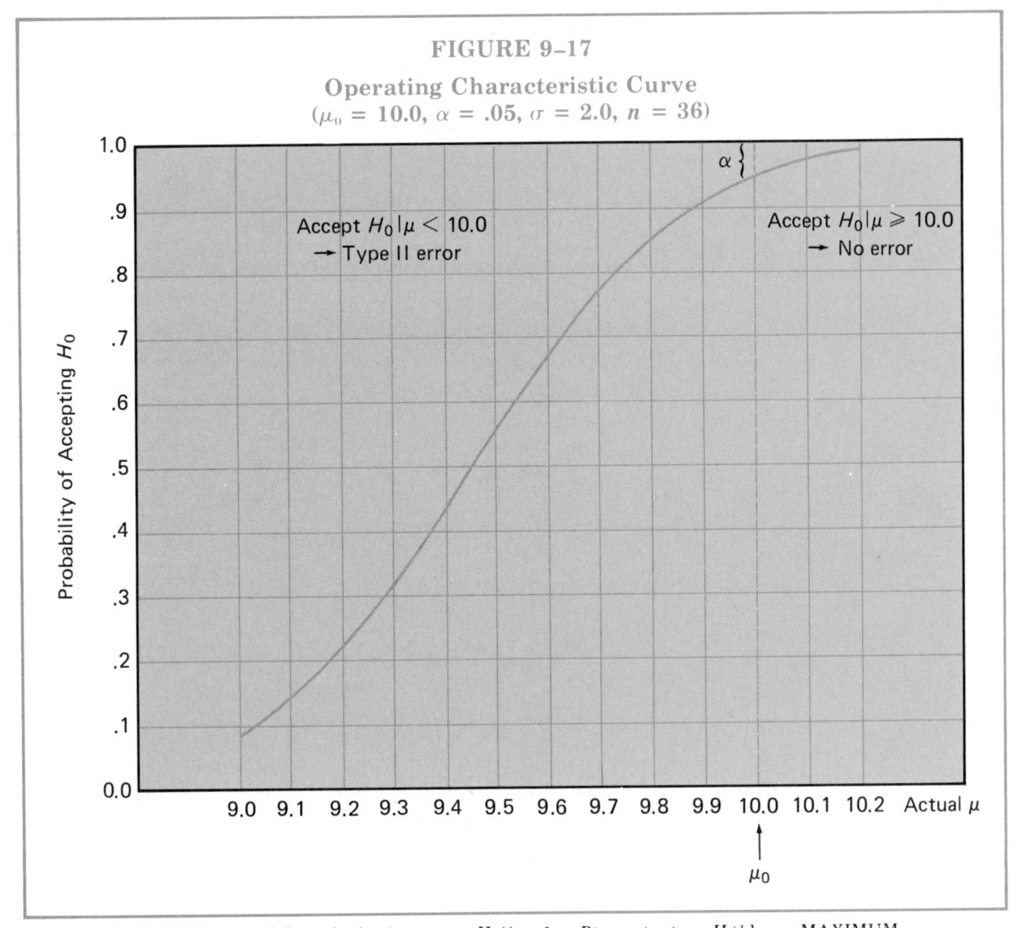

FIGURE 9–17
Operating Characteristic Curve
($\mu_0 = 10.0$, $\alpha = .05$, $\sigma = 2.0$, $n = 36$)

Note: The probability of rejecting a true H_0 [i.e., $1 - P(\text{accept a true } H_0)$] has a MAXIMUM value of .05 at μ_0. By contrast, $P(\text{reject a true } H_0)$ falls to $1 - .9884 = .0116$ at $\mu = 10.2$.

and $\mu_0 = 10.0$. Suppose that the standard deviation of the forecasts is always close to 2.0 percent. Using $\sigma = 2.0$, the writer would accept H_0 provided that

$$\overline{X} > \mu_0 - Z^*(\sigma/\sqrt{n})$$

$$\overline{X} > 10.0 - (1.64)(2.0/\sqrt{36}) = 9.45$$

Figure 9–17 is the operating characteristic curve that arises from calculating $P(\overline{X} > 9.45|\mu)$; for example, two points are $\mu = 9.6$, $\beta = .6736$, where

$$Z = \frac{9.45 - 9.6}{\frac{2.0}{\sqrt{36}}} = -.45,$$

and $\mu = 10.2$, $P(\text{accept } H_0|\mu = 10.2) = .9884$, where

$$Z = \frac{9.45 - 10.2}{\frac{2.0}{\sqrt{36}}} = -2.27$$

Notice that .9884 isn't β or the probability of a Type II error. This is because H_0 is true for any $\mu \geq 10.0$.

We can also conceive of situations in which very low, or left-tail, values of $\overline{X}$ wouldn't make a researcher feel uncomfortable about H_0. Consider the following example.

Educators have witnessed progressive declines in S.A.T. verbal scores. The national average dropped from 466 in the academic year 1966–1967 to 434 in 1974–1975. The principal of J. T. High School believes that students do poorly on S.A.T.s because they don't read much. Specifically, he claims that his students read six hours or less per week. To test this contention with an α of .01, he questions a random sample of 64 students and determines that $\overline{X} = 7$ hours and 40 minutes (i.e., 460 minutes) of reading time and $s = 1$ hour and 10 minutes (i.e., 70 minutes) of reading time.

The principal forms the hypotheses

$$H_0\colon \mu \leq 6 \text{ hours (i.e., 360 minutes)}$$

$$H_1\colon \mu > 6 \text{ hours}$$

He will be suspicious of $\mu \leq 6$ hours only if $\overline{X}$ is large. Hence, we'll place α in the *right* tail (see Figure 9–18).

FIGURE 9–18

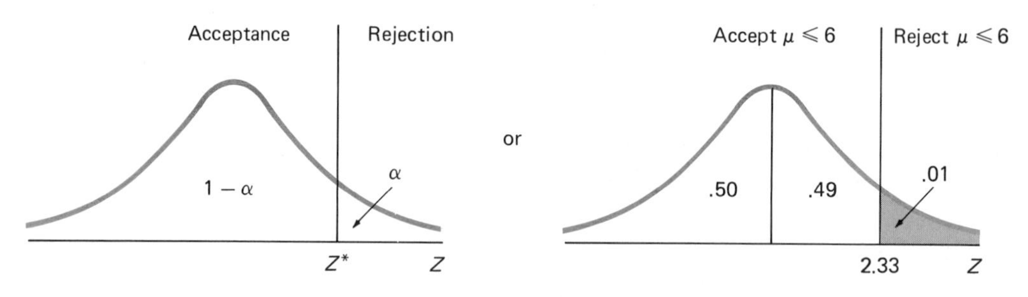

The null hypothesis is a series of μs. By letting $\mu = \mu_0 =$ the *largest* value in H_0, we can reject $\mu < \mu_0$ when we reject $\mu = \mu_0$. Moreover, P(Type I error) will be at a maximum. The computed Z is

$$Z = \frac{460 - 360}{\frac{70}{\sqrt{64}}} = 11.43$$

Since 11.43 > 2.33, the principal rejects H_0 and concludes that students read more than six hours per week.

In every one-tailed test, the "=" sign appears in the null hypothesis. Because of this, H_0 isn't always the belief held by the researcher. If an investigator felt that μ was greater than 15, he would form

$$H_0: \mu \leq 15 \text{ hours}$$

$$H_1: \mu > 15 \text{ hours}$$

SUMMARY

We have noted that estimation and hypothesis testing are closely related. Both of these procedures derive from the characteristics of the sampling distribution. Since the population is unknown, we can't be certain that accepting H_0 or rejecting H_0 is the correct action to take. The researcher's conclusion is simply consistent with the test criterion, α. Irrespective of how convincing the evidence is, hypothesis testing doesn't prove the truthfulness or falsity of the null hypothesis. Rather, this technique serves as a way to assess whether a particular belief seems sensible. In planning a test, the researcher must find an α, n pair that provides a satisfactory balance between P(Type I error) and P(Type II error). After a conclusion is reached, the possibility of one or the other form of error is always present.

EXERCISES

1. Determine the level of significance for a two-tailed hypothesis test if the critical values of Z are:

   a. $Z = -1.34$ and $Z = 1.34$
   b. $Z = -2.08$ and $Z = 2.08$
   c. $Z = -2.41$ and $Z = 2.41$
   d. $Z = -2.63$ and $Z = 2.63$
   e. $Z = -2.87$ and $Z = 2.87$

2. Identify the critical value of Z if a researcher tests

$$H_0: \mu \geq \$8{,}000$$
$$H_1: \mu < \$8{,}000$$

using (a) $\alpha = .02$, (b) $\alpha = .04$, and (c) $\alpha = .08$.

3. Identify the critical value of Z if a researcher tests

$$H_0: \mu \leq \$20{,}000$$
$$H_1: \mu > \$20{,}000$$

using (a) $\alpha = .03$, (b) $\alpha = .06$, and (c) $\alpha = .09$.

4. Given a population with $\sigma = 3$ and a sample of 64 observations, a researcher will accept H_0: $\mu = 18$ as long as an observed $\overline{X}$ is between 17.46 and 18.54. What is the level of significance?

5. Given a population with $\sigma = 7$ and a sample of 81 observations, a researcher will reject H_0: $\mu \geq 18$ as long as an observed $\overline{X}$ is less than 16.54. What is the level of significance?

6. Letting $\sigma = 4$, construct the operating characteristic curve when a sample of 49 observations is used to test

$$H_0: \mu = 26$$
$$H_1: \mu \neq 26$$

at the .05 level of significance.

7. Construct the power curve associated with the information in question 6.

8. Letting $\sigma = 7$, construct the operating characteristic curve when a sample of 81 observations is used to test

$$H_0: \mu \geq 50$$
$$H_1: \mu < 50$$

at the .01 level of significance.

9. Construct the power curve associated with the information in question 8.

10. *TH Magazine* is preparing a brochure in the hope of attracting more advertisers. The magazine believes that many of its readers are "young, sophisticated, and affluent." Using $\alpha = .05$, management would like to determine whether the average income of all subscribers is greater than \$30,000. Forty-nine subscribers are chosen at random. If $\overline{X} = \$33{,}280$ and $s = \$2{,}560$, what will management conclude?

11. If calculated according to the Federal Trade Commission approach, the average amount of tar in brand W cigarettes is said to be 2.3 milligrams. Suppose 36 W cigarettes are randomly selected. For this sample, $\overline{X} = 2.5$ milligrams and $s = 0.3$ milligrams. Test

$$H_0: \mu = 2.3 \text{ milligrams}$$
$$H_1: \mu \neq 2.3 \text{ milligrams}$$

at the .05 level of significance.

12. Because the executives at Corporation Q have to look at thousands of documents and memos each year, the president wishes to introduce an "in-house" speed-reading course. She believes that Q's executives, on the average, read less than 600 words per minute. A random sample of 64 executives is chosen. Test her belief at the .05 level of significance if $\overline{X} = 573$ words per minute and $s = 82$ words per minute.

13. If the equipment at NR Electronics is working properly, the length of a special part is 3.42 centimeters.

a. Management selects a random sample of 81 parts and computes $\overline{X} = 3.37$ centimeters and $s = 0.16$ centimeters. Given a two-tailed test and $\alpha = .05$, should production continue?
b. Using $\sigma = 0.25$ centimeters, $\alpha = .05$, and $n = 81$, what is the probability that management will accept H_0: $\mu = 3.42$ centimeters when the actual μ is 3.45 centimeters?
c. Using $\sigma = 0.25$ centimeters, $\alpha = .01$, and $n = 81$, what is the probability that management will accept H_0: $\mu = 3.42$ centimeters when the actual μ is 3.45 centimeters?
d. Using $\sigma = 0.25$ centimeters, $\alpha = .05$, and $n = 121$, what is the probability that management will accept H_0: $\mu = 3.42$ centimeters when the actual μ is 3.45 centimeters?

14. In connection with question 13, suppose $\sigma = 0.25$ centimeters, $\alpha = .05$, and $n = 64$. Compute the following probabilities:

a. $P(\text{accept } H_0|\mu = 3.40)$
b. $P(\text{reject } H_0|\mu = 3.46)$
c. $P(\text{reject } H_0|\mu = 3.41)$
d. $P(\text{accept } H_0|\mu = 3.38)$

15. One measure of how well a firm is doing is after-tax profits divided by sales, called the "profit margin." A financial analyst believes that all companies in industry L had an average profit margin of less than 4.0 percent in 1979. She chooses 36 companies at random and calculates $\overline{X} = 3.7\%$ and $s = 0.2\%$. Test her belief at the .01 level of significance.

16. The owner of a downtown parking lot is changing his rates. He feels that cars, on the average, are parked for at least 120 minutes. He selects 49 of his receipts at random. Using $\alpha = .01$, what will the owner conclude if $\overline{X} = 100$ minutes and $s = 14$ minutes?

17. The makers of Good Morning cereal market their product in boxes supposedly weighing 300 grams. A public-interest organization wishes to test the weight at the .01 level of significance. Given

$$H_0\text{: } \mu \geq 300 \text{ grams}$$
$$H_1\text{: } \mu < 300 \text{ grams,}$$

what would the organization conclude if a random sample of 64 boxes had a mean of 294 grams and a standard deviation of 5 grams?

18. The president of the W. M. Corporation recently viewed a TV program about the energy crisis. It mentioned that motorists still prefer driving their cars to work in spite of the current price of gasoline. Since the president believes that W. M.'s employees don't use more than 10 gallons of gasoline, on the average, to commute, he feels that the price of gasoline won't deter them from driving either. To test his belief at the .05 level of significance, the president selects a random sample of 49 employees who drive to W. M. What will he conclude if $\overline{X} = 11.9$ gallons and $s = 1.2$ gallons?

19. The Jones family has always enjoyed going to the movies. Given the jump in ticket prices over the years, Mr. Jones is disturbed by his belief that "most movies are finished before you can eat all of your popcorn." In fact, Jones feels that the movies released between 1973 and 1979 run, on the average, no more than 80 minutes. To test this claim, he randomly selects 49 movies from the book *A Description of Recent Films* and calculates $\overline{X} = 93$ minutes and $s = 12$ minutes. Given $\alpha = .05$, what will Jones conclude?

10

The t Distribution

SUBSTITUTING FOR σ WHEN n IS SMALL

The inferences in Chapters Eight and Nine arose from large samples. We have noted two consequences of a large n: (1) the sampling distribution of $\overline{X}$ is normal, and (2) the sample standard deviation will be a good approximation for σ. Because of the first condition, $(\overline{X} - \mu)/(\sigma/\sqrt{n})$ has a standard normal distribution. If s is substituted for σ, the second condition suggests that the distribution of $(\overline{X} - \mu)/(s/\sqrt{n})$ will be almost the same as the distribution of $(\overline{X} - \mu)/(\sigma/\sqrt{n})$. Given these two conditions, setting $(\overline{X} - \mu)/(s/\sqrt{n})$ equal to Z is justified.

When the sampling distribution of $\overline{X}$ is normal, we are able to identify the *actual* distribution of $(\overline{X} - \mu)/(s/\sqrt{n})$. As long as n is large, we can construct approximately the same confidence interval estimates of μ using either the standard normal distribution or the true distribution of the above ratio. If n is small, however, researchers must work with the actual distribution. Since the sampling distribution has to be normal, we will assume that these small samples are selected from a normal population of Xs.

In repeated random sampling, $(\overline{X} - \mu)/(\sigma/\sqrt{n})$ will vary because $\overline{X}$ varies. On the other hand, $(\overline{X} - \mu)/(s/\sqrt{n})$ will vary because *both* $\overline{X}$ and s vary (i.e., $\overline{X}$ and s are random variables whose values represent outcomes of the experiment "select a sample at random").

We shall state that $\overline{X}$ and s are independent whenever sampling is from a normal population. Two samples having equal means (i.e., $\overline{X}_1 = \overline{X}_2$) can have different standard deviations (i.e., $s_1 \neq s_2$). Then, even though

$$\frac{\overline{X}_1 - \mu}{\frac{\sigma}{\sqrt{n}}} = \frac{\overline{X}_2 - \mu}{\frac{\sigma}{\sqrt{n}}},$$

we see that

$$\frac{\overline{X}_1 - \mu}{\frac{s_1}{\sqrt{n}}} \neq \frac{\overline{X}_2 - \mu}{\frac{s_2}{\sqrt{n}}}$$

CHARACTERISTICS OF THE t DISTRIBUTION

The ratio $(\overline{X} - \mu)/(s/\sqrt{n})$ is called a *t value*. We say that t possesses a t (or Student's t) *distribution* with $n - 1$ *degrees of freedom,* or

$$t_{(n-1)} = \frac{\overline{X} - \mu}{\frac{s}{\sqrt{n}}} \qquad (10\text{–}1)$$

The density function for the t distribution is too complicated to reproduce here. Nevertheless, it still yields a curve enclosing a total area of 1.0. As Figure 10–1 indicates, this curve is symmetrical about $t = 0$ and approaches the horizontal axis asymptotically.

FIGURE 10–1

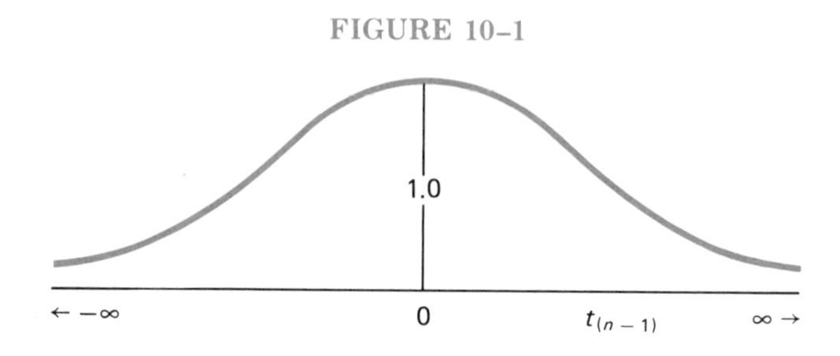

Due to the fact that the degrees of freedom enter the density function, each sample size determines a *different* t distribution. We can speak of *the* standard normal curve, but the t distribution is a *family* of distributions. Since t varies more than Z, we should expect that the t distribution is not as "pinched in" as the standard normal curve. Figure 10–2 confirms the relation between t and Z.

Where "degrees of freedom" is mentioned in the text, we will be studying a sample *sum of squared deviations,* such as $\sum^{n}(X - \overline{X})^2$. Although $\sum^{n}(X - \overline{X})^2$ can be any value, a restriction is placed on the deviations; namely, they must always sum to zero [i.e., $\sum^{n}(X - \overline{X}) = 0$].

FIGURE 10–2

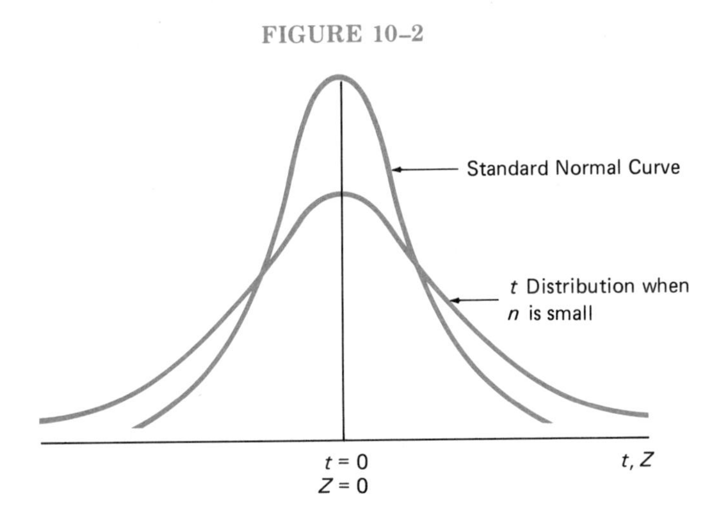

Suppose that $\overline{X} = 7$ and four of five sample observations are 2, 5, 8, and 9. We, then, know

$$(2 - 7) + (5 - 7) + (8 - 7) + (9 - 7) + (X_5 - 7) = 0$$
$$(-5) + (-2) + (1) + (2) + (X_5 - 7) = 0$$
$$X_5 - 7 = 4$$
$$X_5 = 11$$

The last deviation, $X_5 - 7$, must be a number that will make the sum of the n deviations equal to zero. Hence, while $n - 1$ deviations are "free," or independent, the nth deviation is not "free." Because the degrees of freedom identify how many of the deviations in the sum of squared deviations are independent, $\Sigma(X - \overline{X})^2$ represents $n - 1$ degrees of freedom.

If t^* and t^{**} are two values of t, $P(t^* < t < t^{**})$ will change each time n changes. Instead of developing a complete table of areas for n_1, n_2, and so on, we have put parts of many t distributions into one table (see the Appendix). The left column lists the degrees of freedom, and the top row shows what area lies to the right of ts appearing in the body of the table. For a t distribution with 17 degrees of freedom (i.e., $n = 18$), $P(t > 1.740) = .05$. Due to symmetry, $P(t < -1.740)$ is also .05 (see Figure 10–3).

As n increases, the t distribution (1) becomes more concentrated around $t = 0$ and (2) approaches the standard normal curve. This occurs because s improves as an estimator of σ. In Table 10–1, we present values of t^* and Z^* which make $P(t > t^*) = .05$ and $P(Z > Z^*) = .05$ given various ns. Notice that t^* moves closer to Z^* when n increases.

FIGURE 10–3

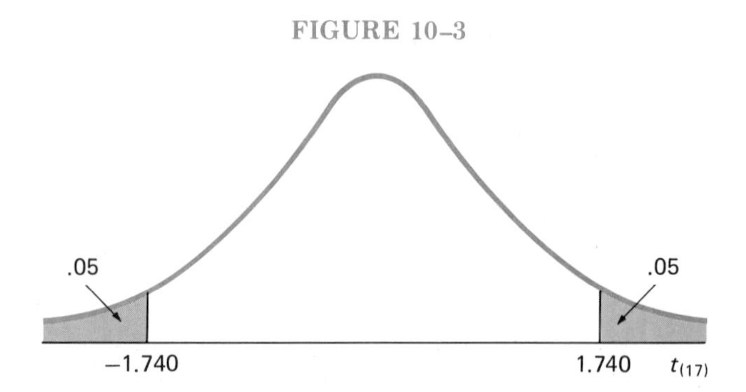

We will follow the tradition of equating $(\overline{X} - \mu)/(s/\sqrt{n})$ with Z if $n \geq 31$. In research, however, the t distribution should be used even if $n = 100$. Although $t = 1.697$ is not much different from $Z = 1.64$, a difference of $1.697 - 1.64 = .057$ could be important. Using 1.64 in place of 1.697 will make $100(1 - \alpha)$ percent confidence intervals and acceptance regions narrower than they actually are. Hence, the probability of rejecting a true null hypothesis will actually be greater than α.

TABLE 10–1

n	t^*	Z^*
11	1.812	1.64
21	1.725	1.64
31	1.697	1.64
61	1.671	1.64
121	1.658	1.64

SMALL-SAMPLE INFERENCES INVOLVING THE SAMPLE MEAN

To reinforce the similarities between interval estimation with t and Z, consider the following example.

A consumer safety agency would like to determine how long a particular electric drill can operate before it overheats. Since the drills are expensive and overheating ruins them, the agency can only test a random sample of ten drills. The agency calculated $\overline{X} = 29$ minutes and $s = 6$ minutes.

Assuming that the population of overheating times is normal, the sampling distribution of the sample average number of minutes would look like Figure 10–4. Moreover, $(\overline{X} - \mu)/(s/\sqrt{n})$ will have a t distribution with 9 degrees of freedom (see Figure 10–5).

FIGURE 10–4

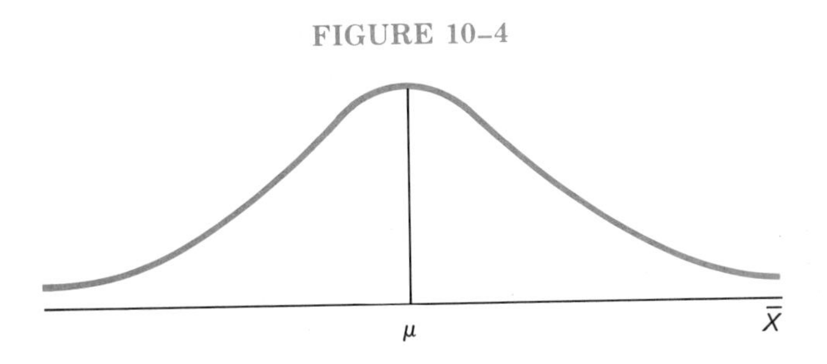

FIGURE 10–5

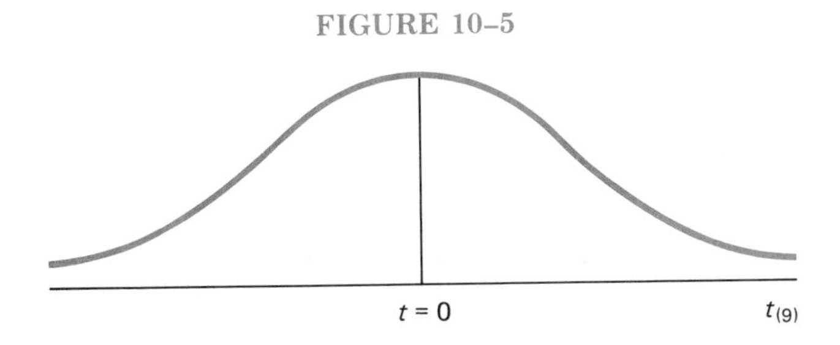

Given

$$P[-t^*_{(n-1)} < t_{(n-1)} < t^*_{(n-1)}] = 1 - \alpha$$

or

$$P\left[-t^*_{(n-1)} < \frac{\overline{X} - \mu}{\frac{s}{\sqrt{n}}} < t^*_{(n-1)}\right] = 1 - \alpha,$$

the $100(1 - \alpha)$ percent confidence interval estimator of μ is

$$\overline{X} - t^*_{(n-1)}(s/\sqrt{n}) < \mu < \overline{X} + t^*_{(n-1)}(s/\sqrt{n}) \qquad (10\text{–}2)$$

Suppose the agency wants a 95 percent confidence interval estimate. With $\alpha = .05$ and t having 9 degrees of freedom, $t^* = 2.262$ makes $P[-t_{(9)} < -t^*] = .025$ and $P[t_{(9)} > t^*] = .025$. We find

$$29 - (2.262)(6/\sqrt{10}) < \mu < 29 + (2.262)(6/\sqrt{10})$$

$$24.7 \text{ minutes} < \mu < 33.3 \text{ minutes}$$

This interval could contain μ or μ could lie outside of it. The agency will, nonetheless, act as though $24.7 < \mu < 33.3$ were true. Consumers will be warned that the drill is not to be used for big jobs (i.e., those needing a lot of time).

We can also conduct hypothesis tests when n is small. The next example illustrates this technique.

A regional automobile club just finished a hectic day. Because of chilling temperatures, over 1,000 batteries had to be recharged. The club contends that the time between a call to its headquarters and the arrival of a service truck averaged no more than 35 minutes. If so, $\mu \leq 35$ minutes will be publicized in future membership campaigns.

The club intends to test

$$H_0: \mu \leq 35 \text{ minutes}$$

$$H_1: \mu > 35 \text{ minutes}$$

at the .01 level of significance. It randomly phones 25 of the motorists who received service, and computes $\overline{X} = 41$ minutes and $s = 13$ minutes.

If the population of waiting times is normal, $(\overline{X} - \mu_0)/(s/\sqrt{n})$ will have a t distribution with 24 degrees of freedom. Figure 10–6 indicates the acceptance and rejection regions when $\alpha = .01$. Inasmuch as

$$t = \frac{41 - 35}{\frac{13}{\sqrt{25}}} = 2.31,$$

the null hypothesis will be accepted. The club won't be reluctant to advertise $\mu \leq 35$ minutes. A Type II error, however, is possible in this situation.

FIGURE 10–6

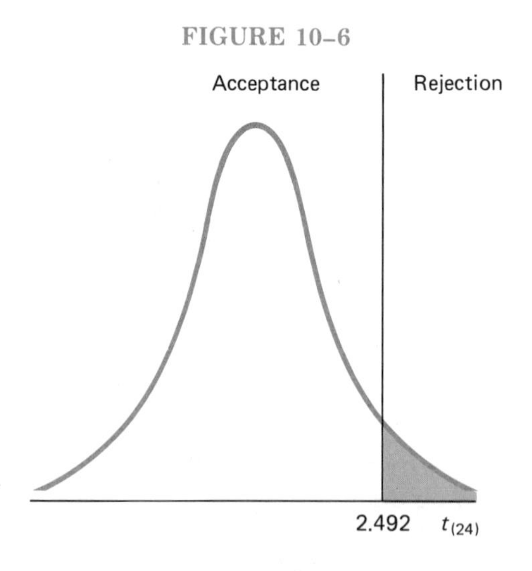

SUMMARY

The t distribution is necessary for making inferences about μ whenever a small sample is drawn from a normal population with an unknown σ. We have seen that the t distribution approaches the standard normal curve as n increases. Because of this, we can equate $(\overline{X} - \mu)/(s/\sqrt{n})$ with Z if n is "large" (i.e., $n \geq 31$). The last section of the chapter substantiated the fact that small-sample inferences are derived the same way as large-sample inferences except that t, not Z, is used. Table 10–2 reviews the circumstances in which t and Z are appropriate.

TABLE 10–2

Population	σ	*Sample Size*	*Basis for Inferences*
Normal	Known	Large	$Z = (\overline{X} - \mu)/(\sigma/\sqrt{n})$
Not normal	Known	Large	$Z = (\overline{X} - \mu)/(\sigma/\sqrt{n})$
Normal	Known	Small	$Z = (\overline{X} - \mu)/(\sigma/\sqrt{n})$
Normal	Unknown	Large	t or $Z = (\overline{X} - \mu)/(s/\sqrt{n})$
Not normal	Unknown	Large	t or $Z = (\overline{X} - \mu)/(s/\sqrt{n})$
Normal	Unknown	Small	$t = (\overline{X} - \mu)/(s/\sqrt{n})$
Not normal	Known	Small	Nonparametric methods
Not normal	Unknown	Small	Nonparametric methods

EXERCISES

1. Determine t^* if:

   a. $P(t_{(13)} > t^*) = .10$
   b. $P(t_{(19)} > t^*) = .025$
   c. $P(t_{(24)} > t^*) = .05$

2. What are the following probabilities?

   a. $P(-1.476 < t_{(5)} < 1.476)$
   b. $P(-2.552 < t_{(18)} < 2.552)$
   c. $P(-2.064 < t_{(24)} < 2.064)$

3. Identify the following probabilities:

   a. $P(t_{(7)} > 2.365)$
   b. $P(t_{(26)} > 2.779)$
   c. $P(t_{(11)} < 2.201)$

4. A researcher chooses 14 observations at random from a normal population. Given $\overline{X} = \$2{,}150$ and $s = \$104$, construct the following confidence interval estimates of μ:

a. 90 percent
b. 95 percent
c. 99 percent

5. Inflation has forced yearly increases in the tuition charges at many colleges and universities. An investigator feels the average tuition at private colleges in 1979 was \$3,200 or more. She selects a random sample of 20 colleges and computes $\overline{X} = \$3{,}050$ and $s = \$246$. Test her contention using $\alpha = .05$ (assume the population is normal).

6. A certain manufacturer produces a component for stereo systems. The part should weigh 1.6 ounces. Suppose a quality-control engineer chooses 12 parts at random. For this sample, $\overline{X} = 1.54$ ounces and $s = 0.035$ ounces. Test

$$H_0\colon \mu = 1.6 \text{ ounces}$$
$$H_1\colon \mu \neq 1.6 \text{ ounces}$$

using $\alpha = .10$ (assume a normal population).

7. The S. W. Company sells a lamp designed for people who read extensively. S. W. is mailing ads to professors since it feels that college instructors, on the average, read 15 or more hours a week. The company selects a random sample of 18 professors. If $\overline{X} = 14.3$ hours and $s = 2.6$ hours, would S. W. accept $H_0\colon \mu \geq 15$ hours at the .05 level of significance (assuming a normal population)?

8. A particular fast-food organization has 1,500 franchises. A researcher would like to determine the average level of sales for these franchises in 1978. He chooses 17 franchises at random. Given $\overline{X} = \$278{,}000$ and $s = \$17{,}200$, and assuming the population is normally distributed, calculate the following confidence interval estimates of μ:

a. 90 percent
b. 95 percent
c. 99 percent

9. The M. Q. Company must decide whether to develop television or radio commercials. Because M. Q. provides services that appeal to doctors, the vice-president of marketing is in favor of advertising on the radio. He believes that doctors can easily listen to a radio in their offices, in their cars, or while they read, but are too busy to watch TV. To determine the average amount of time doctors listen to the radio per week, the vice-president selects a random sample

of 23 physicians. He computes $\overline{X} = 276$ minutes and $s = 28$ minutes. Assuming a normal population, calculate the following confidence interval estimates of μ:

a. 90 percent
b. 95 percent
c. 99 percent

10. The debt-to-equity ratio is a measure of the relationship between the funds borrowed by a company and the funds supplied by stockholders. A business economist believes that the average debt-to-equity ratio for all firms in industry R is no more than 1.25. He chooses 21 firms at random. Suppose $\overline{X} = 1.31$ and $s = .02$. Test his belief using a .05 level of significance (assume the population is normal).

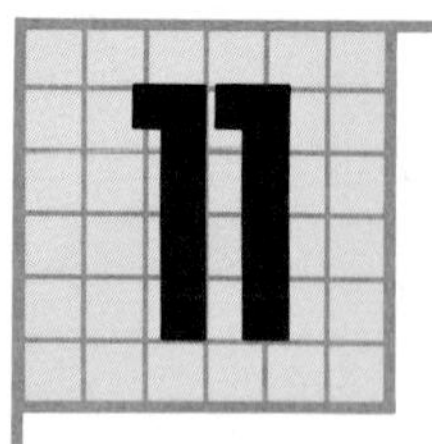

Statistical Inferences Involving the Sample Proportion—Large Samples

THE SAMPLING DISTRIBUTION OF THE SAMPLE PROPORTION

We now want to broaden the discussion of inference by introducing the *sample proportion.* Among other things, this statistic is associated with the Gallup Poll. The following two statements are examples of the way Gallup might phrase its conclusions:

1. If the presidential election were held today, only two-tenths of the electorate would vote for Candidate G.
2. Three-fifths of all adults rank inflation as the top domestic problem.

In each instance, a sample proportion (i.e., .2 and .6, respectively) has been used as a point estimate of the population proportion.

To identify the relationship between the sample proportion,

$$\frac{r}{n} = \frac{\text{number of sample observations having a special attribute}}{\text{sample size}}, \qquad (11\text{–}1)$$

and the population proportion, π, we must derive the *sampling distribution of r/n.* Even before establishing this distribution, however, we can show that r/n is a good estimator of π.

Suppose that a population of Xs consists of observations that are either "successes" or "failures." As in Chapter Five, let's set X equal to 1 if an observation is a "success" and equal to 0 if an observation is a "failure." Table 11–1 presents the population. Notice that the population average, $\mu = \Sigma X/N = 6/10 = .6$, is also the population proportion of "successes."

TABLE 11–1

Attribute	X
S	1
S	1
F	0
S	1
F	0
F	0
S	1
S	1
S	1
F	0
	6

A sample from this population is listed in Table 11–2. The sample proportion of "successes" (i.e., .4) is the same as the sample average, $\overline{X} = \Sigma X/n = 2/5 = .4$. Given our study of $\overline{X}$, we can, thus, anticipate that r/n is an unbiased and consistent estimator of π. Moreover, because of the central limit theorem, its sampling distribution will be normal when n is large. We will use the following example to develop the sampling distribution of r/n.

TABLE 11–2

Attribute	X
F	0
S	1
S	1
F	0
F	0
	2

In the aftermath of the recessions of the 1970s, more and more college students have been attracted to "career" courses. Not surprisingly, the number of majors in business and economics (or BE) has risen substantially. Of the 10,000 undergraduates at University L, 4,000 are BEs.

If we were to select at random 50 students without replacement, the probability of choosing a BE on the first draw would be 4,000/10,000 = .4000. This probability is identical to π, the population proportion of BEs. If 49 BEs are consecutively chosen, the conditional probability P(BE selected on the 50th draw|49 BEs have been selected previously) is 3,951/9,951 = .3970. The draws are virtually independent, and the conditional probability of a "success" is close to π. Each selection is the outcome of a Bernoulli trial.

The binomial distribution details the *cumulative* number of "successes" in n independent Bernoulli trials. When the random variable r is "the number of BEs in a sample of 50 students," its values range from 0, 1, 2, 3 up to 50. We can compute

$$P(r) = \frac{n!}{r!(n-r)!}\pi^r(1-\pi)^{n-r}$$

$$= \frac{50!}{r!(50-r)!}(.40)^r(.60)^{50-r}$$

After computing $P(r)$ for each possible value of r, we could identify the *binomial probability* distribution of r. This distribution would be the same as the *sampling* distribution of r. Since we know that $\mu_r = n\pi$ and $\sigma_r = \sqrt{n\pi(1-\pi)}$, the parameters of the sampling distribution are $\mu_r = (50)(.40) = 20$ BEs and $\sigma_r = \sqrt{(50)(.40)(.60)} = 3.46$ BEs. In repeated random sampling, the average number of BEs will be 20.

Because $n\pi = (50)(.40) = 20 > 5$ and $n(1-\pi) = (50)(.60) = 30 > 5$, we can use the normal distribution to approximate the binomial sampling distribution of r (see Figure 11–1). Using the continuity correction, the probability that r will be between, say, 15 and 28 BEs is $P(14.5 < r < 28.5)$.

FIGURE 11–1

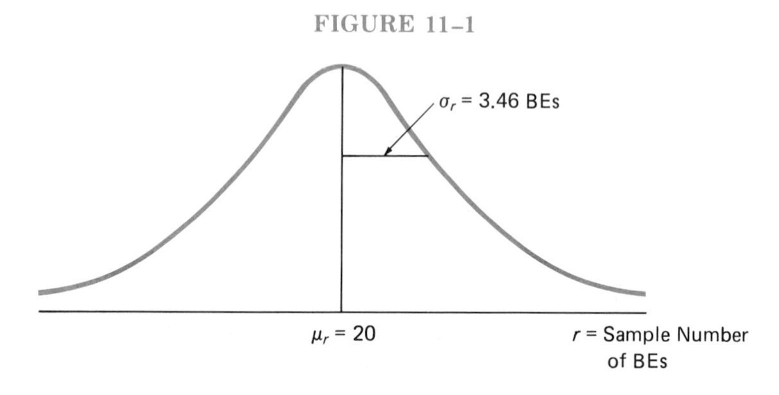

Given Equation 6–10, that is,

$$Z = \frac{r - n\pi}{\sqrt{n\pi(1-\pi)}},$$

this probability, or $P[(14.5 - 20)/3.46 < Z < (28.5 - 20)/3.46]$, is represented by the shaded area in Figure 11–2.

Besides mentioning that the binomial distribution of r is the sampling distribution of r, we have said nothing new. Where inferences are made as a consequence of Bernoulli trials, researchers are

FIGURE 11–2

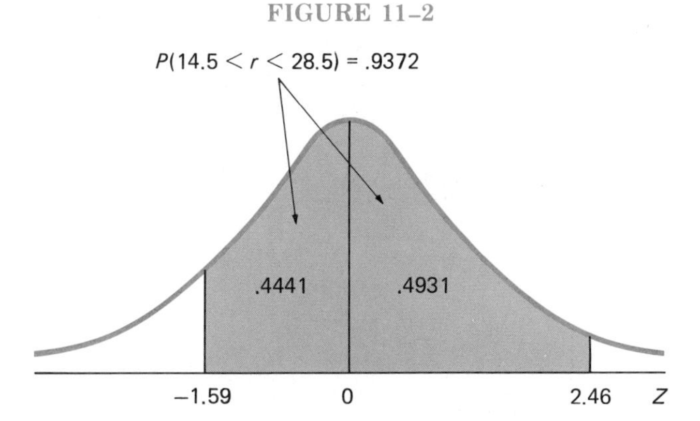

often more concerned about the population proportion of "successes" than about the number of "successes." We will now show how the sampling distribution of r/n can be constructed from the sampling distribution of r.

Every trial of "select 50 students at random" *simultaneously* yields values for both r and r/n. For instance, the outcome $r/n = 12/50$ will occur *only* if $r = 12$ occurs. We can, therefore, state

$$P(r/n = 12/50) = P(r = 12) = \frac{50!}{12!(50-12)!}(.40)^{12}(.60)^{38}$$

The graphs of the sampling distributions of r and r/n will be identical except for the fact that the axes are $P(r)$ and r in the former case and $P(r/n)$ and r/n in the latter. Hence, the sampling distribution of r/n also becomes more symmetrical as the sample size increases.

Recall that the only values of X are 1 and 0, and $r = X_1 + X_2 + \ldots + X_n$. Since

$$E(r/n) \text{ or } \mu_{r/n} = (1/n)E(r), \qquad \text{By virtue of Equation 5–7}$$

$$E(r/n) = (1/n)(n\pi) = \pi \qquad (11\text{–}2)$$

We see that r/n is an unbiased estimator of π.

Next let's find the variance and the standard deviation:

$$V(r/n) = (1/n)^2 V(r) \qquad \text{By virtue of Equation 5–11}$$

$$= (1/n)^2[n\pi(1-\pi)]$$

$$= \pi(1-\pi)/n$$

and

$$\sigma_{r/n} = \sqrt{\pi(1-\pi)/n} \qquad (11\text{–}3)$$

The sample proportion is, therefore, a consistent estimator.

Finally, because r/n is a linear function of r, that is, $r/n = (1/n)r$, r/n will be normal when r is normal. Figure 11–3 is the sampling distribution of the proportion of BEs. We will now confirm that $P(14.5/50 < r/n < 28.5/50) = P(14.5 < r < 28.5)$.

FIGURE 11–3

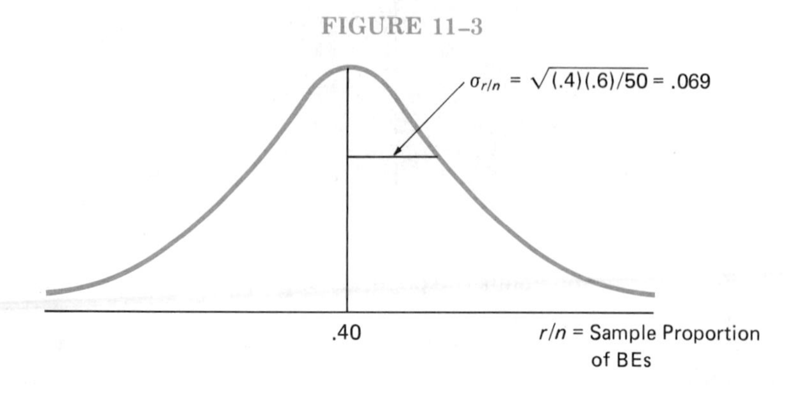

After substituting Equations 11–2 and 11–3 into the expression

$$Z = \frac{\text{value of the random variable} - \text{mean of the probability distribution}}{\text{standard deviation of the probability distribution}},$$

we have

$$Z = \frac{r/n - \pi}{\sqrt{\pi(1 - \pi)/n}} \tag{11–4}$$

The area between $r/n = 14.5/50 = .29$ and $28.5/50 = .57$ is equal to the area between $Z = (.29 - .40)/.069$ and $(.57 - .40)/.069$ (see Figure 11–4). Thus, $P(.29 < r/n < .57) = .9372$, which is the value we derived (see Figure 11–2) for $P(14.5 < r < 28.5)$.

FIGURE 11–4

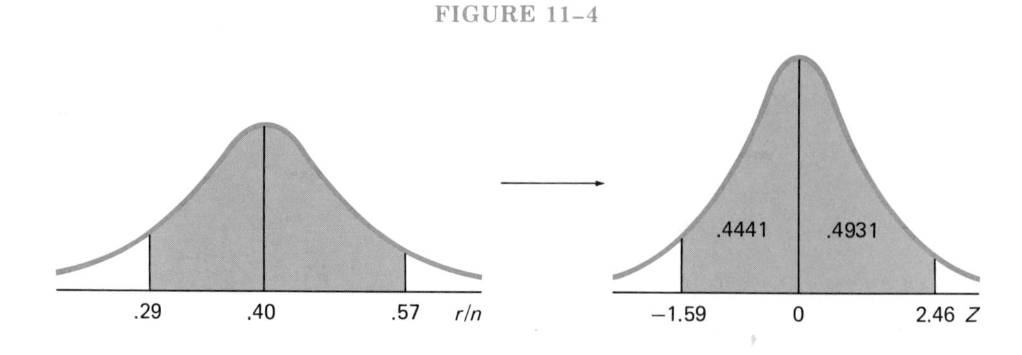

Table 11–3 reviews some of the conclusions of this section.

TABLE 11–3
Summary of the Relationship Between a Population, a Sample, and the Sampling Distribution of the Sample Proportion

Distribution	*Proportion*	*Expected Value*	*Standard Deviation*
Population	π	$\mu_x = \pi$ (where the values of X are 1 and 0)	$\sigma_x = \sqrt{\pi(1 - \pi)}$
Sample	r/n		
Sampling distribution of the sample proportion		$\mu_{r/n} = \pi$	$\sigma_{r/n} = \sqrt{\pi(1 - \pi)/n}$

ESTIMATION WITH THE SAMPLE PROPORTION

If $P(-Z^* < Z < Z^*) = 1 - \alpha$, we can also state the following equalities:

$$P\left(-Z^* < \frac{r/n - \pi}{\sigma_{r/n}} < Z^*\right) = 1 - \alpha$$

and

$$P[-Z^*(\sigma_{r/n}) < (r/n - \pi) < Z^*(\sigma_{r/n})] = 1 - \alpha \qquad (11\text{–}5)$$

The latter expression suggests that the probability of getting a sampling error, $|r/n - \pi|$, less than $Z^*\ (\sigma_{r/n})$ is $1 - \alpha$ (see Figure 11–5). Moreover, we can rearrange the bracketed part of Equation 11–5 to derive the $100(1 - \alpha)$ percent confidence interval estimator of π:

$$r/n - Z^*(\sigma_{r/n}) < \pi < r/n + Z^*(\sigma_{r/n}) \qquad (11\text{–}6)$$

FIGURE 11–5

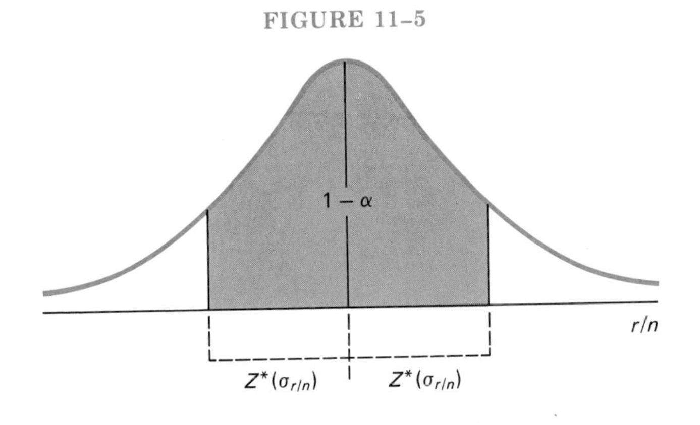

The following example illustrates estimation.

A congressional act changed the structure of the U.S. Postal System in 1970. Even with the change, the Postal Service has continued to operate in the "red." The deficit in 1976 totaled over $1 billion. Stopping Saturday deliveries is one possible cost-cutting step. Congressman W wants to find out how his constituents feel about this plan by questioning a random sample. He would like to be 95 percent confident that an observed r/n (the sample proportion of people opposed to the Saturday recommendation) is no more than .03 from π (the proportion of all people in W's district opposed to the Saturday recommendation).

Since the maximum error, e, is

$$e = Z^*(\sigma_{r/n}) = Z^*\sqrt{\pi(1 - \pi)/n}, \qquad (11\text{–}7)$$

we see that

$$e^2 = (Z^*)^2[\pi(1 - \pi)/n]$$

$$n(e^2) = (Z^*)^2\pi(1 - \pi)$$

$$n = \frac{(Z^*)^2\pi(1 - \pi)}{e^2} \qquad (11\text{–}8)$$

The congressman must choose a sample of size

$$n = \frac{(1.96)^2\pi(1 - \pi)}{(.03)^2}$$

At the presampling stage, Congressman W has no value of r/n to substitute for π. He could estimate π on the basis of (1) a pilot study or (2) his intuition. Another approach to solving Equation 11–8 arises from the fact that the product of π and $1 - \pi$ can *never* be larger than .25 (see Table 11–4).

TABLE 11–4

π	$1 - \pi$	$\pi(1 - \pi)$
.1	.9	.09
.2	.8	.16
.3	.7	.21
.4	.6	.24
.45	.55	.248
.49	.51	.2499
.5	.5	.25
.51	.49	.2499
.55	.45	.248
.6	.4	.24
.7	.3	.21
.8	.2	.16
.9	.1	.09

If we use the value of π that produces the largest possible product of π and $1 - \pi$ (and thus the largest n), we find

$$n = \frac{(Z^*)^2(.5)(.5)}{e^2} \qquad (11\text{–}9)$$

$$= \frac{(1.96)^2(.5)(.5)}{(.03)^2} = 1{,}067$$

When $n = 1{,}067$, the probability associated with $|(r/n - \pi)| \leq .03$ is *at least* .95. For an actual π of .10, $P[|(r/n - \pi)| \leq .03] = .95$ as long as

$$n = \frac{(1.96)^2(.10)(.90)}{(.03)^2} = 384$$

Thus, if π is not close to .5, the n derived by Equation 11–9 will be substantially greater than the required n (Equation 11–8).

The congressman ultimately selects a random sample of 1,000 people and finds 300 opposed to the Saturday recommendation; thus, $r/n = 300/1{,}000 = .30$. Since n is large, r/n is a good estimator of π. He can, then, be 95 percent confident that the sampling error is no more than

$$1.96\sqrt{(r/n)(1 - r/n)/n} = 1.96\sqrt{(.30)(.70)/1{,}000} = .028$$

The corresponding 95 percent confidence interval estimate of π is

$$.30 - 1.96\sqrt{(.30)(.70)/1{,}000} < \pi < .30 + 1.96\sqrt{(.30)(.70)/1{,}000}$$

$$.27 < \pi < .33$$

The congressman can't be certain that π is somewhere in the interval .27 to .33. However, he can feel 95 percent confident that it is and thus that a majority of his constituents won't be upset if Saturday deliveries are stopped.

HYPOTHESIS TESTS INVOLVING THE SAMPLE PROPORTION

Because the value of π enters into the calculation of $\sigma_{r/n}$, any null hypothesis, $\pi = \pi_0$, simultaneously creates an assumption about the standard deviation of the sampling distribution of r/n, that is, $\sigma_{r/n} = \sqrt{\pi_0(1 - \pi_0)/n}$. Whenever

$$Z = \frac{r/n - \pi_0}{\sqrt{\pi_0(1 - \pi_0)/n}} \qquad (11\text{–}10)$$

lies in the rejection region, we would view an observed r/n as "possible but improbable" given $\pi = \pi_0$. Consider the following example.

In a 1964 report, the U.S. Surgeon General announced that people who smoked cigarettes increased their chances of getting certain diseases. Shortly thereafter, a researcher found that 20 percent of the students at University L were smokers.

Despite the warning "Cigarette smoking is dangerous to your health" and the TV ban on cigarette advertising, sales of cigarettes have not declined. The researcher decides to investigate University L again. She believes that .20 is still the proportion of smokers. In a random sample of 100 students, she finds 15 smokers or $r/n = 15/100 = .15$.

FIGURE 11–6

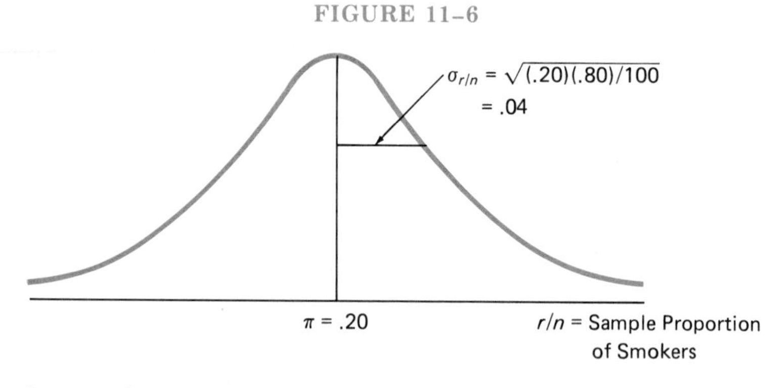

Inasmuch as n is large and $\pi_0 = .20$, the assumed sampling distribution of r/n looks like Figure 11–6. If α is .01, the critical values for testing

$$H_0: \pi = .20$$

$$H_1: \pi \neq .20$$

become $Z = -2.58$ and $Z = 2.58$ (see Figure 11–7).

FIGURE 11–7

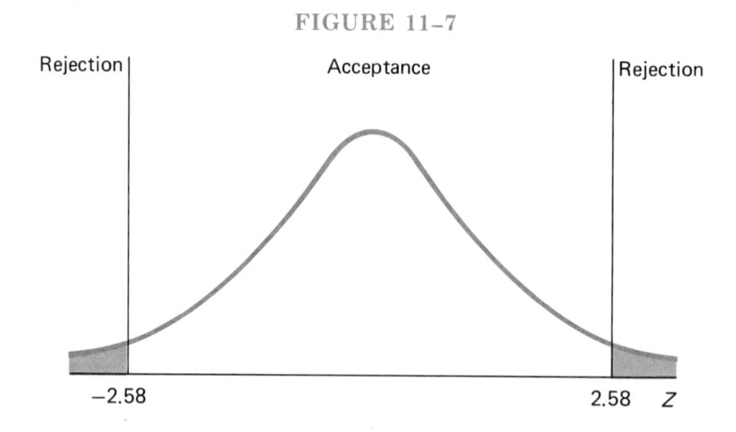

The researcher derives

$$Z = \frac{.15 - .20}{.04} = -1.25$$

Hence, the null hypothesis seems reasonable. Unless π truly is .20, however, accepting H_0 will be a Type II error.

SUMMARY

When X is a random variable having values of only 1 and 0, $\overline{X} = r/n$. Like $\overline{X}$, r/n is an unbiased and consistent estimator. By regarding the selection of a sample as a series of n Bernoulli trials, we were able to reintroduce the binomial distribution. We saw that this distribution is identical to the sampling distribution of r. We can approximate the sampling distributions of r and r/n with a normal curve if n is large. We are able to construct confidence intervals for π and test hypotheses about this parameter. In Chapter Twelve, we will compare the parameters of two populations (e.g., Is the mean of Population 1 equal to the mean of Population 2?).

EXERCISES

1. In a particular population, $\pi = .40$. If a random sample of 64 observations is selected, what are the following probabilities?

   a. $P(.37 < r/n < .41)$
   b. $P(r/n > .39)$
   c. $P(r/n < .42)$
   d. $P(r/n > .43)$
   e. $P(.35 < r/n < .38)$

2. In a particular population, $\pi = .60$. If a random sample of 100 observations is chosen, what are the following probabilities?

   a. $P(r/n > .55)$
   b. $P(r/n < .62)$
   c. $P(.59 < r/n < .63)$
   d. $P(.61 < r/n < .64)$
   e. $P(r/n < .58)$

3. Suppose the sampling distribution of r/n is constructed when $n = 36$ and $\pi = .30$.

   a. What is the expected value of this distribution?
   b. What is the variance?

4. Given $\pi = .70$, a researcher finds that $\sigma_{r/n} = .045$. What is the value of n?

5. Imagine $\pi = .80$. If $n = 64$, what are the following probabilities?

a. $P(|r/n - \pi| < .02)$
b. $P(|r/n - \pi| < .05)$
c. $P(|r/n - \pi| < .09)$
d. $P(|r/n - \pi| < .075)$
e. $P(|r/n - \pi| < .063)$

6. Calculate the probabilities in question 5 if $n = 36$.

7. Mergers are classified as horizontal if they involve two competitors (e.g., two detergent manufacturers), vertical if they involve two companies operating in different phases of the production and distribution of a commodity (e.g., a coal producer and a manufacturer of steel), and conglomerate if they involve two companies in unrelated businesses. A researcher wants to estimate π = the proportion of mergers between 1970 and 1979 which were conglomerate. He selects a random sample of 36 mergers and computes $r/n = .80$.

a. Construct a 90 percent confidence interval estimate of π.
b. Construct a 95 percent confidence interval estimate of π.
c. Construct a 99 percent confidence interval estimate of π.

8. A researcher sets $\pi(1 - \pi)$ equal to .25.

a. What is the sample size for which $P(|r/n - \pi| < .02)$ is at least .90?
b. What is the sample size for which $P(|r/n - \pi| < .04)$ is at least .95?

9. Ever since the United Nations was established, the United States has contributed heavily to its financing. Senator H claims that more than half of all Americans question the effectiveness of this organization. In a random sample of 900 people, 500 did not feel that the U.N. was effective. Test the senator's claim at the .01 level of significance.

10. One method for computing depreciation is referred to as "straight-line." Suppose a $10,000 machine is expected to last five years and it won't have any salvage value. Under the straight-line technique, an accountant would depreciate the machine at a rate of $10,000/5 = $2,000 a year. A person investigating accounting practices feels that no more than 30 percent of the companies in industry W use straight-line depreciation. Among 81 companies chosen at random, 37 applied this method. Test "no more than 30 percent" at the .05 level of significance.

11. A researcher believes that 40 percent or more of the TV commercials aired nationally during "prime time" have sexual over-

tones. From a collection of all commercials broadcast during prime time in 1979, she selects a random sample of 49. If 17 have sexual overtones, is her claim justified at the .05 level of significance?

12. Congress passed a huge increase in employee and employer Social Security taxes in 1977 in response to the projected bankruptcy of the Social Security system. Several factors can be suggested for such a bankruptcy (e.g., a general increase in benefit levels and a growing number of retirees). Some people are in favor of funding the Social Security system solely through income tax revenues. Given π = the proportion of the population in favor of the proposal and $\pi(1 - \pi) = .25$, calculate the sample size for which an investigator could state that $P(|r/n - \pi| < .03)$ is at least .90.

13. Every year a controversy arises over whether the team listed #1 in the AP or UPI national college football rankings is actually the top team. It is repeatedly suggested that play-off games should be arranged at the end of the football season. Imagine that 64 coaches are surveyed at random. Using $\pi(1 - \pi) = .25$, what is the probability that the sample proportion of replies favorable to this idea won't differ from π by more than .02?

14. A highway speed limit of 55 mph was adopted in the 1970s as a strategy for conserving fuel. This limit has also reduced traffic accidents. Of 600 Americans chosen at random, 224 felt that the speed limit should be increased. Given π = the proportion of the population in favor of an increase in the speed limit,

a. Compute a 95 percent confidence interval estimate of π.
b. Compute a 99 percent confidence interval estimate of π.

15. "Zero-base" budgeting requires department heads to justify the total volume of funds they request rather than only increases over their current allocations. A researcher claims that 20 percent or less of the 2,000 largest companies in the United States have adopted zero-base budgeting. In a random sample of 150 companies, 36 reported that they used such a procedure. Given $\alpha = .05$, test the researcher's belief.

Statistical Inferences Involving Two Sample Means or Two Sample Proportions

THE DIFFERENCE BETWEEN TWO SAMPLE MEANS: INDEPENDENT SAMPLES

We have previously tested the null hypotheses $\mu = \mu_0$ and $\pi = \pi_0$. Our inferences were restricted to one population. Inference techniques can also be used to *compare* the parameters of two populations. The following example illustrates a comparative study.

Public health officials are convinced that tooth decay can be reduced by adding a certain amount of fluoride to city water systems. According to the National Institute of Dental Research, however, about half of the people of the United States live in communities lacking fluoridated water. To determine the effectiveness of a fluoride mouthwash, the institute started a special program in 1976.

Assume that two samples of 36 children were randomly selected from "non-fluoride" communities. One sample was supplied with the mouthwash, whereas the other was not. We can view the rinsers (mouthwash users) as part of a hypothetical population of children who (a) live in non-fluoride communities and (b) were given a fluoride mouthwash, and the nonrinsers as coming from a population of children who (a) live in non-fluoride communities and (b) were not given a fluoride mouthwash. We'll call these children Population 1 and Population 2, respectively.

After Sample 1 had rinsed for two years, the following statistics were computed:

Sample 1	*Sample 2*
$\overline{X}_1 = 4.5$ new cavities since 1976	$\overline{X}_2 = 4.8$ new cavities since 1976
$s_1^2 = 0.6$	$s_2^2 = 0.4$

The mouthwash will be recommended for all non-fluoride communities only if μ_1 = the average number of new cavities since 1976 in Population 1 is less than μ_2 = the average number of new cavities since 1976 in Population 2. If $\mu_1 \geq \mu_2$, the mouthwash won't be approved for use.

Whenever two populations have the same means, $\mu_1 = \mu_2$ or $\mu_1 - \mu_2 = 0$. We will test H_0: $\mu_1 - \mu_2 = k$ (where k is a constant, e.g., $k = 0$) by examining $\overline{X}_1 - \overline{X}_2$. The procedure we choose will be influenced by both (1) the size of the samples and (2) whether the samples are selected dependently or independently. When the samples are independent, the observations in Sample 1 do not determine the observations in Sample 2. (This, clearly, is the case in the fluoride mouthwash example.) Under these circumstances, the sample means will be independent as well. We discuss *dependent sampling* later in the chapter. With dependence, the members of each sample are related. Suppose, for instance, that twenty college seniors are randomly chosen and given the Graduate Management Admission Test. Thereafter, these students receive special coaching for such tests. A researcher might be interested in whether "coaching makes a difference." The performance of the before-coaching and after-coaching students could be compared. Both samples would contain the same students.

The Sampling Distribution of $\overline{X}_1 - \overline{X}_2$: Large and Independent Samples

Since $\overline{X}_1$ will vary around μ_1, and $\overline{X}_2$ will vary around μ_2, $\overline{X}_1 - \overline{X}_2$ won't equal zero even if $\mu_1 - \mu_2 = 0$. We now want to identify the *sampling distribution of* $\overline{X}_1 - \overline{X}_2$ when the samples are large and independent. The individual sampling distributions of $\overline{X}_1$ and $\overline{X}_2$ in our example are presented in Figure 12–1.

Inasmuch as $\overline{X}_1 - \overline{X}_2$ is a linear combination of two normal random variables, the sampling distribution of $\overline{X}_1 - \overline{X}_2$ will also be normal, as in Figure 12–2.

We know that the expected value of a difference is equal to the difference of the expected values. Therefore,

$$\begin{aligned} E(\overline{X}_1 - \overline{X}_2) &= E(\overline{X}_1) - E(\overline{X}_2) \\ &= \mu_1 - \mu_2 \end{aligned} \tag{12–1}$$

FIGURE 12–1

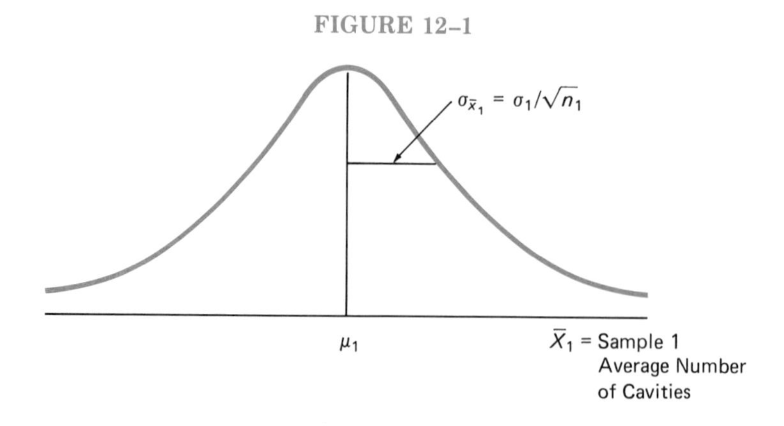

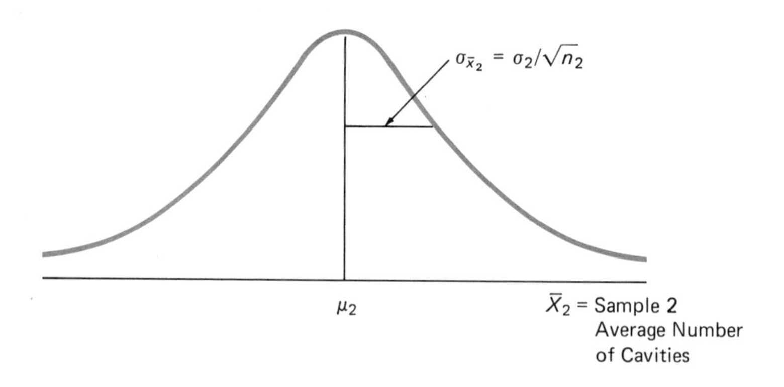

FIGURE 12–2

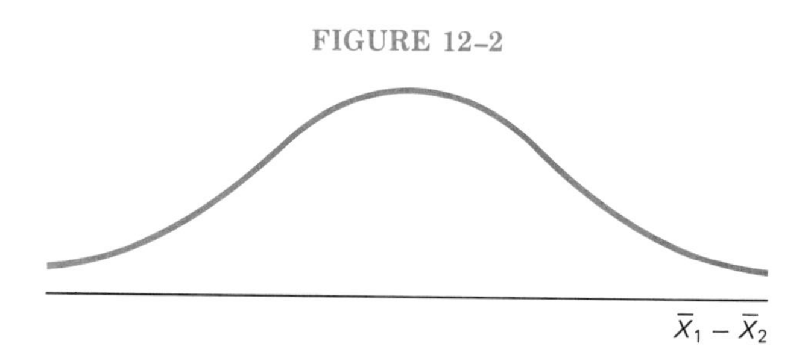

We see that $\overline{X}_1 - \overline{X}_2$ is an unbiased estimator of the difference between two population averages.

If random variables are independent, the variance of a difference equals the sum of the variances. Because of this,

$$\begin{aligned} V(\overline{X}_1 - \overline{X}_2) &= V(\overline{X}_1) + V(\overline{X}_2) \\ &= \sigma_1^2/n_1 + \sigma_2^2/n_2 \end{aligned}$$

and

$$\sigma_{\bar{x}_1 - \bar{x}_2} = \sqrt{\sigma_1^2/n_1 + \sigma_2^2/n_2} \tag{12–2}$$

Notice that $\overline{X}_1 - \overline{X}_2$ is a consistent estimator of $\mu_1 - \mu_2$. The standard deviation of the sampling distribution will decrease as n_1 and n_2 increase.

Hypothesis Tests Involving $\overline{X}_1 - \overline{X}_2$: Large Samples

Once the sampling distribution of $\overline{X}_1 - \overline{X}_2$ has been established, we can determine whether some observed difference in the sample averages is *significantly different* from $\mu_1 - \mu_2$. To evaluate $(\overline{X}_1 - \overline{X}_2) - (\mu_1 - \mu_2)$, we'll replace Figure 12–2 with the standard normal curve and write

$$Z = \frac{(\overline{X}_1 - \overline{X}_2) - (\mu_1 - \mu_2)}{\sigma_{\bar{x}_1 - \bar{x}_2}} \tag{12–3}$$

Because the health officials in our example will recommend the mouthwash if μ_1 is less than μ_2, we'll test

$$H_0\text{: } \mu_1 - \mu_2 \geq 0 \quad \text{(i.e., } \mu_1 \geq \mu_2\text{)}$$

$$H_1\text{: } \mu_1 - \mu_2 < 0 \quad \text{(i.e., } \mu_1 < \mu_2\text{)}$$

Another approach to the problem might account for the benefits *and* the cost of the mouthwash. The cost might be justified only if the difference between μ_1 and μ_2 is more than .5 cavities. Although we will not explore this situation, the hypotheses would appear as

$$H_0\text{: } \mu_1 - \mu_2 \geq -0.5 \quad \text{(i.e., } \mu_1 + 0.5 \geq \mu_2\text{)}$$

$$H_1\text{: } \mu_1 - \mu_2 < -0.5 \quad \text{(i.e., } \mu_1 + 0.5 < \mu_2\text{)}$$

Given a one-tailed test and a level of significance of .01, the acceptance and rejection regions are as shown in Figure 12–3. Since zero is the smallest value in the range $\mu_1 - \mu_2 \geq 0$, the null hypothesis will be rejected as long as

$$Z = \frac{(\overline{X}_1 - \overline{X}_2) - 0}{\sigma_{\bar{x}_1 - \bar{x}_2}}$$

is in the shaded area.

Recall that the data are

$n_1 = 36$ $\quad$ $n_2 = 36$
$\overline{X}_1 = 4.5$ cavities $\quad$ $\overline{X}_2 = 4.8$ cavities
$s_1^2 = 0.6$ $\quad$ $s_2^2 = 0.4$

FIGURE 12–3

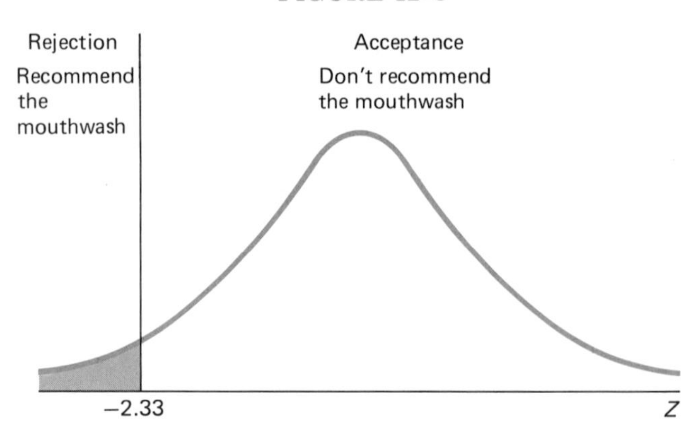

We don't know σ_1^2 and σ_2^2, but large samples make s_1^2 and s_2^2 good estimators. Hence, we can substitute

$$s_{\bar{x}_1 - \bar{x}_2} = \sqrt{s_1^2/n_1 + s_2^2/n_2} \tag{12–4}$$

for Equation 12–2. The computed Z is, therefore,

$$Z = \frac{(4.5 - 4.8) - 0}{\sqrt{.6/36 + .4/36}} = -1.80$$

Since $-1.80 > -2.33$, the dental research institute concludes that a *sample* difference of -0.3 cavities is *not* significantly different from zero. The mouthwash will not be recommended. By accepting H_0, the institute may commit a Type II error. If such an error has been made (i.e., $\mu_1 - \mu_2 < 0$ is actually true), an effective decay-preventing mouthwash will have been withheld from the public.

When sampling is independent, researchers can't "control" all of the factors that influence the outcomes they observe. Some children in Sample 1, for instance, might consume excessive quantities of candy relative to Sample 2 children, or the Sample 1 children might be extremely careless in their brushing habits. These children are likely to have many cavities even if the mouthwash is effective. Therefore, the benefits of the mouthwash could be hidden by the impact of poor diets and/or poor dental hygiene.

Hypothesis Tests Involving $\bar{X}_1 - \bar{X}_2$: Small Samples

We have said that $(\bar{X} - \mu)/(s/\sqrt{n})$ will not be approximately normal when n is small. If either n_1 or n_2 is small,

$$\frac{(\bar{X}_1 - \bar{X}_2) - (\mu_1 - \mu_2)}{\sqrt{s_1^2/n_1 + s_2^2/n_2}}$$

will not be approximately normal either. The small-samples test that we will describe is valid as long as (1) Population 1 and Population 2 are normal and (2) $\sigma_1^2 = \sigma_2^2$. (Testing the "equality of variances" is discussed in Chapter Fifteen.) Two such populations are diagramed in Figure 12–4. We have let $\sigma_1^2 = \sigma_2^2 = \sigma^2$. The parameters μ_1 and μ_2 may or may not be equal. Let's look at an example.

FIGURE 12–4

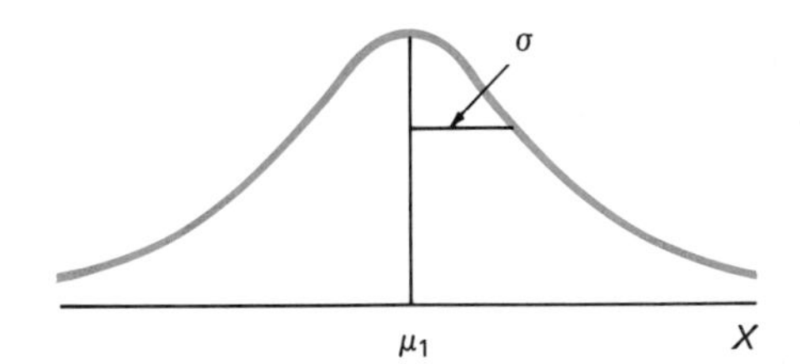

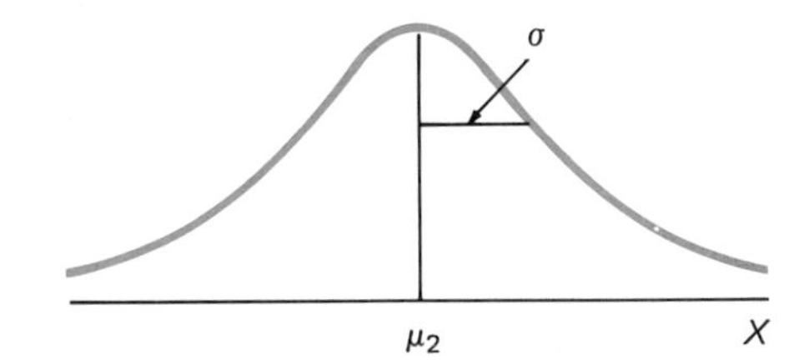

The employees of Company W assemble electronic components for eight hours a day. A worker is paid a fixed rate per finished unit. Management believes that the monotony of the job causes workers to slow down relatively early in the day. Since fatigue reduces output, management would like to change the plant environment.

A supervisor has asked the company to install a stereo system. She feels that music will refresh the workers and improve their spirits. Management decided to try a radio first.

Before the plan was put into action, a random sample of ten employees was selected. The following statistics were computed for May 7, 1979:

$$\overline{X}_1 = 5.2 \text{ finished units}$$

$$s_1^2 = 0.7$$

The next day, the company began playing music on the radio. On May 9, 1979, management chose a random sample of 13 workers and derived

$$\overline{X}_2 = 5.6 \text{ finished units}$$

$$s_2^2 = 0.8$$

If music does increase output, μ_1 = the average number of finished units when no music is played will be less than μ_2 = the average number of finished units when music is played. In testing

$$H_0: \mu_1 - \mu_2 \geq 0$$

$$H_1: \mu_1 - \mu_2 < 0$$

we will assume that both the "before" and "after" distributions are normal and that the two population variances are the same. Given the latter condition, Equation 12–2 can be written as

$$\sigma_{\bar{x}_1 - \bar{x}_2} = \sqrt{(\sigma^2)(1/n_1) + (\sigma^2)(1/n_2)}$$
$$= \sqrt{\sigma^2(1/n_1 + 1/n_2)} \qquad (12\text{–}5)$$

Since s_1^2 *and* s_2^2 supply information about σ^2, we'll estimate σ^2 by calculating an average value for the sample variances. Inasmuch as s^2 improves as an estimator of σ^2 when n increases, more "weight" will be assigned to the s^2 associated with the larger sample. The resulting "pooled" sample variance, s_p^2, is

$$s_p^2 = \frac{(n_1 - 1)s_1^2 + (n_2 - 1)s_2^2}{n_1 + n_2 - 2} \qquad (12\text{–}6)$$

We note that s_1^2 is weighted by $(n_1 - 1)/(n_1 + n_2 - 2)$, while s_2^2 is weighted by $(n_2 - 1)/(n_1 + n_2 - 2)$. Like s_1^2 and s_2^2, s_p^2 is an unbiased estimator of σ^2. If n_1 is larger than n_2, for instance, s_1^2 will have a larger weight than s_2^2.

The two sums of squares included in s_p^2, that is, $\Sigma(X - \overline{X}_1)^2$ and $\Sigma(X - \overline{X}_2)^2$, represent $n_1 - 1$ and $n_2 - 1$ degrees of freedom, respectively. Thus, s_p^2 has $n_1 + n_2 - 2$ degrees of freedom. Additionally, the ratio on the right side of Equation 12–7 possesses a t distribution with $n_1 + n_2 - 2$ degrees of freedom:

$$t_{(n_1 + n_2 - 2)} = \frac{(\overline{X}_1 - \overline{X}_2) - (\mu_1 - \mu_2)}{\sqrt{s_p^2(1/n_1 + 1/n_2)}} \qquad (12\text{–}7)$$

This equality is the basis for converting the sampling distribution of $\overline{X}_1 - \overline{X}_2$ into the t distribution. The null hypothesis will be accepted or rejected according to the value of this ratio.

Returning to our example, Figure 12–5 shows the acceptance and rejection regions if $\alpha = .05$. Management calculates

$$s_p^2 = \frac{(10 - 1)(0.7) + (13 - 1)(0.8)}{10 + 13 - 2} = 0.757$$

and

$$t = \frac{(5.2 - 5.6) - 0}{\sqrt{(0.757)(1/10 + 1/13)}} = -1.093$$

Because -1.093 lies in the acceptance region, management concludes that music doesn't increase output. It will have to find other

FIGURE 12–5

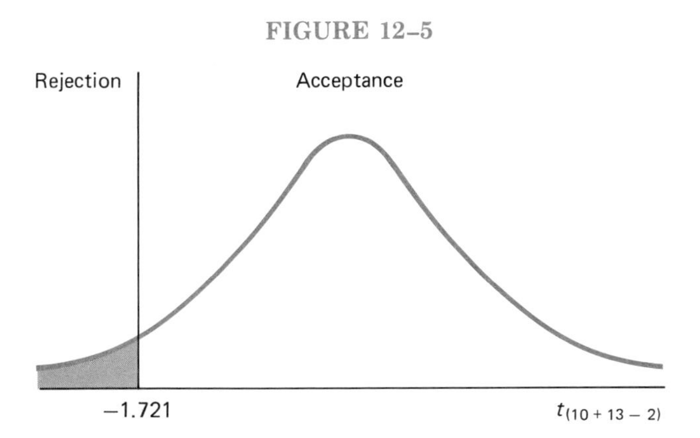

ways to reduce the monotony of assembling the electronic components.

HYPOTHESIS TESTS INVOLVING TWO DEPENDENT SAMPLES

In the two examples thus far, a particular factor (i.e., mouthwash or music) was investigated. By accepting the null hypothesis, a researcher would conclude that such a factor didn't affect X. Neither of our tests, however, eliminated "outside" influences. Thus, H_0 might have been accepted in the last section because the company happened to select very good workers in Sample 1 and very poor workers in Sample 2. (Not intentionally, of course. Remember that the samples were chosen randomly.) Assuming that music actually does increase output, the Sample 2 workers would produce more units than usual. But since $\overline{X}_1$ was high, the difference between $\overline{X}_1$ and $\overline{X}_2$ would still be small.

If each worker in Sample 1 could be paired with an equally skilled worker (who would, then, enter Sample 2), any difference in sample averages would be mostly the result of the music. The variation in $\overline{X}_1 - \overline{X}_2$ would then come from two sources (i.e., music and chance) instead of from three (i.e., music, skill, and chance).

We shall now construct the sampling distribution of $\overline{X}_1 - \overline{X}_2$ when the observations in the second sample are related to those in the first. The criterion (e.g., skill) that determines the pairs has to influence X (e.g., the number of finished units). Matching workers by eye color, for instance, would be inappropriate. The matching, moreover, is done *before* the start of the experiment.

Equation 12–1, $E(\overline{X}_1 - \overline{X}_2) = \mu_1 - \mu_2$, is not changed by the dependence between $\overline{X}_1$ and $\overline{X}_2$. The standard deviation of the difference between the sample averages, however, is now

$$\sigma_{\bar{x}_1 - \bar{x}_2} = \sqrt{V(\overline{X}_1) - 2\text{Cov}(\overline{X}_1,\overline{X}_2) + V(\overline{X}_2)}$$
$$= \sqrt{\sigma_1^2/n_1 + \sigma_2^2/n_2 - 2\text{Cov}(\overline{X}_1,\overline{X}_2)} \qquad (12\text{–}8)$$

This standard deviation will be less than that of Equation 12–2 only if $\text{Cov}(\overline{X}_1,\overline{X}_2)$ is positive. The covariance will be positive whenever good workers in Population 1 are also good workers in Population 2. If good workers in Population 1 became distracted by the radio and poor workers became more motivated, $\text{Cov}(\overline{X}_1,\overline{X}_2)$ would be negative, and σ derived by Equation 12–8 would be greater than that of Equation 12–2.

To estimate $\sigma_{\bar{x}_1 - \bar{x}_2}$ when the samples are dependent, we must estimate σ_1^2, σ_2^2, and $\text{Cov}(\overline{X}_1,\overline{X}_2)$. We can avoid these calculations by adapting the t test presented in Chapter Ten to the case of dependent sampling.

When X_{i1} and X_{i2} are the members of a matched pair, their difference, or d_i, is

$$d_i = X_{i1} - X_{i2} \qquad (12\text{–}9)$$

The average difference is, thus,

$$\bar{d} = \Sigma(X_{i1} - X_{i2})/n = \Sigma d/n \qquad (12\text{–}10)$$

where $n_1 = n_2 = n$. We can rewrite Equation 12–10 as

$$\bar{d} = (1/n)(\Sigma X_{i1} - \Sigma X_{i2})$$
$$= (\Sigma X_{i1}/n) - (\Sigma X_{i2}/n) = \overline{X}_1 - \overline{X}_2 \qquad (12\text{–}11)$$

Hence,

$$E(\bar{d}) = E(\overline{X}_1 - \overline{X}_2) = \mu_1 - \mu_2 = D \qquad (12\text{–}12)$$

FIGURE 12–6

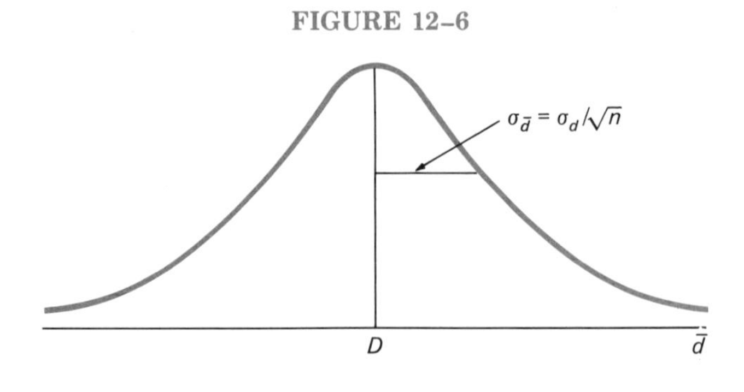

The *average difference* is (1) equal to the difference in averages and (2) an unbiased estimator of $\mu_1 - \mu_2$, or D.

We will treat dependent sampling as a technique for selecting n differences from a population of ds. If this population is normal, the sampling distribution of $\overline{d}$ will look like Figure 12–6.

The sample standard deviation,

$$s_d = \sqrt{\Sigma(d - \overline{d})^2/(n - 1)}, \qquad (12\text{–}13)$$

is an unbiased estimator of σ_d. Because $\Sigma(d - \overline{d}) = 0$, s_d has $n - 1$ degrees of freedom, and

$$t_{(n-1)} = \frac{\overline{d} - D}{\dfrac{s_d}{\sqrt{n}}} \qquad (12\text{–}14)$$

will possess a t distribution with $n - 1$ degrees of freedom. The variance of the X_1 population and the variance of the X_2 population do not have to be identical. To clarify our new test, let's make some changes in the last example.

The management of Company W recognizes that experience plays a major role in determining how many units an employee can assemble per day. Older workers at the plant generally produce more than younger workers. To prevent "experience" from hiding the benefits of music (if they exist), management selected twenty workers at random. These workers were then organized into ten pairs, each pair having employees with similar lengths of service with Company W. The company recorded the output of one member of each pair for a day in which a radio was not played. After a radio was introduced, the productivity of the other member was noted. Table 12–1 indicates the finished units in the "before" and "after" samples.

TABLE 12–1

Employee Pairs	X_{i1}	X_{i2}	d_i
A	2.3	2.6	2.3 − 2.6 = −0.3
B	3.1	3.2	3.1 − 3.2 = −0.1
C	1.0	1.0	1.0 − 1.0 = 0.0
D	4.2	4.4	4.2 − 4.4 = −0.2
E	5.1	5.2	5.1 − 5.2 = −0.1
F	2.2	2.4	2.2 − 2.4 = −0.2
G	4.2	4.4	4.2 − 4.4 = −0.2
H	5.0	5.0	5.0 − 5.0 = 0.0
I	3.1	3.2	3.1 − 3.2 = −0.1
J	4.8	5.6	4.8 − 5.6 = −0.8
			−2.0

Inasmuch as $\bar{d} = -2.0/10 = -0.2$, we can compute

$$s_d = \sqrt{0.48/(10 - 1)} = 0.23$$

from Table 12–2. In terms of D, the original hypothesis about $\mu_1 - \mu_2$ appears as

$$H_0\colon D \geq 0$$

$$H_1\colon D < 0$$

TABLE 12–2

d	$\bar{d}$	$d - \bar{d}$	$(d - \bar{d})^2$
−0.3	−0.2	−0.1	0.01
−0.1	−0.2	0.1	0.01
0.0	−0.2	0.2	0.04
−0.2	−0.2	0.0	0.00
−0.1	−0.2	0.1	0.01
−0.2	−0.2	0.0	0.00
−0.2	−0.2	0.0	0.00
0.0	−0.2	0.2	0.04
−0.1	−0.2	0.1	0.01
−0.8	−0.2	−0.6	0.36
			0.48

We will continue to assume that $\alpha = .05$. The acceptance and rejection regions are shown in Figure 12–7.

Since

$$t = \frac{-0.2 - 0}{\frac{0.23}{\sqrt{10}}} = -2.75,$$

FIGURE 12–7

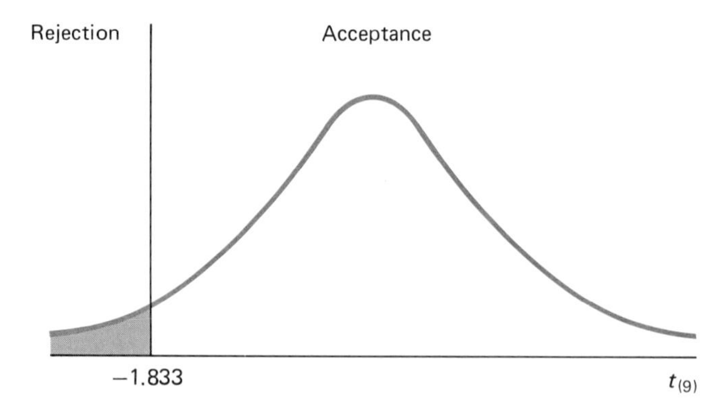

management rejects the null hypothesis and decides that music increases output. If μ_2 is not actually greater than μ_1, a Type I error would be made.

THE DIFFERENCE BETWEEN TWO SAMPLE PROPORTIONS: LARGE AND INDEPENDENT SAMPLES

Researchers can compare population proportions (say, π_1 and π_2) as well as population means. To simplify our symbols, let's write the sample proportion (i.e., r_1/n_1) from Population 1 as p_1 and the sample proportion (i.e., r_2/n_2) from Population 2 as p_2.

The Sampling Distribution of $p_1 - p_2$

The *sampling distribution of* $p_1 - p_2$ will be developed within the context of the following example.

Mary Dunn is vice-president of E. L. Imports. She is going to advertise either in Magazine A or in Magazine B. A and B are sold only by subscription, and Dunn has access to both mailing lists. Because of the nature of her ad, she will choose the magazine having the higher proportion of college-enrolled subscribers. If the proportions are the same, she will advertise in A inasmuch as another E. L. ad was previously run in it.

A random sample of 50 people is drawn from each list. We will view the A subscribers as Population 1 and the B subscribers as Population 2. Dunn collected the following statistics:

$$p_1 = \frac{29 \text{ college students}}{50 \text{ A subscribers}} = .58$$

$$p_2 = \frac{32 \text{ college students}}{50 \text{ B subscribers}} = .64$$

Figure 12–8 displays the sampling distributions of p_1 and p_2 when n_1 and n_2 are large and independent. Since $p_1 - p_2$ is a linear combination of two normal random variables, its sampling distribution will be normal (see Figure 12–9).

We note that

$$\begin{aligned} E(p_1 - p_2) &= E(p_1) - E(p_2) \\ &= \pi_1 - \pi_2 \end{aligned} \tag{12–15}$$

Given independence,

$$V(p_1 - p_2) = V(p_1) + V(p_2)$$

and

$$\sigma_{p_1 - p_2} = \sqrt{(\pi_1)(1 - \pi_1)/n_1 + (\pi_2)(1 - \pi_2)/n_2} \qquad (12\text{–}16)$$

The difference between sample proportions is, thus, an unbiased and consistent estimator of $\pi_1 - \pi_2$.

FIGURE 12–8

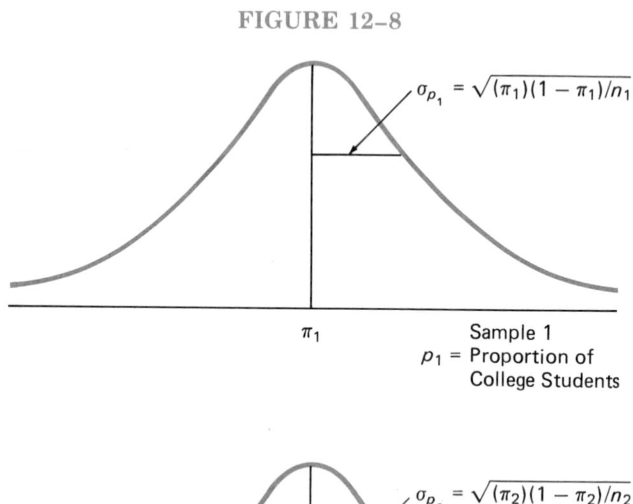

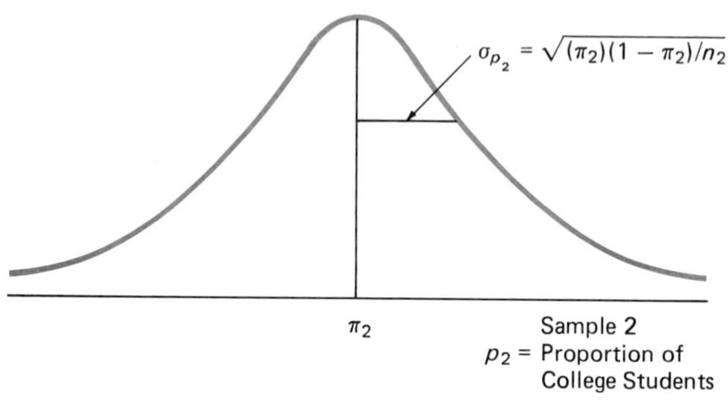

FIGURE 12–9

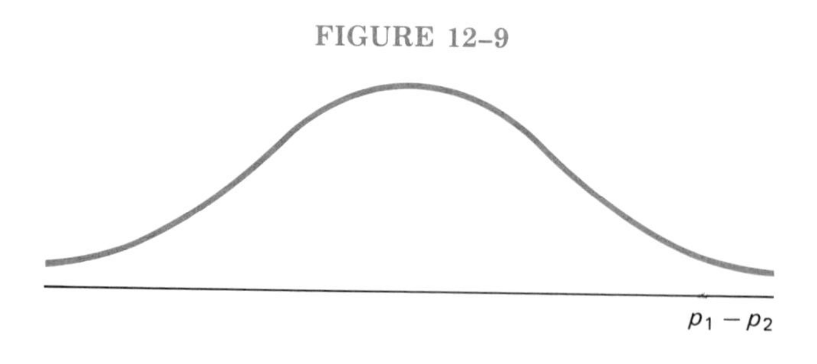

Hypothesis Tests Involving $p_1 - p_2$

An observed $p_1 - p_2$ is examined in relation to

$$Z = \frac{\text{value of the random variable} - \text{mean of the probability distribution}}{\text{standard deviation of the probability distribution}}$$

$$= \frac{(p_1 - p_2) - (\pi_1 - \pi_2)}{\sigma_{p_1 - p_2}} \quad (12\text{–}17)$$

Mary Dunn will test

$$H_0: \pi_1 - \pi_2 \geq 0$$

$$H_1: \pi_1 - \pi_2 < 0$$

because A will be chosen if π_1 is at least as large as π_2.

The hypothesis $\pi_1 - \pi_2 = 0$ is the same as $\pi_1 = \pi_2$. Suppose we let π equal π_1 and π_2. The denominator of Equation 12–17 would, then, be

$$\sqrt{(\pi)(1 - \pi)/n_1 + (\pi)(1 - \pi)/n_2} = \sqrt{(\pi)(1 - \pi)(1/n_1 + 1/n_2)}$$

Recall the logic we associated with the construction of s_p^2. Similarly, an unbiased "pooled" estimator of π is

$$p^* = \frac{n_1 p_1 + n_2 p_2}{n_1 + n_2} \quad (12\text{–}18)$$

Thus, for the *special case* in which the null hypothesis is $\pi_1 - \pi_2 = 0$, we can write Equation 12–17 as

$$Z = \frac{(p_1 - p_2) - 0}{\sqrt{(p^*)(1 - p^*)(1/n_1 + 1/n_2)}} \quad (12\text{–}19)$$

Figure 12–10 shows the acceptance and rejection regions for our example when $\alpha = .05$. Inasmuch as $n_1 = 50$, $n_2 = 50$, $p_1 = .58$, and $p_2 = .64$, Mary Dunn calculates

$$p^* = \frac{(50)(.58) + (50)(.64)}{50 + 50} = .61$$

and

$$Z = \frac{(.58 - .64) - 0}{\sqrt{(.61)(.39)(1/50 + 1/50)}}$$

$$= \frac{-.06}{.0975} = -0.62$$

FIGURE 12–10

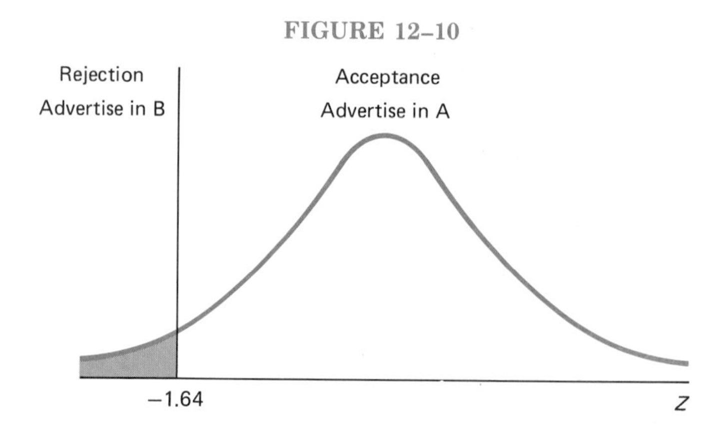

Since $Z = -0.62$ lies in the acceptance region, she will advertise in A. She doesn't believe that an observed $p_1 - p_2$ of $-.06$ is improbable for a sampling distribution with an assumed mean of zero and an estimated standard deviation of .0975.

SUMMARY

Many researchers are interested in comparing one population with another. We have seen that several tests of $\mu_1 - \mu_2 = k$ are possible. The proper test is determined by the nature of the sampling process (e.g., selecting large and independent samples). Using the characteristics of random variables, we confirmed that $\overline{X}_1 - \overline{X}_2$ and $p_1 - p_2$ are unbiased and consistent estimators. In Chapters Fourteen and Fifteen, we will show that inferences can be extended to more than two populations.

EXERCISES

1. Show that the pooled sample variance, s_p^2, is an unbiased estimator of σ^2.

2. Show that the pooled sample proportion, p^*, is an unbiased estimator of π.

3. Corporation G is building three factories in country W. Until the "bugs" have been eliminated, the corporation will operate these factories with American workers, who will be given instruction in W's language. Eighteen workers are selected at random and paired ac-

cording to I.Q. One member of every pair is randomly assigned to the group that will be taught W's language by method A. The other member of the pair will be taught by method B. Following four weeks of intensive study, the employees are given a 100-point test on W's language. The outcomes are listed below. Using $\alpha = .05$, and assuming that the population of score differences is normal, determine whether the average of the hypothetical population of scores produced by method A is the same as the average of the hypothetical population of scores produced by method B.

	Employee Scores, by Method	
Pair	Method A	Method B
#1	63	72
2	84	93
3	72	58
4	95	83
5	46	53
6	71	87
7	59	66
8	84	81
9	93	88

4. An energy analyst selected two independent and random samples of recently manufactured model H and model W cars. After driving the automobiles on a highway, he calculated the number of miles per gallon each car attained. The outcomes are listed below. Using $\alpha = .05$, and assuming normal populations as well as $\sigma_1^2 = \sigma_2^2$, test whether the average number of miles per gallon that can be attained with model H is at most one mile greater than the average for model W.

Sample of Model H Cars	*Sample of Model W Cars*
23.5 mpg	18.3 mpg
22.6	20.4
24.7	19.1
25.2	20.8
23.1	18.4
22.0	19.2

5. Suppose independent and random samples of six observations are chosen from normal populations having equal variances. Given Equation 12–7 and the fact that $P(-2.228 < t_{(10)} < 2.228) = .95$, what is the 95 percent confidence interval estimator of $\mu_1 - \mu_2$?

6. From the information in question 4, construct a 95 percent confidence interval estimate of $\mu_1 - \mu_2$.

7. In every presidential election year, the debate over whether to retain the electoral college or decide the election by popular vote is renewed. A researcher believes that π_1 = the proportion of political scientists favoring the electoral college is greater than π_2 = the proportion of non-political scientists favoring the electoral college. He selects independent and random samples of 64 political scientists and 81 non-political scientists. Thirty-three of the political scientists and 37 of the other people want to retain the electoral college. Test the researcher's contention at the .05 level of significance.

8. Despite the fact that many viewers only wish to learn about tomorrow's high temperature and the probability of precipitation, television newscasters spend a large amount of time commenting on such weather-related concepts as "isobars" and "low-pressure systems." A communications specialist is convinced that TV stations, on the average, currently allot more time to weather reporting than they did ten years ago. She chooses 36 stations at random and computes $\overline{X}_1$ = the sample average time allotted to weather reporting on the nightly newscast in 1969 = 112 seconds, and s_1 = 11 seconds. Based on a second random sample of 49 stations, chosen independently of the first, the specialist calculates $\overline{X}_2$ = the sample average time allotted to weather reporting on the nightly newscast in 1979 = 156 seconds, and s_2 = 9 seconds. Test her belief at the .05 level of significance.

9. An investigator contends that the average time required to assemble a certain electrical component will increase by over 100 seconds after workers drink 12 ounces of beer. He selects seven workers at random and records their assembly times before the beer is served. The workers then drink 12 ounces of beer and immediately assemble another component. The results are given below. Using $\alpha = .01$, and assuming that the population of time differences is normal, test the investigator's contention.

Worker	*Assembly Time Before Drinking Beer (in seconds)*	*Assembly Time After Drinking Beer (in seconds)*
#1	343	408
2	315	457
3	379	526
4	368	574
5	394	461
6	423	533
7	386	427

10. The heavy volume of traffic on the highways into and out of cities during "rush hours" is due to the similarity in the opening

and closing times of most businesses. Congestion could be diminished if some firms were willing to change their hours of operation. A researcher selects two independent and random samples of "large" businesses (i.e., ten employees or more) from City A and City B and notes the following:

n_1	= 100 firms from A	n_2	= 150 firms from B
p_1	= proportion of large firms from A willing to change their times = .22	p_2	= proportion of large firms from B willing to change their times = .28

Given $\alpha = .05$, are the population proportions identical?

11. The board of trustees at a famous women's college is considering whether to admit men to the school. People in favor of such a policy argue that women should learn to compete with men in an academic environment since they will ultimately have to compete with them in the job world. A questionnaire is sent to independent and random samples of 150 "recent" and 200 "older" alumnae. An investigator feels that a larger proportion of the recent alumnae will approve male admissions. If 63 percent of the recent and 54 percent of the older alumnae in the samples approve the change in policy, is the investigator's belief supported at the .05 level of significance?

12. The William O. Miller Manufacturing Company can buy a certain part from either supplier A or supplier B. B contends that his parts, on the average, last more than 100 hours longer than A's. Miller Manufacturing chooses independent and random samples of 21 parts from each supplier and computes

$\overline{X}_1$	= sample A's average time until failure = 762 hours	$\overline{X}_2$	= sample B's average time until failure = 883 hours
s_1	= 46 hours	s_2	= 55 hours

Test B's contention at the .05 level of significance.

13. From the information in question 12, construct a 90 percent confidence interval estimate of $\mu_1 - \mu_2$.

14. A farmer selects independent and random samples of a particular fruit from two fields. Using the following information

n_1	= 100	n_2	= 81
$\overline{X}_1$	= 5.9 ounces	$\overline{X}_2$	= 6.3 ounces
s_1	= .9 ounces	s_2	= .7 ounces

and $\alpha = .05$, test the hypothesis "The average weights of the fruit in each field are equal."

15. Given Equation 12–3 and large ns, identify the 99 percent confidence interval estimator of $\mu_1 - \mu_2$ when two independent and random samples are chosen.

16. Substitute s_1^2 and s_2^2 for σ_1^2 and σ_2^2, respectively, and construct a 99 percent confidence interval estimate of $\mu_1 - \mu_2$ from the information in question 14.

Bayesian Decision Theory

THE FRAMEWORK OF BAYESIAN DECISION MAKING

An expanding body of statistical procedures has recently evolved under the heading of *Bayesian statistics* or *Bayesian decision theory*. The Bayesian approach to decision making differs from the so-called classical techniques that we have been examining. To present some of the contrasts, let's return to the example of the jar-lid manufacturer.

Suppose that 10,000 lids are produced each hour and that 6 percent of the lids will be defective even when the machinery is properly adjusted. If a malfunction occurs, the percentage of defectives will be greater than 6 percent.

Before the next hour's cutting begins, the firm must choose between "continue to produce" and "shut down to check machines." To this end, the manufacturer selects a random sample of 100 lids and calculates the proportion of defectives: $r/n = 11/100 = .11$.

Given $\alpha = .01$, the acceptance and rejection regions for

$$H_0: \pi \leq .06$$

$$H_1: \pi > .06$$

are shown in Figure 13–1. Inasmuch as the computed Z is

$$Z = \frac{r/n - \pi_0}{\sigma_{r/n}} = \frac{.11 - .06}{\sqrt{(.06)(.94)/100}} = 2.08,$$

the firm will continue to produce. By accepting H_0, however, it may incur a Type II error.

FIGURE 13–1

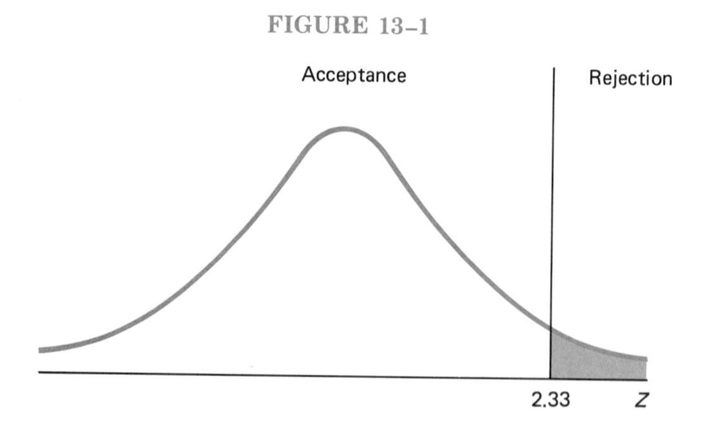

Bayesians would criticize this test on several grounds:

1. The entrepreneur's previous experiences are not *explicitly* introduced.

   The "track record" of the equipment should influence the choice of α. If the machines rarely malfunction, α should be set very low. The classical school doesn't provide an objective way to identify the "proper" α. The usual selection of $\alpha = .05$ or $\alpha = .01$ is arbitrary.
2. In looking for a satisfactory α, n combination, the entrepreneur has to consider the consequences of his actions. Yet he isn't forced to *quantify* these consequences.

   The lid manufacturer hopes that the probability of a Type II error is "low." He knows that producing when $\pi > .06$ will hurt him financially. Despite this, he never answers the question "*How much* will I lose?"

In the lid test, four possible outcomes are evident (see Table 13–1). Situations B and C represent Type I and Type II errors, respectively. The decision maker can control his actions, but he can't control the environment, or *state of nature.*

Bayesians would want the entrepreneur to (1) list all relevant actions and states of nature and (2) determine the dollar value, or *payoff,* associated with each action-state pair. These steps are particularly difficult when a problem arises only once. For example, should Company G move its headquarters from New York City to

TABLE 13–1

		Actions	
		Continue to Produce	Shut Down
Manufacturing Environment	No Malfunction	A	B
	Malfunction	C	D

Connecticut? We can better justify Bayesian analysis if we examine its long-run implications. Therefore, we'll investigate the "continue to produce–shut down" dilemma in detail since the lid manufacturer must choose between these options repeatedly.

We shall adopt the following notation:

a_1 = action 1: continue to produce
a_2 = action 2: shut down
S_1 = state of nature 1: no malfunction
S_2 = state of nature 2: malfunction
b_{ij} = the payoff that corresponds to the ith state of nature and the jth action

In general, any action is an

$$a_j \qquad j = 1, \ldots, m$$

and any state is an

$$S_i \qquad i = 1, \ldots, k$$

Table 13–2 identifies the possible payoffs given $m = 2$ and $k = 2$, as in our example.

TABLE 13–2

		Actions	
		a_1	a_2
State of Nature	S_1	b_{11}	b_{12}
	S_2	b_{21}	b_{22}

To illustrate the process of finding b_{ij}, let's assume the following:

1. π equals .06 when the machines are properly adjusted and .12 otherwise.
2. Due to labor, material, and overhead, it costs 3¢ to produce a lid.
3. The lids are sold for 5¢ each.
4. Management will replace all defective lids after being notified by the customer.
5. Each production run is shipped to only one customer (i.e., one customer gets all of the defective lids).
6. 10,000 lids can be produced in an hour.

Computations of the payoffs for the situations identified in Table 13–1 are as follows:

CASE A

Without defectives, the profits on 10,000 lids would be (5¢)(10,000) − (3¢)(10,000) = \$200. Inasmuch as 6 percent of 10,000 = 600 lids will be exchanged when the machines are working properly, the company will have an additional cost of (3¢)(600) = \$18. Profits will, then, decline to

$$\$200 - \$18 = \$182 = b_{11}$$

CASE B

A 12-minute inspection will convince the entrepreneur that he incorrectly shut down. During this time, (12/60)(10,000) = 2,000 lids aren't produced. Profits on the lids cut in the rest of the hour will be (5¢)(8,000) − (3¢)(8,000) − replacement costs, or

$$\$160 - (3¢)(.06)(8{,}000) = \$145.60 = b_{12}$$

CASE C

If the machines malfunction, the manufacturer will have to replace (.12)(10,000) = 1,200 lids at a cost of (3¢)(1,200) = \$36. A customer having to return 12 percent of his shipment will be extremely annoyed. To pacify these people, the company will grant a \$50 rebate. Hence, profits will be

$$\$200 - \$36 - \$50 = \$114 = b_{21}$$

CASE D

When the manufacturer shuts down and finds that the equipment has to be readjusted, production is halted for 18 minutes. Of the 7,000 lids that can be cut in the remaining 42 minutes, 6 percent, or 420, will be defective, and profits will be

$$\$140 - (3¢)(420) = \$127.40 = b_{22}$$

Table 13–3 is the *payoff table* for the lid problem. We intend to examine three decision-making contexts:

1. *Decision making under certainty.* The S_i is known before an a_j is taken.
2. *Decision making under uncertainty.* The S_i is not known before an a_j is taken.
3. *Decision making under risk.* The S_i is not known before an a_j is taken, but the experimenter estimates $P(S_i)$.

TABLE 13–3

	a_1	a_2
S_1	\$182	\$145.60
S_2	\$114	\$127.40

Imagine that a red light goes on whenever the machines malfunction. The firm would, therefore, select a_1 if no light flashes and a_2 if the light appears. Such actions are pursued because $182 is the largest payoff given $\pi = .06$, and $127.40 is the largest payoff given $\pi = .12$. Since S_i is always known, this scenario illustrates decision making under certainty.

Suppose, however, that S_i is not known—a case of decision making under uncertainty. If the entrepreneur picks a_1, he could earn either $182 or $114. By choosing a_2, he could earn either $145.60 or $127.40. Neither of the actions is clearly preferable (e.g., a_1 yields both the largest *and* the smallest payoff). We must develop a decision criterion.

If the entrepreneur is cautious and pessimistic, he will choose a_2. The worst that can happen with a_2 is better than the worst with a_1. Pessimists use a strategy called *maximin.* In effect, they locate the smallest payoff in each column and pick the a_j having the maximum column minimum (see Table 13–4).

TABLE 13–4

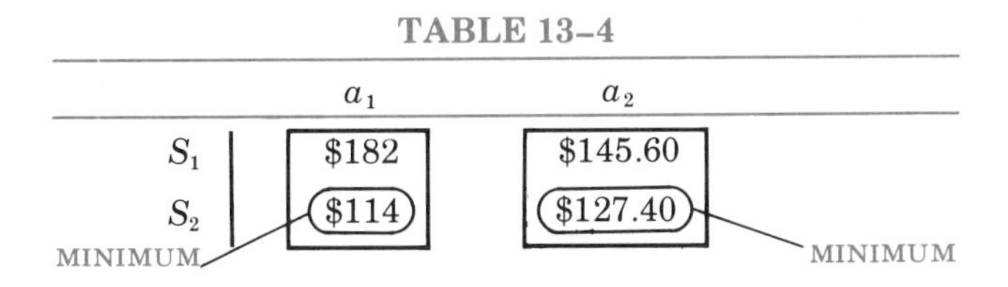

	a_1	a_2
S_1	$182	$145.60
S_2	($114) MINIMUM	($127.40) MINIMUM

On the other hand, if the manufacturer is aggressive and optimistic, he will choose a_1. It has the potential for generating the greatest b_{ij}. Optimists apply a *maximax* strategy. After determining the largest payoff in each column, they pick the a_j having the maximum column maximum (see Table 13–5).

TABLE 13–5

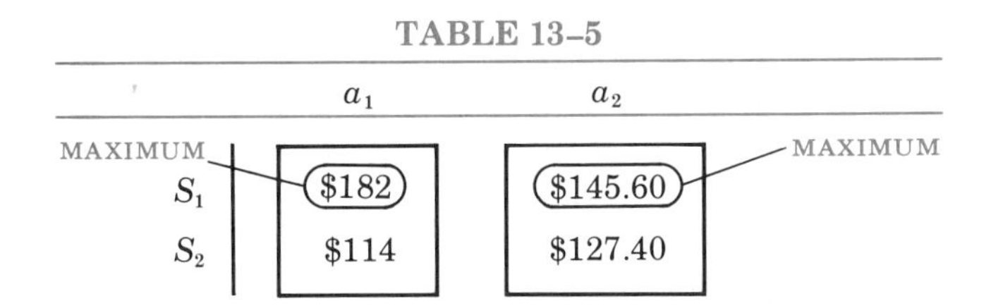

	a_1	a_2
S_1	($182) MAXIMUM	($145.60) MAXIMUM
S_2	$114	$127.40

We can imagine an entrepreneur who will be disgusted with himself if his action is incorrect. If he selects a_2 but $\pi = .06$, the decision will "cost" him $36.40. This "cost" is the difference between the most he could get by knowing $\pi = .06$ (i.e., $182) and the payoff from a_2 (i.e., $145.60). We will say that $36.40 is the *opportunity loss,* or *OL,* of a_2 given $\pi = .06$. The *OL* of a_1 given $\pi = .06$ is $0.

Formally,

$$\frac{OL \text{ of } a_j}{\text{given } S_i} = \frac{\text{largest payoff}}{\text{in the } S_i \text{ row}} - \frac{\text{payoff for}}{a_j \text{ given } S_i} \qquad (13\text{–}1)$$

Table 13–6 displays all of the opportunity losses for our example. A decision maker wanting to minimize his disappointment, or *OL*, would select a_1. The "cost" of a wrong decision would be only $13.40 instead of $36.40.

TABLE 13–6

	a_1	a_2
S_1	182 – 182 = $0	182 – 145.60 = $36.40
S_2	127.40 – 114 = $13.40	127.40 – 127.40 = $0

In decision making under uncertainty, strategies (e.g., maximin) reflect the outlook of the experimenter. The heart of Bayesian theory, however, is decision making under risk. Bayesians assign probabilities to the states of nature. We will identify the "optimal" action, or *Bayes' act,* once these probabilities have been introduced.

CALCULATING EXPECTED PAYOFFS

Expected Monetary Value

Bayesians believe that decision makers should be able to determine $P(S_i)$. When the problem is unique, such probabilities will be subjective. But when the problem is recurrent, we can use past experience as a guide.

Suppose that the lid manufacturer previously inspected his equipment after each run. Eighty percent of the time nothing was wrong. We can, thus, estimate $P(S_1)$ and $P(S_2)$ as .80 and .20, respectively. We will use the information in Table 13–7 to calculate the *expected monetary value,* or *EMV,* of a_1 (continue to produce) and a_2 (shut down).

TABLE 13–7

$P(S_i)$	S_i	a_1	a_2
.80	S_1	$182	$145.60
.20	S_2	$114	$127.40

Consider the following long-run outcomes:

1. If the lid manufacturer always chooses a_1, he will make $182 eighty percent of the time and $114 twenty percent of the time. The average payoff will be

$$(182)(.80) + (114)(.20) = b_{11}P(S_1) + b_{21}P(S_2) = \$168.40$$

2. If the lid manufacturer always chooses a_2, he will make \$145.60 eighty percent of the time and \$127.40 twenty percent of the time. The average payoff will be

$$(145.60)(.80) + (127.40)(.20) = b_{12}P(S_1) + b_{22}P(S_2) = \$141.96$$

The expected monetary value of act a_j is defined as

$$EMV_j = \sum_{i=1}^{k} b_{ij}P(S_i) \tag{13-2}$$

The *Bayes' act* is the a_j with the largest *EMV*. Because the average payoff from a_1 is greater than the average payoff from a_2, "continue to produce" is the Bayes' act. Unlike maximin, maximax, or minimum *OL*, the *EMV* strategy allows the decision maker to inject his prior experiences, via $P(S_i)$, into the search for the optimal a_j.

Notice that a different estimate of $P(S_i)$ might change the preferred act. To identify the circumstances under which a_1 would no longer be chosen, we will imagine that $p = P(S_1)$ and $1 - p = P(S_2)$. The *EMV*s of a_1 and a_2 will be the same when

$$\overbrace{(182)(p) + (114)(1 - p)}^{EMV_1} = \overbrace{(145.60)(p) + (127.40)(1 - p)}^{EMV_2}$$

$$182p + 114 - 114p = 145.60p + 127.40 - 127.40p$$

$$p = .269$$

$$1 - p = .731$$

For any $P(S_1) > .269$, EMV_1 will be greater than EMV_2. Even if the entrepreneur isn't confident about setting $P(S_1)$ equal to .80, he now sees that a_1 would be optimal for a true $P(S_1)$ as low as .27.

Remember that *EMV* is an *average* payoff. An *actual* payoff may differ from it considerably. When a particular b_{ij} could ruin a business, Bayesians reformulate the payoff table. The following example includes a disastrous consequence.

Mr. Roe is going to invest some of his capital in a foreign country. He must choose between investment #1 and investment #2. The current social and political climate in this country is stable. Nonetheless, a revolution could occur. If there is a revolution, investment #2 will suffer far more than investment #1. Roe has established the payoffs and probability estimates in Table 13–8.

TABLE 13–8

$P(S_i)$	S_i	a_1 *Invest in #1*	a_2 *Invest in #2*
.90	No revolution	\$1,000,000	\$2,000,000
.10	Revolution	−\$50,000	−\$500,000

We find that $EMV_1 = (1{,}000{,}000)(.90) + (-50{,}000)(.10) =$ \$895,000, whereas $EMV_2 = (2{,}000{,}000)(.90) + (-500{,}000)(.10) =$ \$1,750,000. Since \$1,750,000 > \$895,000, a_2 is the Bayes' act. With a_2, however, the decision maker could experience a \$500,000 loss. If he is a gambler, the prospect of a \$2 million gain may outweigh the risk of such a loss. A non-gambler, though, would feel uncomfortable about using the *EMV* criterion. Bayesians can account for different attitudes toward risk by converting monetary payoffs into measures called *utilities*. Under these circumstances, the optimal act maximizes *expected utility*. Because calculating utilities is difficult, we won't study the procedure here.

Expected Opportunity Loss

When a decision maker chooses an incorrect action, he will have an opportunity loss. The a_j that maximizes *EMV* also minimizes the *expected opportunity loss*, or *EOL*.

Recall that the opportunity losses for the lid manufacturer are as shown in Table 13–9. We compute the *EOL* from a_j as

$$EOL_j = \sum_{i=1}^{k} OL_{ij}P(S_i) \tag{13–3}$$

Thus, $EOL_1 = (0)(.80) + (13.40)(.20) =$ \$2.68, and $EOL_2 = (36.40)(.80) + (0)(.20) =$ \$29.12. Because \$2.68 < \$29.12, a_1 minimizes *EOL*.

TABLE 13–9

$P(S_i)$	S_i	a_1	a_2
.80	S_1	\$0	\$36.40
.20	S_2	\$13.40	\$0

With perfect information (i.e., knowledge of S_i), the manufacturer would choose a_1 80 percent of the time and a_2 20 percent of the time (see Table 13–10). The most he could average, then, is

$$(182)(.80) + (127.40)(.20) = \$171.08$$

$$= \text{expected payoff with perfect information } (EPPI)$$

If a_1 were always chosen, the average payoff would be \$168.40. We see that EOL_j is the difference between *EPPI* and EMV_j; for example,

$$EOL_1 = \$171.08 - \$168.40 = \$2.68$$

Inasmuch as the entrepreneur could average \$2.68 more per decision by knowing the true state of nature, the *EOL* of the Bayes'

TABLE 13–10

$P(S_i)$	S_i	a_1	a_2
.80	S_1	\$182	\$145.60
.20	S_2	\$114	\$127.40

MAXIMUM PAYOFF WHEN S_1 OCCURS (\$182)
MAXIMUM PAYOFF WHEN S_2 OCCURS (\$127.40)

act indicates the *cost of uncertainty*. Moreover, it establishes the highest price that a person should be willing to pay for perfect information. Bayesians refer to the cost of uncertainty as the *expected value of perfect information*, or *EVPI*.

Imagine that a red-light signaling device would cost \$2.00 a production run. Since this device would correctly identify S_i, the manufacturer could average

$$\$171.08 - \underset{\text{cost of the signal}}{\underset{\uparrow}{\$2.00}} = \$169.08$$

On the other hand, if the cost of the signal were \$3.00 a run, the average payoff would be \$171.08 − \$3.00 = \$168.08. In the latter case, perfect information is too expensive. The entrepreneur would have no incentive to install the device (i.e., \$168.08 < \$168.40).

Up to this point, a decision has been made without obtaining a sample of lids. For the classical approach, of course, sampling is essential. Due to the fact that sample information is imperfect, the manufacturer would pay less than $EVPI$ = \$2.68 for it. We will now show how Bayesians combine before-sampling and after-sampling information.

BAYES' THEOREM

As new information becomes available, decision makers revise their plans. Bayesians integrate $P(S_i)$, b_{ij}, and sample outcomes by applying *Bayes' theorem*. Consider the following example.

The R. K. Company sells scientific instruments. The marketing department prefers employees who possess both a technical background and graduate training in business. The personnel from this department can be sorted into the categories listed in Table 13–11.

Suppose that an employee is selected at random. We find that

$$P(A) = 75/100 = .75$$
$$P(B) = 70/100 = .70$$
$$P(C) = 25/100 = .25$$
$$P(D) = 30/100 = .30$$

TABLE 13–11

	A *MBA*	*C* *Non-MBA*	*Row Total*
B *Engineer*	50	20	70
D *Non-Engineer*	25	5	30
Column Total	75	25	

These marginal probabilities are said to be *prior probabilities* because they are computed before additional information about an employee is known.

We have previously confirmed that

$$P(A \text{ and } B) = P(B|A)P(A)$$
$$= P(A|B)P(B)$$

Since $P(B|A)P(A) = P(A|B)P(B)$,

$$P(B|A) = \frac{P(A|B)P(B)}{P(A)}$$

In terms of Table 13–11,

$$P(B|A) = \frac{(50/70)(70/100)}{(75/100)} = 50/75 = .67$$

Let's put $P(A)$ in an alternative form. Inasmuch as the A event includes A and B employees as well as A and D employees,

$$P(A) = P(A \text{ and } B) + P(A \text{ and } D)$$
$$= (50/100) + (25/100) = 75/100 = .75$$

Therefore,

$$P(B|A) = \frac{P(A|B)P(B)}{\underbrace{P(A|B)P(B)}_{P(A \text{ and } B)} + \underbrace{P(A|D)P(D)}_{P(A \text{ and } D)}} \qquad (13\text{–}4)$$

If we identify B and D as "causes" and A and C as "effects," Equation 13–4 relates the probability of cause B given effect A to the probability of effect A given cause B and the probability of effect A given cause D.

In Bayesian analysis, sampling generates "effects" (e.g., 11 defective lids). Such effects are "caused" by the state of nature (e.g., $\pi = .06$). We can observe the effects, but we're uncertain as to the causes. When E is an effect and S_i is a cause, Equation 13–4 looks like

$$P(S_i|E) = \frac{P(E|S_i)P(S_i)}{\sum_{i=1}^{k} P(E|S_i)P(S_i)} \tag{13–5}$$

We call this *Bayes' theorem*. With it, we can determine the probability of the ith cause if effect E occurs. The next example illustrates Bayes' theorem.

A particular county has offered Company W a set of tax concessions and loans if W will locate there. The probability that W will accept the proposal is .7. Mr. Allen is currently unemployed. If W accepts, Allen believes the probability of his getting a job somewhere in the county will be .6. If W doesn't accept, the probability of his getting a job will be .2.

Assume that Allen finds a job. What is the probability that Company W accepted the county's proposal? In other words, we want to calculate the probability that W accepted *given* that Allen found a job. The effects are "job" and "no job," while the causes are "W accepts" and "W doesn't accept." We note

$$P(\text{W accepts}) = .7$$

$$P(\text{W doesn't accept}) = 1 - .7 = .3$$

$$P(\text{job}|\text{W accepts}) = .6$$

$$P(\text{job}|\text{W doesn't accept}) = .2$$

Then,

$$P(\text{W accepts}|\text{job}) = \frac{P(\text{job}|\text{W accepts})P(\text{W accepts})}{P(\text{job}|\text{W accepts})P(\text{W accepts}) + P(\text{job}|\text{W doesn't accept})P(\text{W doesn't accept})}$$

$$= \frac{(.6)(.7)}{(.6)(.7) + (.2)(.3)} = .875$$

We are now ready to explore the way in which decision makers revise their prior probabilities [for our lid manufacturer, $P(\pi = .06)$ and $P(\pi = .12)$] after they obtain sample information. The new probabilities, $P(S_1|E)$ and $P(S_2|E)$, are referred to as *posterior probabilities*.

DECISION MAKING WITH POSTERIOR PROBABILITIES

Let's calculate the posterior probabilities for our manufacturer after a sample of 100 lids containing $r = 11$ defectives has been observed. The entrepreneur must add this information to the "old" information [i.e., $P(S_i)$ and b_{ij}]. Due to Bayes' theorem,

$$P(S_i|r = 11) = \frac{P(r = 11|S_i)P(S_i)}{\sum_{i=1}^{k} P(r = 11|S_i)P(S_i)}$$

Since r = the number of defectives is a binomial random variable and $n\pi \geq 5$ and $n(1 - \pi) \geq 5$, whether $\pi = .06$ or $.12$, we can approximate $P(r = 11|\pi = .06)$ and $P(r = 11|\pi = .12)$ by using the normal distribution. As such,

$$\overbrace{P(r = 11|\pi = .06)}^{\text{binomial}} \approx \overbrace{P(10.5 < r < 11.5|\pi = .06)}^{\text{normal}}$$
$$= P(1.89 < Z < 2.32) = .0192$$

and

$$P(r = 11|\pi = .12) \approx P(10.5 < r < 11.5|\pi = .12)$$
$$= P(-.46 < Z < -.15) = .1176$$

The probability of getting 11 defectives when the equipment is functioning properly is only .0192.

Because the prior probabilities are $P(S_1) = .80$ and $P(S_2) = .20$, the posterior probabilities are

$$P(S_1|r = 11) = \frac{(.0192)(.80)}{(.0192)(.80) + (.1176)(.20)} = .40$$

and

$$P(S_2|r = 11) = \frac{(.1176)(.20)}{(.0192)(.80) + (.1176)(.20)} = .60$$

If we compare the posterior to the prior probabilities, we see that the probability of S_1 has been adjusted downward, whereas the probability of S_2 has been adjusted upward. The direction of these changes is sensible. The sample evidence should heighten the manufacturer's confidence in S_2 (see Table 13–12).

TABLE 13–12

$P(S_i \mid r = 11)$	S_i	a_1	a_2
.40	S_1	$182	$145.60
.60	S_2	$114	$127.40

When posterior probabilities have been derived, EMV_j is recomputed:

$$EMV_1 = (182)(.40) + (114)(.60) = \$141.20$$

$$EMV_2 = (145.60)(.40) + (127.40)(.60) = \$134.68$$

The optimal act is still a_1.

SUMMARY

The action taken as a result of classical hypothesis testing is dependent upon an observed sample. Bayesians, however, integrate the decision maker's experiences and sample information. Using *EMV* as a decision criterion is appealing whenever the same problem appears over and over. Nonetheless, Bayesians also apply *EMV* to "one-time-only" situations. Critics of the Bayesian approach argue that calculating $P(S_i)$ and b_{ij} is often impossible. Bayesian analysis forces experimenters to consider all dimensions of a problem explicitly. They can't just arbitrarily establish a level of significance and then collect a sample. In fact, Bayesians may even choose not to use a sample.

EXERCISES

1. Given the following payoff table, indicate the action that will be chosen when the decision maker's objective is (a) maximin, (b) maximax, and (c) minimize maximum opportunity losses.

	a_1	a_2	a_3
S_1	$10	$5	$20
S_2	–$7	$32	–$25
S_3	$18	–$3	$42

2. Given the following payoff table, indicate the action that will be chosen when the decision maker's objective is (a) maximin, (b) maximax, and (c) minimize maximum opportunity losses.

	a_1	a_2	a_3
S_1	−$10	$50	$40
S_2	$20	$5	−$60
S_3	$90	−$40	$70

3. Based on the payoff table below, identify the action that yields the largest expected monetary value.

$P(S_i)$	S_i	a_1	a_2
.60	S_1	$158	$195
.40	S_2	$176	$139

4. Based on the payoff table below, identify the action that yields the largest expected monetary value.

$P(S_i)$	S_i	a_1	a_2	a_3
.30	S_1	$50	$12	$86
.50	S_2	$15	$73	$18
.20	S_3	$65	$41	$23

5. Using the payoff table in question 3, compute the following:

a. the expected opportunity loss associated with each action
b. the expected payoff with perfect information
c. the expected value of perfect information

6. Using the payoff table in question 4, compute the following:

a. the expected opportunity loss associated with each action
b. the expected payoff with perfect information
c. the cost of uncertainty

7. Suppose that 30 percent of all income tax returns are submitted by low-income families, 60 percent are submitted by middle-income families, and 10 percent are submitted by high-income families. Imagine, moreover, that the Internal Revenue Service audits 1 percent of the low-income, 5 percent of the middle-income, and 25 percent of the upper-income returns. A particular return is selected at random. If the return has been audited, what is the probability that a middle-income family submitted it?

8. Consider the information for 1972 in the table below. A researcher selects a firm at random and determines that it is owned

by a woman. What is the probability that the firm is part of the construction industry?

Industry	*Percentage of All U.S. Firms Included in the Industry*	*Percentage of Firms in the Industry Which Are Owned by a Woman*
Construction	12	1.5
Manufacturing	5	1.8
Other	83	4.3

Source: *Statistical Abstract of the United States, 1977,* p. 552.

9. Ralph Wilson is a member of the sales-forecasting unit of a large corporation. He hopes to be appointed to the position of director for this unit. Since he has rarely seen eye-to-eye with the company president, Mr. Oliver, Wilson believes that the probability of his getting the appointment if Oliver continues to head the company is only .20. If Oliver retires, however, Wilson feels that the probability of his getting the appointment is .60. Wilson also believes that .50 represents the probability that Oliver will retire soon. Suppose Wilson is appointed unit director. What is the probability that Oliver has retired?

10. A certain panel of TV critics previews every new series and predicts whether it will become a success or a failure. Of all the series previewed, only 25 percent receive a "success" prediction. When the panel predicts success, the prediction is correct 40 percent of the time. When the panel predicts failure, the prediction is incorrect 20 percent of the time. If a new series is successful, what is the probability that the panel predicted this?

11. For some time the Sonya Leebove Bakery has sold between zero and five large cakes every day. The cakes are baked the same day they are put on sale, and unsold cakes are destroyed at the end of the day. The cakes cost $5 apiece to make and sell for $10. Construct a payoff table for profits, where the a_js indicate the number of cakes that are baked, the S_is are the number of cakes demanded, and the b_{ij}s are the profits corresponding to a particular S_i, a_j combination ($i = 0, 1, \ldots, 5$ and $j = 1, 2, \ldots, 5$).

12. Using the payoff table in question 11, determine the number of cakes that should be baked each day if the Sonya Leebove Bakery's objective is (a) maximin, (b) maximax, and (c) minimize maximum opportunity losses.

13. Suppose that the bakery in question 11 has kept a record of the requests for its large cakes. This information appears below.

Demand for Large Cakes	*Percentage of Days*
0	.10
1	.20
2	.20
3	.30
4	.15
5	.05

a. Using the payoff table for profits, identify the strategy that maximizes expected monetary value.
b. Compute the expected payoff with perfect information.
c. Compute the expected value of perfect information.

14. A toy manufacturer is considering whether to introduce a new electronic game for children between the ages of seven and nine. He believes that the probability of a high sales level for the new game is .6. The payoff table for profits is shown below.

	Introduce the Game	*Don't Introduce the Game*
Low Sales Level	−$50,000	$0
High Sales Level	$90,000	$0

a. Which act maximizes expected monetary value?
b. Compute the expected payoff with perfect information.
c. Compute the expected value of perfect information.
d. What value of *P*(high sales level) makes the *EMV* of each act the same?

15. The company in question 14 always tests its games on a group of ten youngsters. Tests involving other games yielded the following results:

Outcome of the Game	*Percentage of Games Enjoyed by only 3 Children*
High sales	15
Low sales	70

Only three of the youngsters liked the new electronic game.

a. Calculate the posterior probabilities.
b. Using the posterior probabilities, identify the act that maximizes *EMV*.

16. The R. K. Lamp Company has been buying parts from the same supplier for many years. R. K. is willing to accept any shipment containing 4 percent or fewer defectives. It used to inspect 100 percent of the parts before deciding what action to take with a shipment. The outcomes of those inspections are presented in the table below. There are 10,000 parts in a shipment. Suppose the lamp company selects a random sample of 400 parts from a newly arrived shipment and determines that 18 parts are defective. Using the normal-curve approximation for the binomial distribution of defectives, calculate the posterior probability of each state of nature.

Percentage of Defectives in the Shipment	*Percentage of Shipments*
2	60
4	20
6	15
8	5

The Chi-Square Distribution

DEVELOPMENT OF THE CHI-SQUARE DISTRIBUTION

Chapter Fourteen introduces the *chi-square probability distribution*. This distribution can be used to:

1. Test hypotheses about σ^2,
2. Test hypotheses about the way in which a random variable is distributed (e.g., Is X normal?),
3. Test whether two random variables are independent,
4. Test whether three or more populations have the same π.

Before studying these inference situations, we need to identify the characteristics of the chi-square distribution.

Whenever we randomly select an observation from a normal population, we know that $P(-1 < Z < 1) = .68$ and $P(-2 < Z < 2) = .95$. The square of Z,

$$Z^2 = \left(\frac{X - \mu}{\sigma}\right)^2,$$

is also a random variable. If $P(-Z^* < Z < Z^*) = 1 - \alpha$, $P(Z^2 < Z^{*2}) = 1 - \alpha$; for example,

$$P(Z^2 < 1) = .68$$

$$P(Z^2 < 4) = .95$$

As long as the X population is normal, $[(X - \mu)/\sigma]^2$ possesses a chi-square distribution with 1 degree of freedom:

$$\chi^2_{(1)} = \left(\frac{X - \mu}{\sigma}\right)^2 \qquad (14\text{–}1)$$

where χ is the Greek letter *chi*. By looking at the right side of Equation 14–1, we can conclude the following:

1. $\chi^2_{(1)}$ is never negative. It is positive for $X < \mu$ and $X > \mu$, and zero for $X = \mu$.
2. $\chi^2_{(1)}$ can equal infinity. If $Z = \infty$, $(Z = \infty)^2 = \infty$.
3. $\chi^2_{(1)}$ is skewed to the right. Since $P(Z^2 < 1) = .68$ implies $P(\chi^2_{(1)} < 1) = .68$ or $P(0 < \chi^2_{(1)} < 1) = .68$, 68 percent of the distribution lies in the interval 0 to 1.

The chi-square density function yields Figure 14–1, in which all of these characteristics are illustrated.

FIGURE 14–1

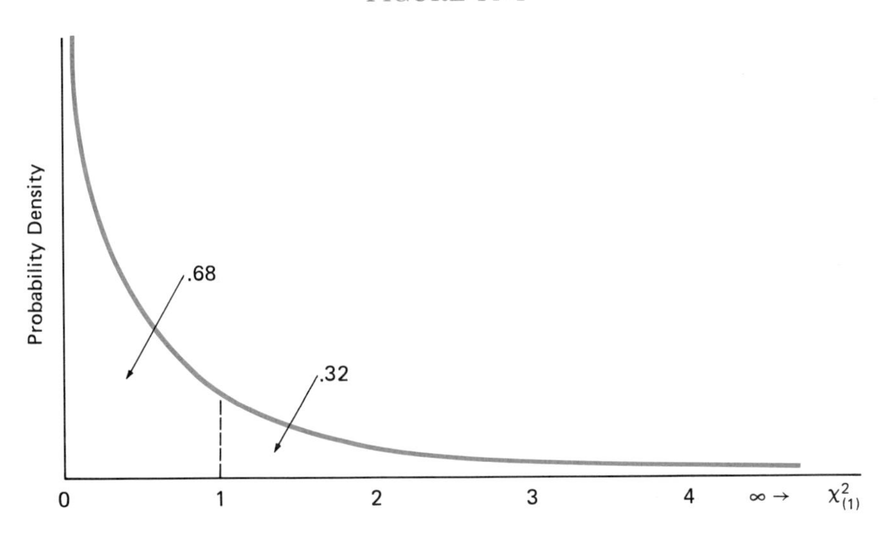

Let's now select n observations independently and square their corresponding Z values. We shall state that the *sum* of independent χ^2 variables has a chi-square distribution with v degrees of freedom, where v = the *sum* of the degrees of freedom for each term. Thus,

$$\chi^2_{(n)} = \sum_{i=1}^{n} Z_i^2 = \sum\left(\frac{X-\mu}{\sigma}\right)^2$$

$$= \frac{\Sigma(X-\mu)^2}{\sigma^2} \qquad (14\text{–}2)$$

Because $\Sigma(X - \mu)^2/\sigma^2$ includes more than one Z^2, the probability of a large value of $\chi^2_{(n)}$ is higher than the probability of a large value of $\chi^2_{(1)}$. Figure 14–2 indicates that the chi-square distribution becomes less skewed as the degrees of freedom increase.

The table of χ^2 values in the Appendix contains parts of many

FIGURE 14–2

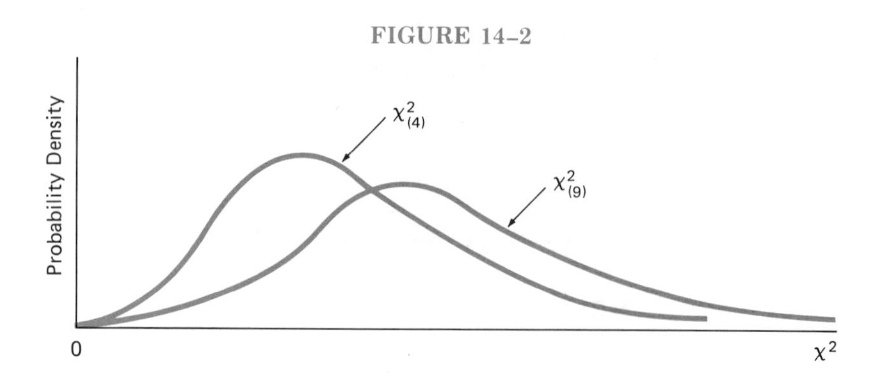

chi-square distributions. Degrees of freedom appear in the left column. The top row lists the probability that χ^2 will be greater than the value in the body of the table. For example, given $\chi^2_{(12)}$,

$$P(\chi^2_{(12)} > 5.23) = .95$$

and

$$P(\chi^2_{(12)} > 26.22) = .01$$

Figure 14–3 illustrates these probabilities.

FIGURE 14–3

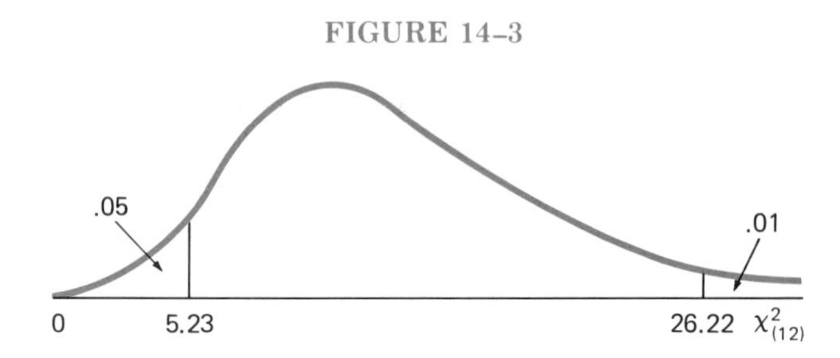

HYPOTHESIS TESTS INVOLVING THE SAMPLE VARIANCE

To establish an inference about σ^2, we have to examine the random variable $(n - 1)s^2/\sigma^2$. If a sample is chosen at random from a normal population, this ratio has a chi-square distribution with $n - 1$ degrees of freedom; that is,

$$\chi^2_{(n-1)} = \frac{(n - 1)s^2}{\sigma^2} \qquad (14\text{–}3)$$

The proof of Equation 14–3 rests on expanding Equation 14–2:

$$\chi^2_{(n)} = \frac{\sum^{n}(X - \mu)^2}{\sigma^2}$$

$$= (1/\sigma^2)\Sigma[(X - \mu) + \overbrace{(\overline{X} - \overline{X})}^{0}]^2$$

$$= (1/\sigma^2)\Sigma[(X - \overline{X}) + (\overline{X} - \mu)]^2$$

$$= \frac{\Sigma(X - \overline{X})^2}{\sigma^2} + \frac{(\overline{X} - \mu)^2}{\frac{\sigma^2}{n}}$$

Since

$$s^2 = \frac{\Sigma(X - \overline{X})^2}{n - 1}$$

and

$$Z^2 = \frac{(\overline{X} - \mu)^2}{\frac{\sigma^2}{n}},$$

the last line is the same as

$$\chi^2_{(n)} = \frac{(n - 1)s^2}{\sigma^2} + \chi^2_{(1)}$$

We noted in Chapter Ten that $\overline{X}$ and s^2 are independent when the population is normal. Thus the two terms on the right are independent. In order for the sum of the terms to equal $\chi^2_{(n)}$, $(n - 1)s^2/\sigma^2$ must equal $\chi^2_{(n-1)}$; that is,

$$\chi^2_{(n)} = \chi^2_{(n-1)} + \chi^2_{(1)}$$

and $v = (n - 1) + 1 = n$.

Suppose we hypothesize that $\sigma^2 = \sigma_0^2$. The sample variance will generally be close to σ_0^2 if H_0 is true. Therefore, we will be suspicious of the null hypothesis when $(n - 1)s^2/\sigma_0^2$ is "very small" or "very large." We determine "large" and "small" in relation to critical values for $\chi^2_{(n-1)}$. These values, $\chi^{2*}_{(n-1)}$ and $\chi^{2**}_{(n-1)}$, are established by the level of significance (see Figure 14–4). Consider the following example.

The jar-lid manufacturer has been informed that a new kind of cutting machine is on the market. The local distributor is willing

FIGURE 14–4

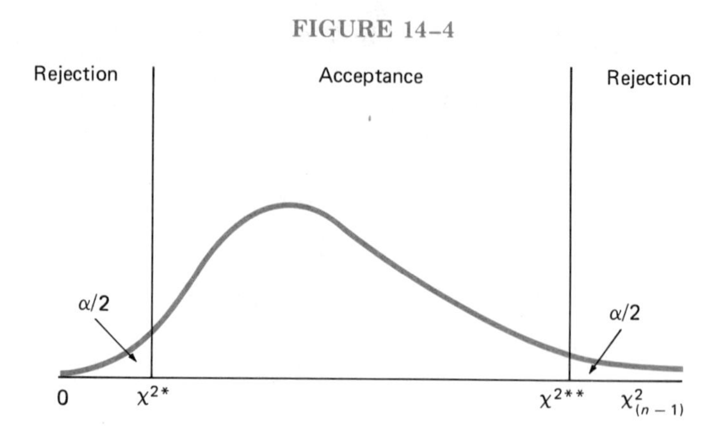

to let him try it for one day. Thereafter, the machine can be returned, leased, or purchased.

Since his old equipment produces lids with diameters having a variance of .25, the entrepreneur won't even discuss a lease-or-buy decision if the variance caused by the new machine is greater than or equal to .25. He decides to conduct a test at the .05 level of significance. Once the new machine is installed, he picks a random sample of 25 lids and measures the diameters. He calculates $s^2 = .20$.

The hypotheses are

$$H_0: \sigma^2 \geq .25$$

$$H_1: \sigma^2 < .25$$

Inasmuch as $\alpha = .05$, $n = 25$, and the test is one-tailed, the critical value of $\chi^2_{(25-1)}$ is 13.85, illustrated in Figure 14–5.

FIGURE 14–5

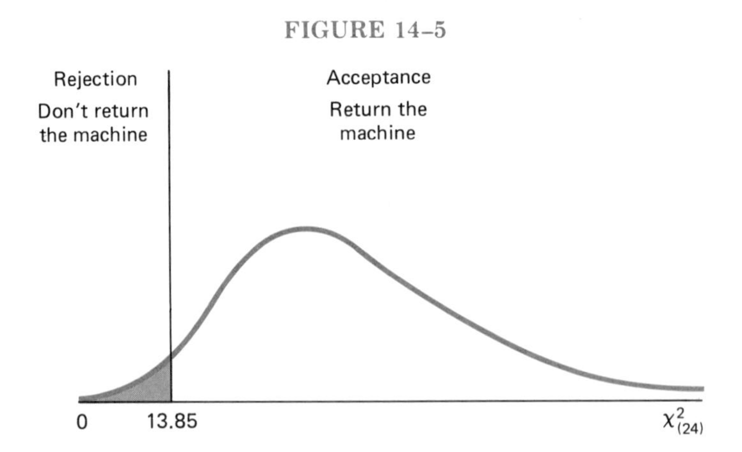

The computed value of χ^2 is

$$\frac{(n - 1)s^2}{\sigma_0^2} = \frac{(25 - 1)(.20)}{.25} = 19.2$$

Hence, the entrepreneur accepts H_0 and returns the machine. He feels that observing $s^2 = .20$ is "possible and probable" when a random sample of 25 lids is selected from a normal population of diameters and $\sigma^2 = .25$.

GOODNESS-OF-FIT TESTS

Researchers may wish to determine whether a random sample has been selected from a population having a particular distribution (e.g., binomial). As in our previous hypothesis-testing illustrations, we first assume that H_0 is true. Then the conditions implied by that hypothesis provide a basis for evaluating an observed outcome. The next example is a case in point.

Company D makes metal clamps and packages them in boxes of three. D has a reputation for selling an inferior product. The company was recently taken over by a conglomerate, which hired a quality-control engineer. He believes that r = the number of defectives per box possesses a binomial distribution with π = the probability that a clamp is defective = .4.

The engineer would like to test this hypothesis so that he can assess improvements in the quality of manufacturing a year from now. He selects a random sample of 100 boxes and records the data in Table 14–1.

TABLE 14–1

r *Number of Defective Clamps*	*Number of Boxes*
0	23
1	31
2	28
3	18
	100

If $\pi = .4$, the probability of finding r defectives among three clamps would be

$$P(r) = \frac{3!}{r!(3 - r)!}(.4)^r(.6)^{3 - r}$$

Therefore,

$$P(0) = .216$$
$$P(1) = .432$$
$$P(2) = .288$$
$$P(3) = .064$$

Imagine a population of boxes. Let's put the boxes in four categories: $r = 0, 1, 2$, and 3. Provided that H_0 is true, $P(r)$ represents the *relative frequency* of each category (e.g., 28.8% of the boxes will contain two defectives). If sample relative frequencies always matched those of the population and we chose 100 boxes, we would expect to see the sample distribution in Table 14–2.

TABLE 14–2

r	*Expected Frequency* $nP(r)$
0	21.6
1	43.2
2	28.8
3	6.4
	100.0

Of course, even when the null hypothesis is correct, the sample and population relative frequencies and, hence, the actual (A) and expected (E) frequencies will differ because of sampling error. H_0 will be questionable if the $A_i - E_i$ (i = category #1, . . . , k) deviations are "large." Our measure of the differences will be

$$\sum_{i=1}^{k} \frac{(A_i - E_i)^2}{E_i}$$

This summation is zero as long as each deviation is zero.

Under a true H_0, the above expression is *approximately* a χ^2 variable with $k - 1 - m$ degrees of freedom; that is,

$$\chi^2_{(k-1-m)} \approx \sum \frac{(A - E)^2}{E} \qquad (14\text{–}4)$$

The k identifies the number of categories, while m indicates the number of parameters that have to be estimated. In our example, a value for π is part of the null hypothesis. Thus, $m = 0$. We can rationalize the remaining $k - 1$ degrees of freedom by noting that $\Sigma(A - E) = 0$. After the deviations for $k - 1$ categories are known, the kth deviation is "fixed."

Formally, the hypotheses are

H_0: The distribution of boxes is binomial with $\pi = .4$

H_1: The distribution of boxes is not binomial with $\pi = .4$ (i.e., the distribution could be binomial with $\pi = .45$, binomial with $\pi = .52$, normal, etc.)

Assuming that $\alpha = .05$, $\chi^{2*}_{(4-1-0)} = 7.81$ (see Figure 14–6).

FIGURE 14–6

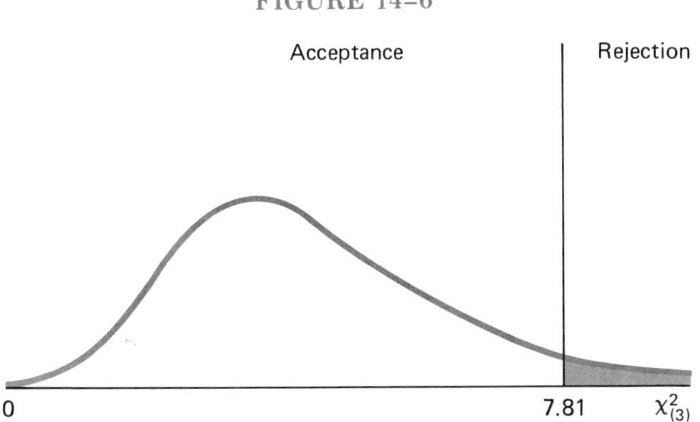

χ^2 is calculated in Table 14–3. Since 24.59 > 7.81, the engineer rejects H_0. If it were true, the probability of observing a summation greater than 7.81 would be only .05. Notice that he is rejecting "binomial with $\pi = .40$" rather than simply "binomial."

TABLE 14–3

A	E	$A - E$	$(A - E)^2$	$\frac{(A - E)^2}{E}$
23	21.6	1.4	1.96	.09
31	43.2	−12.2	148.84	3.45
28	28.8	−.8	.64	.02
18	6.4	11.6	134.56	21.03
100	100.0			24.59

The approximation between the distribution of $\Sigma[(A - E)^2/E]$ and the chi-square distribution improves as n increases. The rule-of-thumb used in applying χ^2 is that none of the expected frequencies should be less than 5. If the engineer had chosen 50 boxes, he would have violated this "rule" (see Table 14–4). When we can't pick a sample large enough to make all E_is 5 or more, the alter-

TABLE 14–4

r	E
0	(.216)(50) = 10.8
1	(.432)(50) = 21.6
2	(.288)(50) = 14.4
3	(.064)(50) = (3.2)
	50.0

native is to combine categories. Table 14–4 would, then, look like Table 14–5.

To investigate a context where $m \neq 0$, imagine that the quality-control engineer feels that the distribution of boxes is binomial, but he won't specify a value for π. Then,

H_0: The distribution is binomial

H_1: The distribution is not binomial

TABLE 14–5

r	E
0	10.8
1	21.6
2 or 3	14.4 + 3.2 = 17.6

Because the expected frequencies depend on π, we first have to estimate π = the probability of a defective clamp. Our estimate will be

$$\frac{\text{number of defective clamps in the sample}}{\text{number of clamps in the sample}} = \frac{(0)(23) + (1)(31) + (2)(28) + (3)(18)}{(3)(100)}$$

$$= 141/300 = .47$$

Using $\pi = .47$,

$$P(r) = \frac{3!}{r!(3-r)!}(.47)^r(.53)^{3-r}$$

The $P(r)$ values lead to Table 14–6.

When H_0 is true and we estimate π, $\Sigma[(A - E)^2/E]$ is approximately a χ^2 variable with $k - 1 - m = 4 - 1 - 1 = 2$ degrees of freedom. Given $\alpha = .05$, the critical value of χ^2 is 5.99 (see Figure 14–7). Since $13.26 > 5.99$, the engineer concludes that the distribution of boxes is not binomial.

TABLE 14–6

r	$P(r)$	A	$E = nP(r)$	$A - E$	$(A - E)^2$	$\frac{(A - E)^2}{E}$
0	.149	23	14.9	8.1	65.61	4.40
1	.396	31	39.6	−8.6	73.96	1.87
2	.351	28	35.1	−7.1	50.41	1.44
3	.104	18	10.4	7.6	57.76	5.55
		100	100.0			13.26

FIGURE 14–7

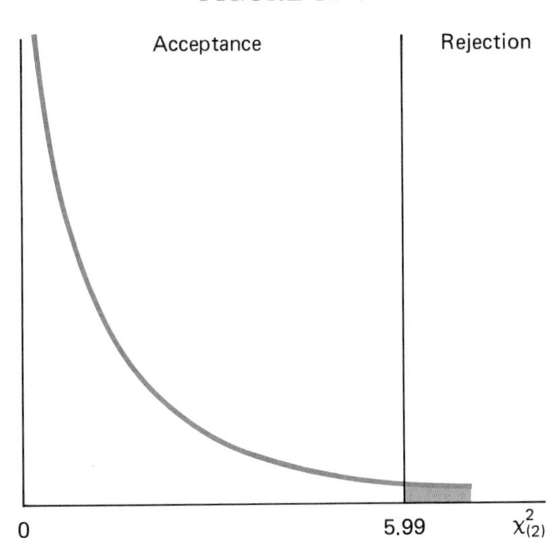

TESTS FOR INDEPENDENCE

In Chapter Five, the population of business administration professors at University L constituted the sample space for the experiment "select a faculty member at random." Given $R = 2$ sex categories and $C = 3$ degree categories, we created a table of joint events (see Table 14–7).

TABLE 14–7

		Degree			
		B.A.	M.A.	Ph.D.	*Row Total*
Sex	Female	1	1	3	5
	Male	3	5	7	15
	Column Total	4	6	10	20 Faculty Members

We will now refer to a table such as 14–7 as an $R \times C$ *contingency table*. Since the joint probabilities are not all equal to the product of the corresponding marginal probabilities, for example,

$$P(\text{female and Ph.D.}) \neq P(\text{female})P(\text{Ph.D.})$$

$$3/20 \neq (5/20)(10/20),$$

"sex" and "degree" are related. Let's look at another example.

If the President of the United States hopes to run for a second term, he must anticipate the political implications of his policies. Unpopular, but possibly effective, actions may not be undertaken. To remove this "political" restriction from the office, certain people have advocated limiting the presidency to one term of six years. A researcher is studying the connection between "sex" and the way adults feel about the proposal. She chose a random sample of 100 adults and constructed a 2×2 contingency table (Table 14–8).

TABLE 14–8

		Attitude		
		For a Six-Year Term	Against a Six-Year Term	*Row Total*
Sex	Female	35	22	57
	Male	19	24	43
	Column Total	54	46	

Suppose that $p_{ij}(i = 1, \ldots, R$, and $j = 1, \ldots, C)$ is the probability that an adult, picked at random, belongs to the joint category in the ith row and jth column. If we had information on the population of adults, we could complete Table 14–9 and confirm independence or dependence between "sex" and "attitude" by noting whether

$$p_{11} = P(\text{female})P(\text{for})$$

$$p_{12} = P(\text{female})P(\text{against})$$

$$p_{21} = P(\text{male})P(\text{for})$$

$$p_{22} = P(\text{male})P(\text{against})$$

Because we possess only a sample, we will have to *infer* independence or dependence. The hypotheses are

H_0: Sex and attitude are independent

H_1: Sex and attitude are not independent

TABLE 14–9

	For	*Against*	*Marginal Probability*
Female	p_{11}	p_{12}	P(female)
Male	p_{21}	p_{22}	P(male)
Marginal Probability	P(for)	P(against)	

Let's use the sample relative frequencies to estimate the marginal probabilities for the population:

$$\text{estimated } P(\text{female}) = 57/100 = .57$$

$$\text{estimated } P(\text{male}) = 43/100 = .43$$

$$\text{estimated } P(\text{for}) = 54/100 = .54$$

$$\text{estimated } P(\text{against}) = 46/100 = .46$$

When the null hypothesis is true, the product of the estimated marginal probabilities will yield estimated joint probabilities (see Table 14–10). We would estimate, for instance, that 19.78 percent of all adults are both "male" and "against."

TABLE 14–10

	For	*Against*	*Marginal Probability*
Female	(.57)(.54) = .3078	(.57)(.46) = .2622	.57
Male	(.43)(.54) = .2322	(.43)(.46) = .1978	.43
Marginal Probability	.54	.46	

If the sample relative frequencies in each joint category were the same as the estimated p_{ij}s, Table 14–11 would identify the expected distribution of outcomes. We wish to find the difference between the actual and expected frequencies. Given a correct H_0, the summation of $(A_{ij} - E_{ij})^2/E_{ij}$ is approximately a χ^2 variable with $k - 1 - m$ degrees of freedom.

TABLE 14–11

	For	*Against*
Female	30.78	26.22
Male	23.22	19.78

The number of categories, k, is equal to RC. In tests for independence, m = the number of probabilities that are "freely" estimated. Notice that the sum of the estimates at the end of the R rows or at the bottom of the C columns in Table 14–10 is 1.0. Once we compute estimated P(female) = .57, the estimate of P(male) is automatically determined; that is, estimated P(male) = 1 − .57 = .43. Moreover, estimated P(against) = 1 − estimated P(for) = .46. Thus, $R - 1$ row estimates and $C - 1$ column estimates are "free." As a consequence,

$$\begin{aligned} k - 1 - m &= RC - 1 - m \\ &= RC - 1 - (R - 1 + C - 1) \\ &= (R - 1)(C - 1) \end{aligned}$$

and

$$\chi^2_{[(R-1)(C-1)]} \approx \sum_{j=1}^{C} \sum_{i=1}^{R} \frac{(A_{ij} - E_{ij})^2}{E_{ij}} \qquad (14\text{–}5)$$

For our example, H_0: Sex and attitude are independent, $(R - 1)(C - 1) = (2 - 1)(2 - 1) = 1$ degree of freedom. Given a level of significance of .05, the acceptance and rejection regions are as shown in Figure 14–8.

FIGURE 14–8

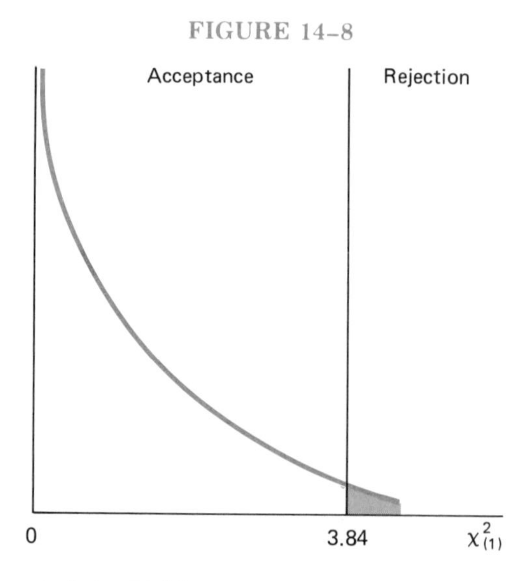

χ^2 is computed in Table 14–12. Because 2.93 < 3.84, the researcher accepts H_0. She concludes that knowing the sex of an adult doesn't provide any insight into that person's opinion of the proposed six-year presidential term.

TABLE 14–12

A	E	$A - E$	$(A - E)^2$	$\frac{(A - E)^2}{E}$
35	30.78	4.22	17.81	0.58
19	23.22	−4.22	17.81	0.77
22	26.22	−4.22	17.81	0.68
24	19.78	4.22	17.81	0.90
100	100.00			2.93

TESTS INVOLVING PROPORTIONS IN MORE THAN TWO POPULATIONS

We explored the test of H_0: $\pi_1 = \pi_2$ in Chapter Twelve. The χ^2 distribution is the framework for testing the equality of three or more population proportions. To clarify this procedure, we'll use the following example.

Although most cars are equipped with seat belts, many drivers look upon belts as uncomfortable and inconvenient. As such, they do not wear them. An analyst for the National Highway Safety Council wants to investigate the proportion of seat-belt wearers in each of three parts of the country. He independently selects three random samples—from New England, the Middle Atlantic, and the Southeast. Table 14–13 presents the results.

TABLE 14–13

		Location			
		New England	Middle Atlantic	Southeast	*Row Total*
Behavior	Motorist Wears a Belt	20	16	24	60
	Motorist Doesn't Wear a Belt	80	84	76	240
	Column Total	100	100	100	

In contrast to the researcher studying attitudes toward a six-year presidential term (Table 14–8), the analyst did not obtain his results by (1) choosing 300 motorists at random and then (2) classifying them according to "behavior" and "location." Since the selections are from three populations of drivers, the column totals are under his control.

TABLE 14–14

	New England	*Middle Atlantic*	*Southeast*
Wears Seat Belt	(.20)(100) = 20	(.20)(100) = 20	(.20)(100) = 20
Doesn't Wear Seat Belt	(.80)(100) = 80	(.80)(100) = 80	(.80)(100) = 80

We are interested in π_j = the proportion of seat-belt wearers in population j. If the three population proportions are equal (i.e., $\pi_1 = \pi_2 = \pi_3 = \pi$), the individual sample proportions will be unbiased estimators of π. Because we want to use as much information as possible, we'll derive

$$\text{"pooled" estimate of } \pi = \frac{\text{number of seat-belt wearers in the three samples}}{\text{number of motorists in the three samples}} = \frac{60}{300} = .20$$

The "pooled" estimate of $1 - \pi$ is $1 - .20 = .80$.

If $\pi_j = \pi$ is true, we would expect the sample frequencies in Table 14–14. The hypotheses are

$$H_0\colon \pi_1 = \pi_2 = \pi_3$$

$$H_1\colon \text{The } \pi_j\text{s are not all equal}$$

FIGURE 14–9

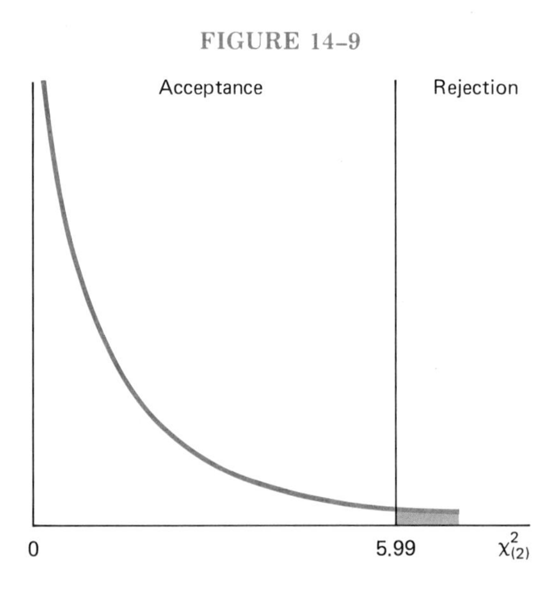

We will apply Equation 14–5 and assume that $\alpha = .05$. The critical value of a χ^2 variable with $(2 - 1)(3 - 1) = 2$ degrees of freedom is 5.99 (see Figure 14–9).

The calculated value of χ^2 appears in Table 14–15.

TABLE 14–15

A	E	$A - E$	$(A - E)^2$	$\frac{(A - E)^2}{E}$
20	20	0	0	0
80	80	0	0	0
16	20	−4	16	.8
84	80	4	16	.2
24	20	4	16	.8
76	80	−4	16	.2
300	300			2.0

Due to the fact that 2.0 lies in the acceptance region, the analyst accepts $\pi_1 = \pi_2 = \pi_3$. He concludes that the proportion of seat-belt wearers doesn't vary by location.

SUMMARY

The χ^2 distribution has a variety of applications. We have seen that the structure of χ^2 tests parallels that of tests we studied earlier. In each testing situation, we established H_0 and then compared the actual outcomes to those expected if H_0 were true. The level of significance separated "possible and probable" from "possible but improbable" occurrences. For tests of goodness-of-fit, independence, and the equality of several population proportions, the χ^2 distribution only approximates the distribution of $\Sigma[(A - E)^2/E]$ or $\Sigma\ \Sigma[(A - E)^2/E]$. The approximation gets better as the sample size increases.

EXERCISES

1. Determine χ^{2*} if

   a. $P(\chi^2_{(16)} > \chi^{2*}) = .05$

   b. $P(\chi^2_{(29)} > \chi^{2*}) = .975$

   c. $P(\chi^2_{(7)} > \chi^{2*}) = .01$

2. Identify the following probabilities:

a. $P(\chi^2_{(24)} > 13.85)$

b. $P(\chi^2_{(11)} > 21.92)$

c. $P(\chi^2_{(18)} > 37.16)$

3. What are the following probabilities?

a. $P(2.70 < \chi^2_{(9)} < 19.02)$

b. $P(1.61 < \chi^2_{(5)} < 9.24)$

c. $P(9.89 < \chi^2_{(24)} < 45.56)$

4. A cigarette manufacturer believes that the amount of tar in his cigarettes is normally distributed with $\mu = 4.5$ milligrams and $\sigma = .27$ milligrams. He selects 400 cigarettes at random and determines the following:

X *Amount of Tar (in milligrams)*	*Number of Cigarettes*
3.99 or less	31
4.0–4.19	58
4.2–4.39	72
4.4–4.59	93
4.6–4.79	67
4.8–4.99	54
5.0 or more	25
	400

Test the manufacturer's belief at the .05 level of significance. (Note: To compute Z, substitute the class boundaries for X.)

5. For the data in question 4, suppose that $\overline{X} = 4.4$ milligrams and $s = .32$ milligrams. Using $\alpha = .05$, test

H_0: The population of the amounts of tar is normal.

H_1: The population of the amounts of tar isn't normal.

[Note: To approximate Z, calculate $Z \approx (X - \overline{X})/s$.]

6. A researcher would like to select a random sample of 28 observations from a normal population. Given that

$$\chi^2_{(n-1)} = \frac{(n-1)s^2}{\sigma^2}$$

and $P(14.57 < \chi^2_{(27)} < 43.19) = .95$, identify the 95 percent confidence interval estimator of σ^2.

7. Suppose an investigator picks a random sample of 18 observations from a normal population and computes $s^2 = 3.7$. Construct a 90 percent confidence interval estimate of σ^2.

8. Mary Smith tosses a coin five times and records the total number of tails. By performing this experiment often, she obtains the following outcomes:

r *Number of Tails*	*Number of Sequences of 5 Tosses*
0	21
1	72
2	148
3	163
4	74
5	22
	500

If the coin is properly balanced, the probability distribution of r will be binomial with $\pi = .5$. Using $\alpha = .05$, test whether r is binomial with $\pi = .5$.

9. A specialist in urban economics selected a random sample of 100 companies that had moved out of a certain large city. The presidents of these companies were asked to indicate whether the move was due principally to taxes, expansion, or other factors (e.g., crime and pollution). The responses are listed below.

X *Principal Factor Behind the Move*	*Number of Companies*
0 (high taxes)	38
1 (expansion requirements)	27
2 (other)	35
	100

Given $\alpha = .05$, test H_0: The probability distribution of X is *uniform* (i.e., the probability associated with each factor is 1/3).

10. A manufacturer of boxes has just completely overhauled his equipment. If the production process is functioning correctly, the weights of the boxes will be normally distributed with $\mu = 12$ ounces and $\sigma = .30$ ounces. He chooses 16 boxes at random and calculates $s = .37$ ounces. Test H_0: $\sigma^2 = .09$ (or $\sigma = .30$) when α is .01.

11. The table below shows the 1972 size distribution of companies involved in retailing. A researcher believes that both the 1972 and the 1979 distributions are identical. He selects a random sample of 1,000 retailers operating in 1979 and sorts them into the categories in the 1972 table. He finds that 298, 105, 223, 232, 61, 43, and 38 firms, respectively, occupy these categories. Test his belief at the .05 level of significance.

Sales	*Proportion of All Retail Firms*
Less than $30,000	.302
$ 30,000–49,999	.110
50,000–99,999	.178
100,000–299,999	.263
300,000–499,999	.062
500,000–999,999	.042
1,000,000 or more	.043

Source: *Statistical Abstract of the United States, 1977*, p. 836.

12. An investigator is studying Americans' knowledge of current affairs. As part of her study, she asks a random sample of 400 adults to indicate the major source of their current affairs information. Based on the following table and an α of .05, test H_0: The probability distribution of X is uniform (i.e., the probability associated with each source is 1/4).

X *Source of Information*	*Number of Adults*
0 (newspapers)	120
1 (television)	135
2 (radio)	87
3 (magazines)	58
	400

13. A researcher is interested in the opinions undergraduates have concerning "big business." He selects four independent and random

samples, each containing 100 undergraduates, from among the freshman, sophomore, junior, and senior classes, and he asks these students, "Do you feel that corporate executives are unscrupulous?" The outcomes are given below. Using $\alpha = .05$, determine whether all four population proportions, where π_j = the population j proportion of undergraduates who believe that corporate executives are unscrupulous, are the same.

		Samples			
		Freshmen	Sophomores	Juniors	Seniors
Responses	Yes	22	34	45	41
	No	78	66	55	59
		100	100	100	100

14. In June 1974 *American Education* magazine reported the following breakdown of fields of college study:

Field	*Percentage of All College Students Choosing the Field as a Major*
Business	13.9
Education	12.1
Humanities and Social Sciences	20.5
Biological and Health Sciences	11.5
Engineering, Mathematics, and Physical Sciences	9.0
Other	33.0

A University Q administrator would like to determine whether this distribution applies to Q as well. She selects 100 students at random and finds that the number of students majoring in each of these fields is 9, 11, 23, 15, 7, and 35, respectively. Given $\alpha = .05$, are the University Q and national distributions identical?

15. A sociologist wants to investigate the relationship between the socioeconomic background of corporation presidents and their involvement in community activities (e.g., speech making or volunteering to help charities). He selects 300 presidents at random and establishes the contingency table displayed below. Using $\alpha = .01$, are "Income of Parents" and "Community Involvement" independent?

		Income of Parents		
		Low	Moderate	High
Community Involvement	Very Active	34	49	33
	Moderately Active	39	30	29
	Not Active	27	32	27

16. An economist would like to identify how company presidents feel about government regulations in business. She chooses 100 presidents at random and categorizes them according to whether their companies are "small" (less than $100,000 in sales) or "large" ($100,000 or more in sales). The economist then constructs the table below. Given $\alpha = .05$, are "Opinion" and "Size of Company" independent?

		President's Opinion about Regulation		
		Too Much	Adequate	Not Enough
President's Company	Small	10	25	5
	Large	30	20	10

15

The F Distribution

CHARACTERISTICS OF THE F DISTRIBUTION

In Chapter Fourteen, we developed a test for the equality of more than two population proportions. Given k populations, we can also investigate H_0: $\mu_1 = \mu_2 = \ldots = \mu_k$. To do so, however, we must employ the F *distribution*. Besides its use in testing the equality of population means, the F distribution enables researchers to infer whether σ_1^2 and σ_2^2 are equal.

Suppose that both Population 1 and Population 2 are normal. If we independently select a random sample from each population, we know that

$$\frac{(n_1 - 1)s_1^2}{\sigma_1^2} = \chi^2_{(n_1 - 1)}$$

and

$$\frac{(n_2 - 1)s_2^2}{\sigma_2^2} = \chi^2_{(n_2 - 1)}$$

Furthermore,

$$\frac{\dfrac{(n_1 - 1)s_1^2}{\sigma_1^2}}{\dfrac{(n_2 - 1)s_2^2}{\sigma_2^2}} = \frac{\chi^2_{(n_1 - 1)}}{\chi^2_{(n_2 - 1)}} \qquad (15\text{–}1)$$

When the population variances are the same, Equation 15–1 reduces to

$$\frac{(n_1 - 1)s_1^2}{(n_2 - 1)s_2^2} = \frac{\chi^2_{(n_1 - 1)}}{\chi^2_{(n_2 - 1)}} \qquad (15\text{–}2)$$

and we can write

$$\frac{s_1^2}{s_2^2} = \frac{\dfrac{\chi^2_{(n_1 - 1)}}{n_1 - 1}}{\dfrac{\chi^2_{(n_2 - 1)}}{n_2 - 1}} \quad \begin{array}{l} \text{A chi-square variable divided by its degrees of freedom} \\ \\ \text{A chi-square variable divided by its degrees of freedom} \end{array} \qquad (15\text{–}3)$$

The ratio of two independent χ^2 variables divided by their degrees of freedom has an F distribution with v_1 and v_2 degrees of freedom, where v_1 = the degrees of freedom in the numerator and v_2 = the degrees of freedom in the denominator. We see, therefore, that s_1^2/s_2^2 will possess an F distribution, that is,

$$F_{(n_1 - 1,\, n_2 - 1)} = \frac{s_1^2}{s_2^2}, \qquad (15\text{–}4)$$

as long as samples are chosen randomly and independently from two *normal* populations having *equal* variances.

Since χ^2 is never less than zero, F is non-negative. An infinite number of F distributions can be constructed. Figure 15–1 illustrates that F is skewed to the right.

FIGURE 15–1

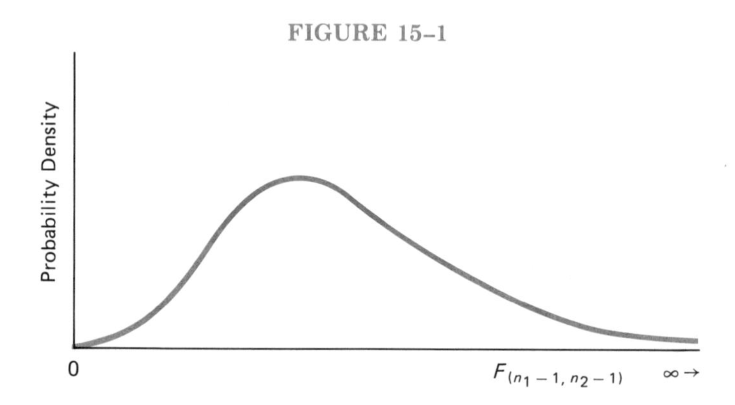

There are two tables of F values in the Appendix. Both contain details for a variety of F distributions. The top row displays v_1, whereas v_2 is presented in the left column. In one table,

$$P\left[F_{(v_1,\, v_2)} > \begin{array}{c}\text{the value in}\\ \text{the body of}\\ \text{the table}\end{array}\right] = .05$$

In the other table, the probability is .01. Thus,

$$P[F_{(7,10)} > 3.14] = .05$$

and

$$P[F_{(7,10)} > 5.20] = .01$$

To find the F value for the lower α area, we proceed as follows:

1. Given a distribution with v_1 and v_2 degrees of freedom, *reverse* the degrees of freedom and determine the value of a for which

$$P[F_{(v_2,\, v_1)} > a] = \alpha$$

2. Compute the *reciprocal* of a (i.e., $1/a$). Then,

$$P\left[F_{(v_1,\, v_2)} < \frac{1}{a}\right] = \alpha$$

For example, when $v_1 = 7$, $v_2 = 10$, and $\alpha = .05$, we can note

$$P[F_{(10,7)} > 3.64] = .05,$$

$$1/3.64 = .27,$$

and

$$P[F_{(7,10)} < .27] = .05$$

Figure 15–2 is illustrative.

FIGURE 15–2

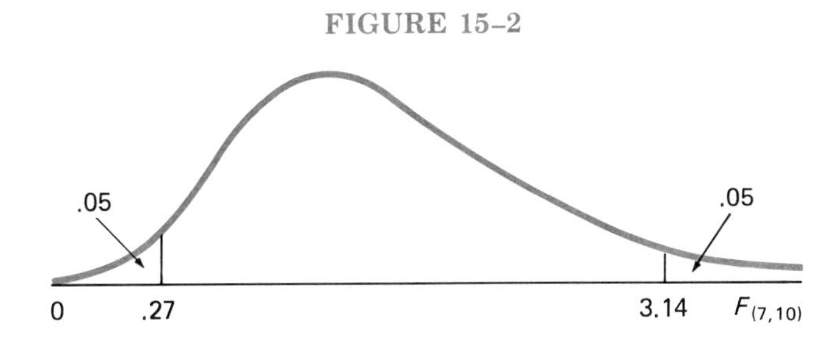

HYPOTHESIS TESTS INVOLVING TWO SAMPLE VARIANCES

As the basis for testing hypotheses about two population variances, let's reconsider a situation from Chapter Twelve.

The management of a company is investigating the relationship between music and productivity. It picks random samples independently from two populations. Population 1 consists of workers who were not exposed to music. Population 2 consists of workers who were exposed to music. The company collects the following information:

$\overline{X}_1 = 5.2$ finished units $\overline{X}_2 = 5.6$ finished units
$s_1^2 = .7$ $s_2^2 = .8$
$n_1 = 10$ $n_2 = 13$

To apply the t test to the hypothesis H_0: $\mu_1 - \mu_2 = 0$ (i.e., music doesn't affect the average level of output), two conditions have to be met:

1. The populations are normal.
2. $\sigma_1^2 = \sigma_2^2$.

We are now in a position to infer the truth or falsity of the second condition.

We'll test

$$H_0: \sigma_1^2 = \sigma_2^2$$

$$H_1: \sigma_1^2 \neq \sigma_2^2$$

Imagine that the level of significance is .10. If H_0 is true and the populations are normal, s_1^2/s_2^2 will have an F distribution with $10 - 1 = 9$ and $13 - 1 = 12$ degrees of freedom. Since the sample variances will be unbiased estimators of the same population variance, s_1^2 should be close to s_2^2. The null hypothesis will, therefore, appear questionable whenever s_1^2/s_2^2 is "very small" or "very large."

Inasmuch as $P[F_{(9,12)} > 2.80] = .05$ and $P[F_{(9,12)} < .326] = .05$, the acceptance and rejection regions are as shown in Figure 15–3. The computed F is

$$\frac{s_1^2}{s_2^2} = \frac{0.7}{0.8} = .875$$

FIGURE 15–3

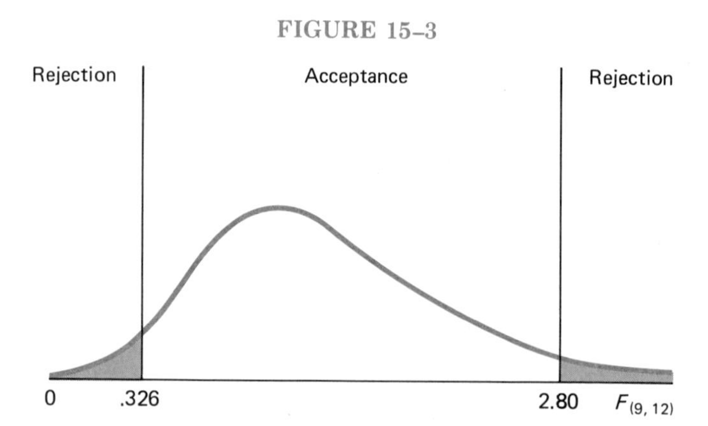

Since .875 lies in the acceptance region, the company concludes that the difference in sample variances occurred by chance. After accepting H_0, management can move on to the t test.

THE ANALYSIS OF VARIANCE

Analysis of variance, or ANOVA, is the technique associated with hypotheses of the form

$$H_0: \mu_1 = \mu_2 = \ldots = \mu_k$$

$$H_1: \text{Not all of the } \mu_j\text{s are equal}$$

We will explore ANOVA in the context of the following example.

Many companies are plagued by high rates of absenteeism among their employees. Such companies are either unable to meet production schedules or they are forced to incur the cost of overtime labor. In addition to disciplining workers who miss frequently, some corporations believe that a ten-hour-a-day, four-day week will reduce absences. Management may also offer bonuses whenever output in a particular period is greater than a specified level. Under the latter arrangement, an employee not only has a financial incentive to show up, but will anger his colleagues if he doesn't.

Company Q employs three shifts of 300 workers each. Q would like to determine whether μ_j = the average number of absences per day for shift j is the same for each shift. If Q concludes that there is a difference in the average number of absences, all workers will be asked to answer a series of questions designed to identify morale problems. The day and night shifts would be administered the questionnaire first.

The payroll office collects information on X_{ij} = the total number of absences for shift j on day i. The company independently selects random samples of five absenteeism reports from the day, evening, and night shifts. Table 15–1 lists the results.

TABLE 15–1

	Shift	
Day X_{i1}	Evening X_{i2}	Night X_{i3}
19	22	25
23	22	27
24	23	28
24	25	30
25	28	30
$\overline{X}_1 = 23$	$\overline{X}_2 = 24$	$\overline{X}_3 = 28$
$s_1^2 = 5.5$	$s_2^2 = 6.5$	$s_3^2 = 4.5$

Each of the shifts constitutes a population. The criterion that distinguishes these populations is called a *factor*. We will refer to the various categories of the factor as *treatments*. Thus, in our example, "shift" is the factor and "day," "evening," and "night" are treatments.

We know that the sample averages will generally differ among themselves even when $\mu_1 = \mu_2 = \mu_3$. Because of this, a researcher will reject a hypothesis of "equal population, or treatment, means" only when $\overline{X}_1$, $\overline{X}_2$, and $\overline{X}_3$ are "very different."

There is a temptation to approach the three-sample test in terms of separate t tests of

$$H_0\colon \mu_1 - \mu_2 = 0$$
$$H_0\colon \mu_1 - \mu_3 = 0$$
$$H_0\colon \mu_2 - \mu_3 = 0$$

When we reject any of those hypotheses, we automatically reject H_0: $\mu_1 = \mu_2 = \mu_3$. Two problems arise, however:

1. If k = the number of populations is large, we might have to conduct many t tests. When $k = 6$, for instance,

$$\frac{6!}{2!4!} = 15$$

   t tests are possible.
2. Suppose that $\alpha = .05$ and the t tests are independent. For 15 tests, the probability of *incorrectly* rejecting at least one H_0 is

$$P(\text{at least one Type I error}) = 1 - P(\text{no Type I error})$$
$$= 1 - (.95)^{15} = .54$$

   Under a true H_0: $\mu_1 = \mu_2 = \mu_3$, therefore, the probability of a Type I error ceases to be α if a series of t tests is performed.

FIGURE 15–4

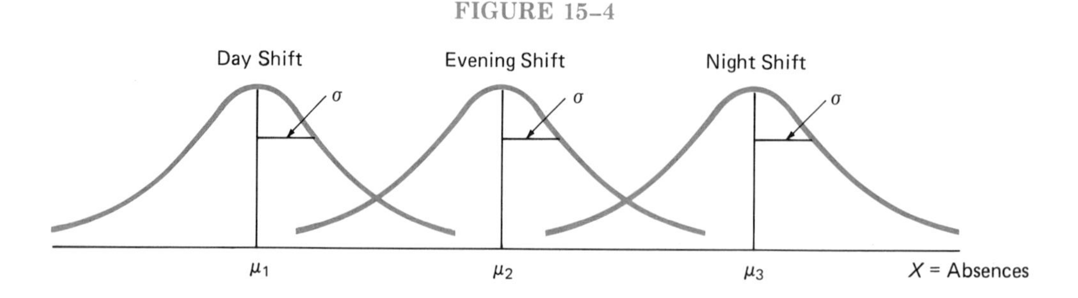

With ANOVA, researchers gain efficiency and the guarantee that their chosen level of significance will, in fact, represent P(Type I error). ANOVA is similar to the t test in that it evolves from assuming *normal* populations and *equal* σ^2s (i.e., $\sigma_1^2 = \sigma_2^2 = \sigma_3^2 = \sigma^2$).

Imagine that every μ_j is different. Figure 15–4 might display the populations. If we selected a sample from each population, we would expect to find a large difference among the $\overline{X}$s. By contrast, when $\mu_1 = \mu_2 = \mu_3$, the populations would appear as in Figure 15–5. Under these circumstances, sampling from each shift would be *equivalent* to drawing three samples from the *same* population.

FIGURE 15–5

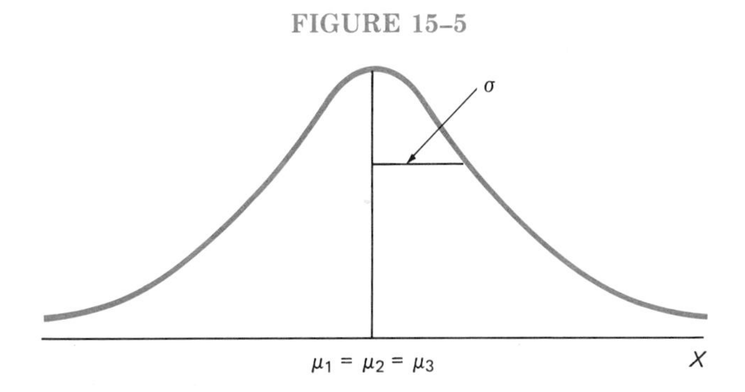

We verified in Chapter Seven that the variance of the sampling distribution of $\overline{X}$ is $\sigma_{\overline{x}}^2 = \sigma^2/n$. We can derive

$$n(\sigma_{\overline{x}}^2) = \sigma^2 \qquad (15\text{–}5)$$

from such a relationship. To measure the difference in the $\overline{X}$s associated with the three samples, we'll calculate

$$s_{\overline{x}}^2 = \begin{matrix}\text{the variance of the}\\ \text{sample averages}\end{matrix} = \frac{\sum_{j=1}^{k} (\overline{X}_j - \overline{\overline{X}})^2}{k-1} \qquad (15\text{–}6)$$

where

$$\overline{\overline{X}} = \begin{matrix}\text{the average of the}\\ \text{sample averages}\end{matrix} = \frac{\sum_{j=1}^{k} \overline{X}_j}{k} \qquad (15\text{–}7)$$

(Note: Equations 15–6 and 15–7 parallel the formulas for the sample variance and the sample mean.)

We shall use $n(s_{\overline{x}}^2)$ as an estimator of σ^2. The rationale for concentrating on σ^2 even though we are testing the equality of means will be explained shortly.

Returning to our example, based on Table 15–1, we find that

$$\bar{\bar{X}} = \frac{\bar{X}_1 + \bar{X}_2 + \bar{X}_3}{3} = \frac{23 + 24 + 28}{3} = 25$$

and

$$s_{\bar{x}}^2 = \frac{(23 - 25)^2 + (24 - 25)^2 + (28 - 25)^2}{3 - 1} = 7.0$$

Given $n_1 = n_2 = n_3 = n$,

$$n(s_{\bar{x}}^2) = (5)(7.0) = 35$$

If 35 were close to σ^2, we would feel comfortable in believing that the samples were picked from identical populations. However, σ^2 is unknown. Thus we have to develop an alternative basis for evaluating $n(s_{\bar{x}}^2)$.

Let's now study the difference within each sample. Since every member of a population is exposed to the same treatment, any deviation between a sample observation and its sample mean must occur by chance. Inasmuch as $\sigma_1^2 = \sigma_2^2 = \sigma_3^2 = \sigma^2$, *each* s_j^2 is an unbiased estimator of σ^2. We can combine the information contained in all of the sample variances, however, and estimate σ^2 by way of the "pooled" variance:

$$\begin{aligned} s_p^2 &= \frac{(n_1 - 1)\, s_1^2 + (n_2 - 1)\, s_2^2 + (n_3 - 1)\, s_3^2}{(n_1 - 1) + (n_2 - 1) + (n_3 - 1)} \\ &= \frac{(5 - 1)(5.5) + (5 - 1)(6.5) + (5 - 1)(4.5)}{(5 - 1) + (5 - 1) + (5 - 1)} = 5.5 \end{aligned} \qquad (15\text{–}8)$$

This estimate of σ^2 is *independent* of the previous estimate derived by using Equation 15–6. Unlike the value of $n(s_{\bar{x}}^2)$, the value of s_p^2 will generally be close to σ^2 regardless of whether H_0 is true or false. We know that $n(s_{\bar{x}}^2)$ will be close to σ^2 and, hence, close to s_p^2 *only* if H_0 is true. We will, therefore, reject H_0 whenever $n(s_{\bar{x}}^2)$ is "much larger" than s_p^2. We may state the following: If k populations are normally distributed, $\sigma_1^2 = \sigma_2^2 = \ldots = \sigma_k^2$, and H_0: $\mu_1 = \mu_2 = \ldots = \mu_k$ is true, $n(s_{\bar{x}}^2)/s_p^2$ will have an F distribution with $k - 1$ and $k(n - 1)$ degrees of freedom; that is,

$$F_{[k - 1,\ k(n - 1)]} = \frac{n(s_{\bar{x}}^2)}{s_p^2} \qquad (15\text{–}9)$$

The numerator has $k - 1$ degrees of freedom because $\Sigma(\bar{X} - \bar{\bar{X}})$ equals zero. Once $k - 1$ deviations are determined, the kth deviation is "fixed." On the other hand, s_p^2 includes k summations,

$$\sum_{i=1}^{n} (X_{i1} - \overline{X}_1)^2, \ldots, \sum_{i=1}^{n} (X_{ik} - \overline{X}_k)^2,$$

each of which has $n - 1$ degrees of freedom. (Note: These summations correspond to the numerators of the s_j^2 terms.)

Company Q will test

$$H_0\colon \mu_1 = \mu_2 = \mu_3$$

$$H_1\colon \text{Not all of the } \mu_j\text{s are equal}$$

Assume that the level of significance is .05. When F has $3 - 1 = 2$ and $3(5 - 1) = 12$ degrees of freedom, its critical value is 3.89 (see Figure 15–6).

FIGURE 15–6

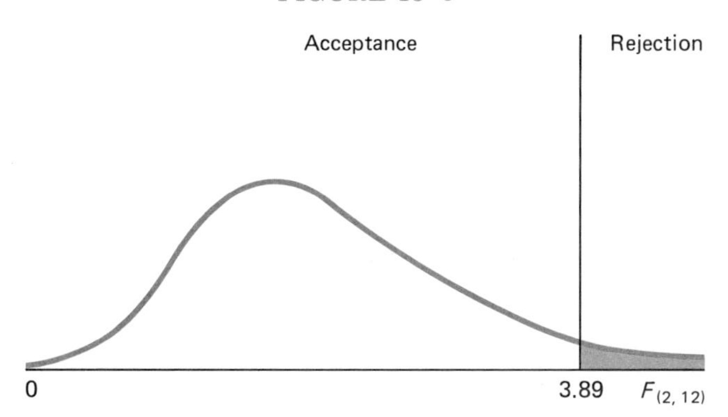

Management computes

$$\frac{n(s_{\overline{X}}^2)}{s_p^2} = \frac{35}{5.5} = 6.36$$

Inasmuch as $6.36 > 3.89$, 35 is "very large" relative to 5.5. The numerator and denominator should be similar when the samples are selected from identical populations. Company Q therefore rejects the hypothesis that each shift has the same average number of absences. Management will now attempt to discover why such shift, or treatment, effects occur.

THE ANOVA TABLE

The results of an analysis-of-variance test are often displayed in tabular form. We will now investigate the structure of such a table.

Suppose we combine the k samples. An observation, X_{ij}, will differ from the over-all mean, $\bar{\bar{X}}$, (1) *entirely* because of chance when there are *no* treatment effects or (2) *partly* because of chance and *partly* because of treatment effects.

Given $X_{ij} - \bar{\bar{X}}$, let's add and subtract $\bar{X}_j$:

$$X_{ij} - \bar{\bar{X}} = X_{ij} - \bar{\bar{X}} + \overbrace{\bar{X}_j - \bar{X}_j}^{0}$$

$$= (X_{ij} - \bar{X}_j) + (\bar{X}_j - \bar{\bar{X}}) \qquad (15\text{–}10)$$

Since $(X_{ij} - \bar{X}_j)$ can only be due to chance, the impact of the treatments (if there is an impact) will be contained in $(\bar{X}_j - \bar{\bar{X}})$.

Although the dispersion in the data could be identified by the variance of the kn observations, we'll restrict our attention to the numerator of the variance,

$$\sum_{j=1}^{k} \sum_{i=1}^{n} (X_{ij} - \bar{\bar{X}})^2$$

This *sum of squares* is called the *total variation*. We can derive it by squaring the right side of Equation 15–10 and adding the kn squares.

We're going to proceed in three stages (continuing to use our example of absenteeism at Company Q):

1. Substitute every X_{ij}, along with $\bar{X}_j$, into Equation 15–10 and calculate

$$[(X_{ij} - \bar{X}_j) + (\bar{X}_j - \bar{\bar{X}})]^2$$

or

$$(X_{ij} - \bar{X}_j)^2 + 2(X_{ij} - \bar{X}_j)(\bar{X}_j - \bar{\bar{X}}) + (\bar{X}_j - \bar{\bar{X}})^2$$

For X_{23} = the second observation in the third sample (Table 15–1) = 27, the last line would be

$$(27 - 28)^2 + 2(27 - 28)(28 - 25) + (28 - 25)^2$$

2. Sum the squares in step 1 for each sample:

$$\sum_{i=1}^{n} (X_{ij} - \bar{X}_j)^2 + 2(\bar{X}_j - \bar{\bar{X}}) \overbrace{\sum_{i=1}^{n} (X_{ij} - \bar{X}_j)}^{0} + n(\bar{X}_j - \bar{\bar{X}})^2$$

or

$$\sum_{i=1}^{n} (X_{ij} - \bar{X}_j)^2 + n(\bar{X}_j - \bar{\bar{X}})^2$$

3. Add the k sums in step 2:

$$\overbrace{\sum_{j=1}^{k}\sum_{i=1}^{n}(X_{ij}-\overline{X}_j)^2}^{\text{within-sample variation}} + \overbrace{n\sum_{j=1}^{k}(\overline{X}_j-\overline{\overline{X}})^2}^{\text{between-sample variation}}$$

We are, thus, able to "decompose" the total variation into a "within-sample" (or unexplained-by-treatments) variation and a "between-sample" (or explained-by-treatments) variation.

Whenever a "sum of squares," SS, is divided by its degrees of freedom, the ratio is referred to as a *mean square* (e.g., the sample variance, s^2, is a mean square). Given $k = 3$,

$$\frac{\text{Within-sample variation}}{\text{Degrees of freedom}} = \frac{SS_W}{k(n-1)} = \frac{\Sigma\Sigma(X_{ij}-\overline{X}_j)^2}{k(n-1)}$$

$$= \frac{(n_1-1)\dfrac{\sum_i (X_{i1}-\overline{X}_1)^2}{(n_1-1)} + (n_2-1)\dfrac{\sum_i (X_{i2}-\overline{X}_2)^2}{(n_2-1)} + (n_3-1)\dfrac{\sum_i (X_{i3}-\overline{X}_3)^2}{(n_3-1)}}{(n_1-1)+(n_2-1)+(n_3-1)}$$

$$= \frac{(n_1-1)s_1^2+(n_2-1)s_2^2+(n_3-1)s_3^2}{(n_1-1)+(n_2-1)+(n_3-1)} = s_p^2$$

Moreover,

$$\frac{\text{Between-sample variation}}{\text{Degrees of freedom}} = \frac{SS_B}{k-1} = \frac{n\sum_j(\overline{X}-\overline{\overline{X}})^2}{k-1}$$

$$= n(s_{\overline{X}}^2)$$

TABLE 15–2

Source of Variation	*Degrees of Freedom*	*Sum of Squares*	*Mean Squares*	*F*
Between samples	$k - 1$	SS_B	$\dfrac{SS_B}{k-1}$	$\dfrac{SS_B/(k-1)}{SS_W/k(n-1)}$
Within samples	$k(n-1)$	SS_W	$\dfrac{SS_W}{k(n-1)}$	
Total	$kn - 1$	SS_T		

After decomposing the total variation and calculating two mean squares, we have all of the information that is vital for an F test.

The general outline of an ANOVA table is presented in Table 15–2. Using our absenteeism example, we can construct Table 15–3.

TABLE 15–3

Source of Variation	*Degrees of Freedom*	*Sum of Squares*	*Mean Squares*	*F*
Between samples	2	70	35	
Within samples	12	66	5.5	35/5.5 = 6.36
Total	14	136		

CONFIDENCE INTERVALS FOR THE DIFFERENCE BETWEEN TWO POPULATION MEANS

Having concluded that the population means are not all equal, Company Q intends to survey the day- and night-shift employees on their attitudes toward work (e.g., "Do you take pride in your job?"). The company would first like to establish a $100(1 - \alpha)$ percent confidence interval estimate of the difference between μ_1 = the average number of absences for the day shift and μ_3 = the average number of absences for the night shift.

Recall our discussion of two-sample t tests in Chapter Twelve. When small random samples are chosen independently from Populations 1 and 3 and $\sigma_1^2 = \sigma_3^2$, Equation 12–7 becomes

$$t_{(n_1 + n_3 - 2)} = \frac{(\overline{X}_1 - \overline{X}_3) - (\mu_1 - \mu_3)}{\sqrt{\underbrace{s_p^2}(1/n_1 + 1/n_3)}}$$

Derived from Equation 12–6

We can improve our estimator of σ^2 by using Equation 15–8. Under these circumstances,

$$\frac{(\overline{X}_1 - \overline{X}_3) - (\mu_1 - \mu_3)}{\sqrt{\underbrace{s_p^2}(1/n_1 + 1/n_3)}}$$

Derived from Equation 15–8

has a t distribution with $n_1 + n_2 + n_3 - 3$ degrees of freedom.

If

$$P\left[-t^* < \frac{(\overline{X}_1 - \overline{X}_3) - (\mu_1 - \mu_3)}{\sqrt{s_p^2(1/n_1 + 1/n_3)}} < t^*\right] = 1 - \alpha,$$

we are able to say

$$P[(\overline{X}_1 - \overline{X}_3) - t^*\sqrt{s_p^2(1/n_1 + 1/n_3)} < (\mu_1 - \mu_3) < (\overline{X}_1 - \overline{X}_3) + t^*\sqrt{s_p^2(1/n_1 + 1/n_3)}] = 1 - \alpha \quad (15\text{–}11)$$

The portion of Equation 15–11 enclosed in brackets is the 100 $(1 - \alpha)$ percent *confidence interval estimator* of $\mu_1 - \mu_3$.

With $n_1 = 5$, $n_2 = 5$, $n_3 = 5$, and $100(1 - \alpha)$ percent = 95%, $t^* = 2.179$. Therefore,

$$(23 - 28) - (2.179)\sqrt{(5.5)(1/5 + 1/5)} < \mu_1 - \mu_3 < (23 - 28) + (2.179)\sqrt{(5.5)(1/5 + 1/5)}$$

$$-8.23 < \mu_1 - \mu_3 < -1.77$$

Management can be 95 percent confident that the day shift averages somewhere between 8.23 and 1.77 fewer absences than the night shift. Two other 95 percent confidence intervals are

$$-4.23 < \mu_1 - \mu_2 < 2.23$$

$$-7.23 < \mu_2 - \mu_3 < -0.77$$

We can be 95 percent confident of *each* of these intervals. However, we can't be 95 percent confident that they are *simultaneously* true.

SUMMARY

Like the χ^2 distribution, the F distribution can be used in situations where samples have been selected from more than two populations. Although the H_0 in ANOVA concerns means, inferences are derived from two independent estimates of σ^2. The ANOVA table is used to display the calculations associated with a test of H_0: $\mu_1 = \mu_2 = \ldots = \mu_k$. Similar tables will be introduced in the chapters on regression.

EXERCISES

1. Determine F^* if

   a. $P(F_{(8,26)} > F^*) = .05$
   b. $P(F_{(10,29)} > F^*) = .05$
   c. $P(F_{(24,18)} > F^*) = .05$
   d. $P(F_{(7,16)} > F^*) = .01$
   e. $P(F_{(12,23)} > F^*) = .01$

2. Determine F^* if

   a. $P(F_{(9,12)} < F^*) = .05$
   b. $P(F_{(20,24)} < F^*) = .05$
   c. $P(F_{(6,9)} < F^*) = .05$
   d. $P(F_{(15,24)} < F^*) = .01$
   e. $P(F_{(4,20)} < F^*) = .01$

3. A researcher selects random samples independently from two normal populations. Given $n_1 = 11$, $s_1^2 = 3.08$, and $n_2 = 25$, $s_2^2 = 2.65$, test H_0: $\sigma_1^2 = \sigma_2^2$ at the .10 level of significance.

4. Suppose three normal populations have the same variance. An investigator independently chooses three random samples from these populations and obtains the observations below. Using $\alpha = .05$, test H_0: $\mu_1 = \mu_2 = \mu_3$.

Sample from Population #1	*Sample from Population #2*	*Sample from Population #3*
5	7	8
8	10	7
11	3	15
7	4	8
9	5	9
4	7	11
6	4	10

5. Given the data in question 4,

   a. Construct an ANOVA table.
   b. Calculate a 95 percent confidence interval estimate of $\mu_2 - \mu_3$.
   c. Calculate a 99 percent confidence interval estimate of $\mu_2 - \mu_3$.

6. A farmer plants the same crop on three acres. Because soil composition differs from one acre to another, he wants to determine whether $\mu_1 = \mu_2 = \mu_3$, where μ_j = the average height of the crop on acre j. He independently and randomly selects seven plants from each acre and records the data below. Test H_0: $\mu_1 = \mu_2 = \mu_3$ at the .05 level of significance.

Sample from Acre #1	*Sample from Acre #2*	*Sample from Acre #3*
8.1 inches	7.6 inches	8.4 inches
7.9	7.4	8.6
8.2	7.7	8.3
8.3	7.3	8.9
8.0	7.2	9.0
8.2	7.3	9.1
8.3	7.4	8.8

7. Given the information in question 6, construct an ANOVA table.

8. A financial analyst believes that the average return on stockholders' equity, a measure of profitability, is the same in industry #1, #2, and #3. She chooses independent and random samples of six companies from each industry. The outcomes are listed below. Test $\mu_1 = \mu_2 = \mu_3$ at the .01 level of significance.

Sample from Industry #1	*Sample from Industry #2*	*Sample from Industry #3*
7.6%	9.4%	10.2%
8.3	9.7	12.6
7.9	10.1	13.3
8.4	9.3	13.1
8.5	9.2	14.5
8.4	9.8	11.4

9. Using the data in question 8, construct (a) an ANOVA table and (b) a 95 percent confidence interval estimate of $\mu_1 - \mu_3$.

10. The machines used by Company W are powered by special batteries, which W can purchase from three suppliers. Although cost is an important consideration, the lifetime of such batteries is also of interest. Company W examined independent and random samples of five batteries produced by each supplier. Based on the following information (battery life in hundreds of hours) and $\alpha = .05$, test H_0: $\mu_1 = \mu_2 = \mu_3$, where μ_j = the average life of supplier j's batteries.

Sample from Supplier #1	*Sample from Supplier #2*	*Sample from Supplier #3*
7.2	8.4	9.2
6.8	7.5	8.6
7.3	7.1	9.3
6.7	8.3	8.9
7.5	6.8	8.6

11. Using the data in question 10, construct (a) a 90 percent confidence interval estimate of $\mu_1 - \mu_3$ and (b) a 95 percent confidence interval estimate of $\mu_1 - \mu_3$.

12. Allen & Co. manufactures product B at two factories. The vice-president wants to know whether the weight of the finished product has the same variance at both factories. Suppose the weights are normally distributed. If the vice-president selects independent and random samples of ten Bs from each factory and finds that $s_1 = .62$ ounces and $s_2 = .55$ ounces, is $H_0: \sigma_1^2 = \sigma_2^2$ acceptable at the .10 level of significance?

13. A consumer-information agency is comparing the effectiveness of three household glues. In one experiment, five independently and randomly selected jars of each glue were used to bind two pieces of wood, and a researcher recorded the time required for the glue to set. Given the results below, test $\mu_1 = \mu_2 = \mu_3$ at the .05 level of significance.

Sample of Glue #1	*Sample of Glue #2*	*Sample of Glue #3*
15 minutes	12 minutes	19 minutes
11	10	20
14	11	22
16	13	18
13	10	21

14. Given the data in question 13, construct (a) an ANOVA table and (b) a 99 percent confidence interval estimate of $\mu_2 - \mu_3$.

15. An investigator is studying the heights of workers on the assembly line at two plants. Based on independent and random samples of 21 workers from plant #1 and 25 workers from plant #2, she computes $s_1 = 0.9$ inches and $s_2 = 1.3$ inches. Assuming that height is normally distributed, can she accept $H_0: \sigma_1^2 = \sigma_2^2$ at the .02 level of significance?

Linear Regression Analysis: *Descriptive Aspects*

THE RATIONALE OF LINEAR REGRESSION

For most variables of interest (e.g., Y), observations will generally differ among themselves. In this chapter, we intend to "explain" why these differences occur.

Policymakers are often concerned with such explanations. For instance, when the United States is suffering through a recession, officials try to stimulate corporate spending. Table 16–1 outlines two possible plans of attack and indicates how the effects generated by the two governmental actions are supposed to work. More information is necessary, however, before the strategies can be implemented. *Linear regression* is an approach to estimating the *numerical relationship* between two variables (e.g., tax credits and corporate spending). Based on regression estimates, economists might decide that raising credits from 10 percent to 11 percent could boost spending by $5 billion. We will use the following example to develop our analysis.

TABLE 16–1

Action	*Consequences*						
1. The Federal Reserve System buys government securities.	Increase in the money supply	→	Reduction in interest rates	→	Increase in corporate spending	→	Increase in GNP
2. Congress raises the investment tax credit.	Reduction in the cost of equipment	→	Increase in corporate spending	→	Increase in GNP		

John Davis is the operator of a newsstand. During his first month in business, he always ordered a large number of papers.

Despite the surpluses on some days, Davis wanted to protect himself from "lost" sales. He wishes to revise his ordering strategy in the future. Sales for the first month are recorded in Table 16–2. Davis believes that the *sequence* of sales (i.e., Y_1, Y_2, etc.) will change every month. He does not believe, however, that the *distribution* of Y will change.

Let's view the data in the table as a population. Since Davis

TABLE 16–2

Day	Y *Number of Newspapers Sold*
#1	35
#2	39
#3	23
#4	29
#5	41
#6	48
#7	28
#8	51
#9	37
#10	45
#11	27
#12	32
#13	43
#14	19
#15	31
#16	42
#17	22
#18	33
#19	25
#20	47
#21	38

expects the distribution to remain the same, he might order $\mu_y =$ 35 papers per day. Papers that he sells for 15¢ cost him 10¢. On days when demand exceeds 35, he would buy additional papers (at 15¢) from the drugstore across the street (to maintain his customers' "good will").

Because the typical value of Y (i.e., 35) won't usually match the actual Y_i, the μ_y order criterion exposes Davis to both (1) the cost of unsold papers and (2) the "cost" of missed profits. These costs will be high if the variance of Y,

$$\sigma_y^2 = \frac{\Sigma(Y - \mu_y)^2}{N},$$

is large.

We want to see whether the ordering strategy can be improved; that is, is there a way of ordering Q papers such that $\Sigma(Y - Q)^2/N$ is less than $\sigma_y^2 = 1{,}694/21 = 80.67$? Our goal will be to find an *independent variable*, X, that varies in a consistent or systematic fashion with the *dependent variable*, Y. In choosing X, we must examine the factors that affect Davis' sales.

The newsstand is located in a mall with five other stores. Inasmuch as more shoppers come to the mall when the stores advertise discounts (e.g., "25% off," "two-for-the-price-of-one," etc.), Davis reasons that his newspaper sales may be affected by X = the number of stores advertising discounts. Table 16–3 pairs the value of Y on day i with the value of X on that day.

TABLE 16–3

Day	Y	X
#1	35	3
#2	39	5
#3	23	1
#4	29	3
#5	41	3
#6	48	5
#7	28	1
#8	51	5
#9	37	3
#10	45	5
#11	27	1
#12	32	3
#13	43	5
#14	19	1
#15	31	1
#16	42	5
#17	22	1
#18	33	3
#19	25	1
#20	47	5
#21	38	3

If we graph Y against X, we obtain Figure 16–1, which is referred to as a *scatter diagram*. The points seem to move upward and to the right. This suggests that higher values of Y are usually paired with higher values of X. Davis' order when $X = 5$, then, should be larger than his order when $X = 3$. As in the tax-credit example, however, there is still the question "*How much* larger?"

We intend to look at the population of Y as a collection of *subpopulations*, or *conditional distributions*, of Y given X (see Table

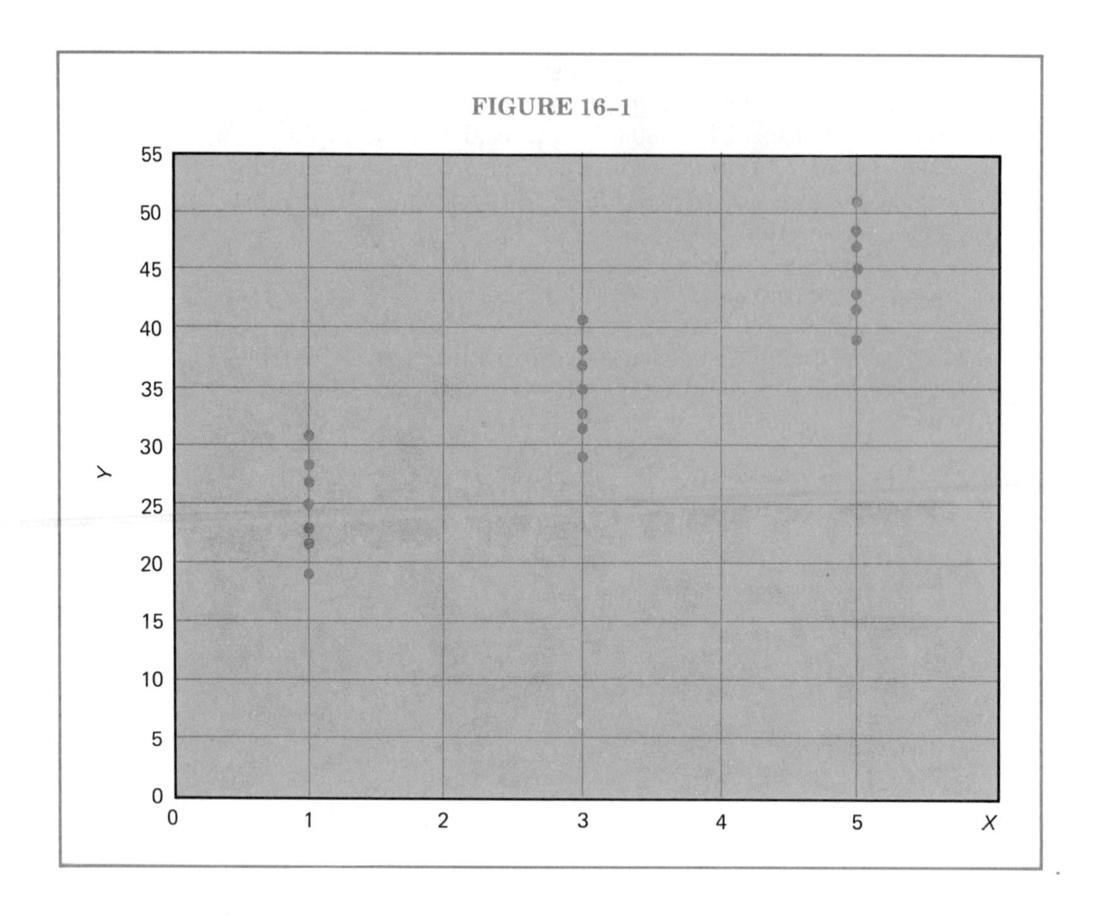

16–4). Each of these subpopulations has a mean, $\mu_{y|x}$, and a variance, $\sigma^2_{y|x}$ (e.g., $\mu_{y|x=1} = 25$ and $\sigma^2_{y|x=1} = 14$). The data illustrate two of the regression assumptions:

1. *Linearity:* The "systematic" relation between Y and X can be written as

$$\mu_{y|x} = A + BX \tag{16–1}$$

where Equation 16–1 is the *population regression line*, A is the *intercept* (i.e., the value of $\mu_{y|x}$ when $X = 0$), and B is the *slope* (i.e., the change in the value of $\mu_{y|x}$ per every one-unit change in X). For the data in Table 16–4, $\mu_{y|x} = 20 + 5X$. The graph of this line appears in Figure 16–2.
2. *Homoscedasticity:* The variances of the subpopulations are the same.

The other regression assumptions will be identified in Chapter Seventeen.

TABLE 16–4

Subpopulation X = 1	*Subpopulation X = 3*	*Subpopulation X = 5*
19	29	39
22	32	42
23	33	43
25	35	45
27	37	47
28	38	48
31	41	51

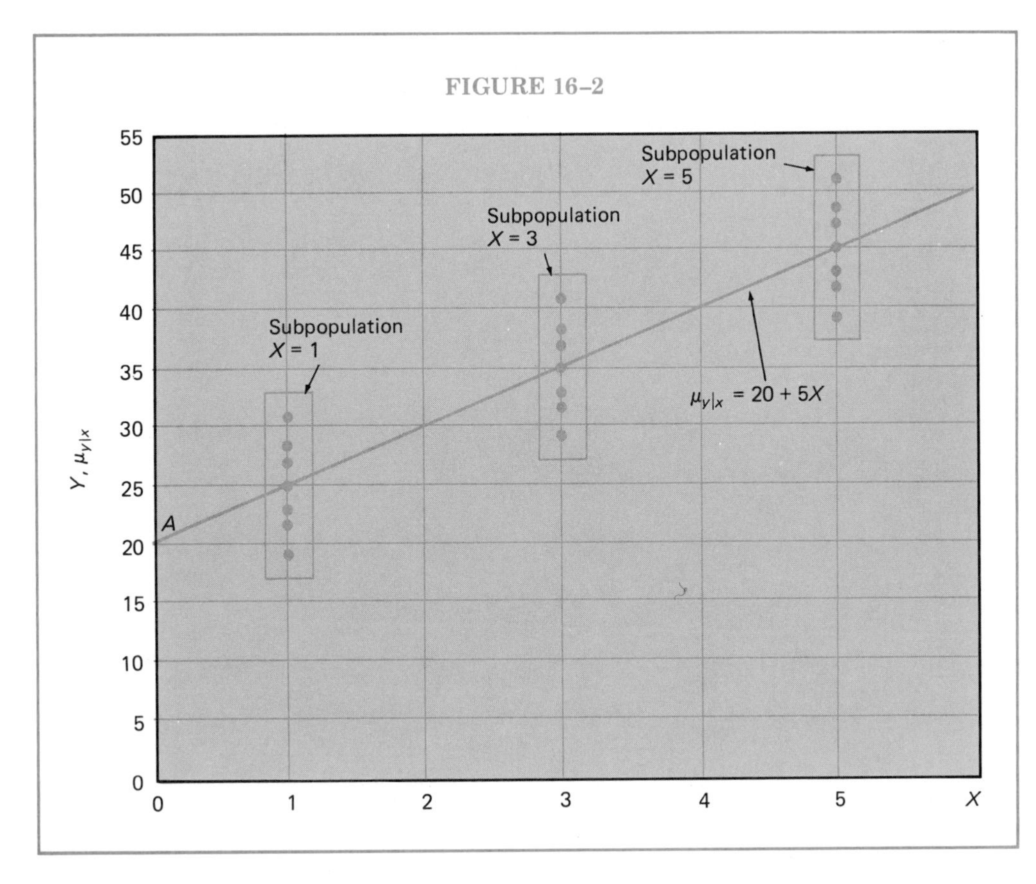

FIGURE 16–2

Suppose the difference between a Y_i and its subpopulation mean is represented by the Greek letter *epsilon*, ϵ. Then,

$$Y_i = \mu_{y|x} + (Y_i - \mu_{y|x}) = \mu_{y|x} + \epsilon_i$$
$$= (A + BX) + \epsilon_i \qquad (16\text{–}2)$$

Given $X = 3$,

$$Y = 41 = [20 + 5(3)] + 6 \quad \epsilon$$

while

$$Y = 33 = [20 + 5(3)] + (-2) \quad \epsilon$$

Other factors in addition to X (e.g., the type of stories in the news) influence Y. Their *cumulative* effect is captured by ϵ. Because the ϵs are *random* and *nonobservable,* Equation 16–2 can't be used to determine the exact value of Y when A, B, and X are known.

The errors push the Y_is above and below the population regression line. The variance of Y around $\mu_{y|x} = A + BX$ is the sum of the N individual $(Y - \mu_{y|x})^2$, or ϵ^2, terms divided by N (i.e., 294/21 = 14). Whenever the subpopulation variances are identical, however, the regression variance equals $\sigma^2_{y|x}$. We will let $\sigma^2_{y|x}$ indicate both (1) a subpopulation variance and (2) the regression variance.

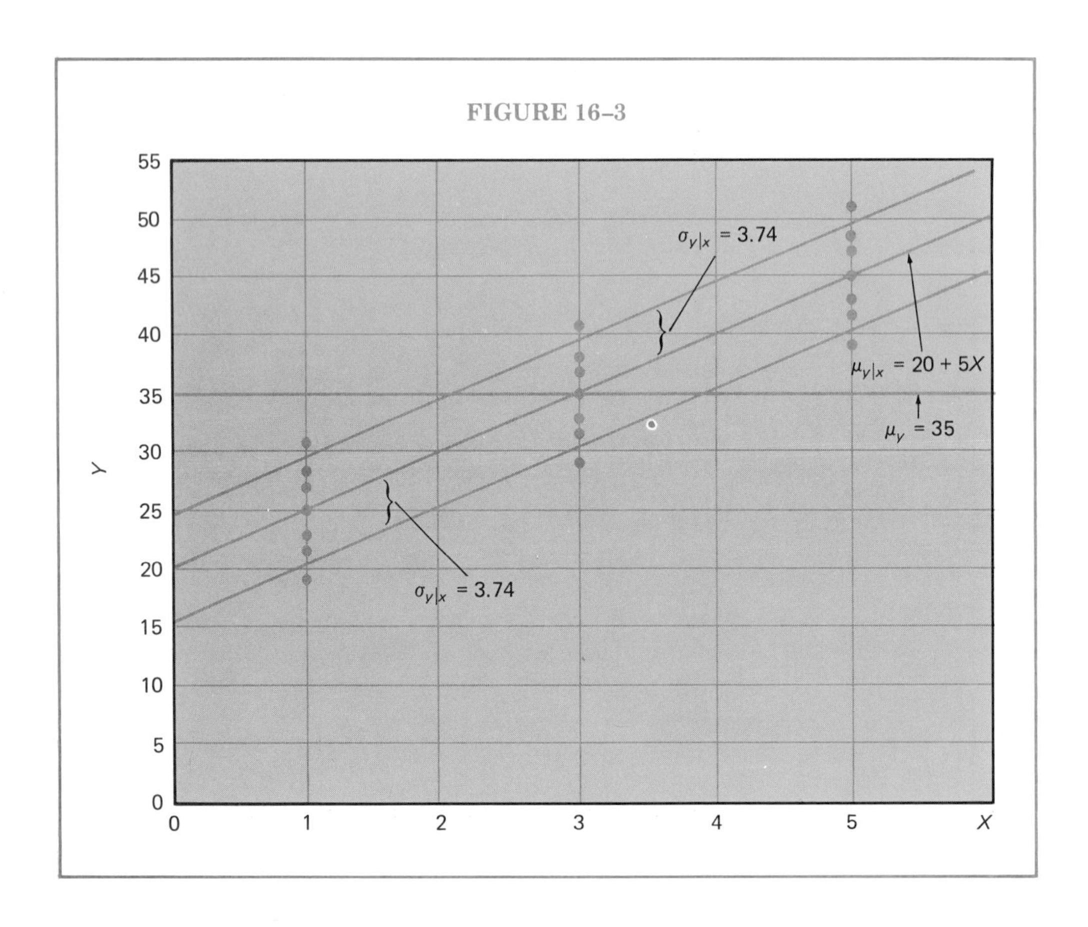

Figure 16–3 shows which Y_is lie within one standard deviation, or $\sigma_{y|x} = \sqrt{14} = 3.74$ papers, of $\mu_{y|x} = 20 + 5X$. The variance of Y around the order quantity falls from 80.67 to 14 if $\mu_{y|x}$, rather than μ_y, is the daily order. This new strategy, therefore, reduces $\Sigma(Y - Q)^2/N$ by $(80.67 - 14)/80.67$, or 83 percent.

DERIVING THE SAMPLE REGRESSION LINE

If a researcher has only a sample of X,Y pairs, he must estimate $\mu_{y|x} = A + BX$ with the *sample regression line*:

$$Y^* = a + bX \tag{16–3}$$

where Y^* is an estimator of $\mu_{y|x}$ and the foundation for predicting a future Y
a is an estimator of A
b is an estimator of B

To obtain the estimates, we will identify particular subpopulations and select Y_is from them at random. The values of X are, thus, controlled, or predetermined, by the experimenter, but Y is generated by chance.

Suppose we pick two Y_is from each subpopulation in Table 16–4. This will yield six X,Y pairs. Although we can't write down the sample Y values until *after* the drawing is completed, we already know the X values (i.e., $X = 1, 1, 3, 3, 5, 5$). Table 16–5 displays one possible set of outcomes. Since there are $7!/2!5! = 21$ ways to choose two Y_is from a subpopulation, and the choices are independent of those from other subpopulations, we could make (21)(21)(21) = 9,261 different tables.

TABLE 16–5

Y	X
23	1
28	1
32	3
37	3
45	5
51	5

The observations in Table 16–5 are presented in a scatter diagram in Figure 16–4. We sense that large Ys are paired with large Xs. To quantify the association between Y and X, we have to compute a and b. Infinitely many lines pass among the sample points.

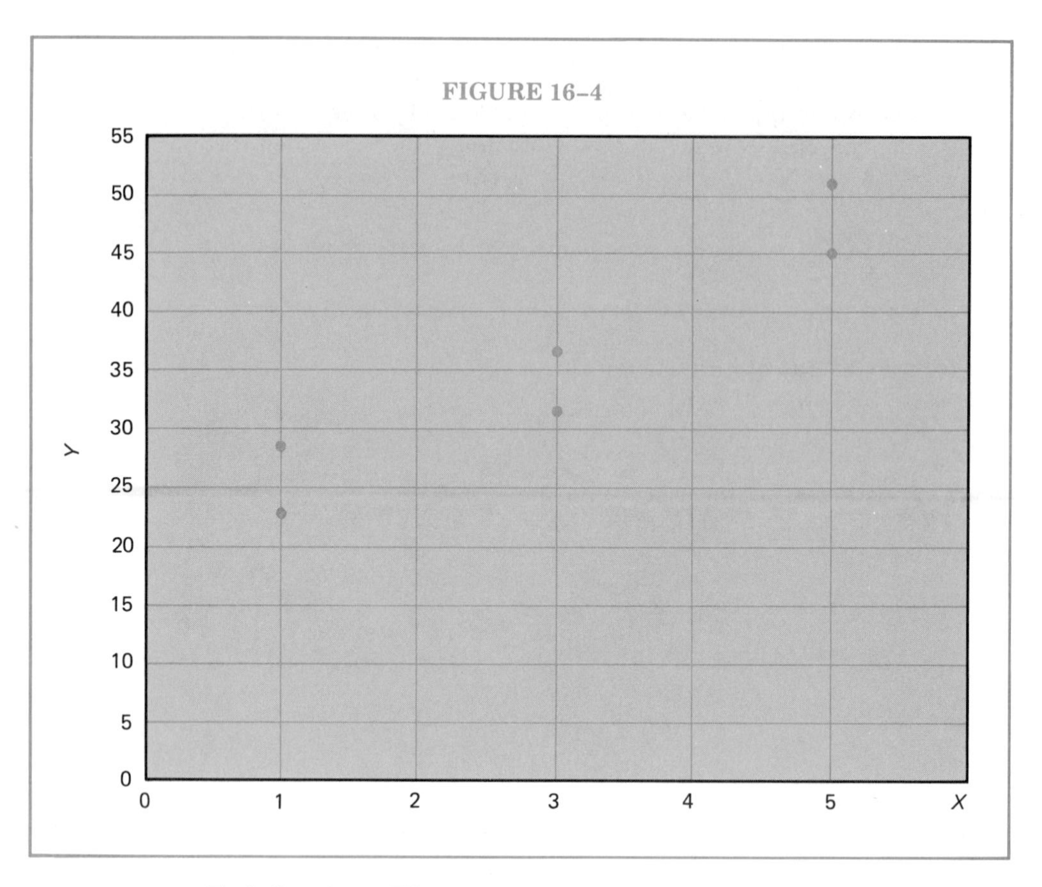

Each line (e.g., $Y^* = 5 + 10X$, $Y^* = 4 + 16X$, or $Y^* = 28 + 0.5X$) represents a specific combination of a and b values. We need a criterion for deciding which line is appropriate.

We want the line to come close to the observed points, that is, a line which provides a *good fit.* The six points in Figure 16–4 can't all lie on the same line. Hence, any line that we ultimately derive will produce *residuals, e*s, or differences between the actual Y when $X = X_i$ and the computed Y^* when $X = X_i$.

The "best" line, based on the regression criterion, is the one for which the *sum of the squared residuals,* $\Sigma(Y - Y^*)^2$ or $\Sigma(Y - a - bX)^2$ or Σe^2, is as small as possible. As we will discuss in Chapter Seventeen, such a line has properties that are important for establishing inferences about $\mu_{y|x}$, Y, A, and B. Using calculus, we could substantiate the fact that the a and b values for the "least-squares" line are the solutions to the following simultaneous equations:

$$\Sigma Y = an + b\Sigma X \tag{16–4}$$

$$\Sigma XY = a\Sigma X + b\Sigma X^2 \tag{16–5}$$

Although Equations 16–4 and 16–5 are called *normal* equations, "normal" isn't a reference to the normal distribution.

From the sample, we can obtain ΣY, ΣX, ΣXY, ΣX^2, and n = the number of X,Y pairs. The summations are shown in Table 16–6. The normal equations become

$$216 = 6a + 18b$$

$$738 = 18a + 70b$$

TABLE 16–6

Y	X	XY	X^2
23	1	23	1
28	1	28	1
32	3	96	9
37	3	111	9
45	5	225	25
51	5	255	25
216	18	738	70

To eliminate a, we'll (1) multiply the first equation by -3 and (2) add the new equation to the second one:

$$\begin{array}{rrcl} & -648 & = & -18a - 54b \\ + & 738 & = & 18a + 70b \\ \hline & 90 & = & 0 + 16b \end{array}$$

If we, then, substitute $b = 90/16 = 5.625$ into $216 = 6a + 18b$, we find

$$216 = 6a + (18)(5.625)$$

$$6a = 216 - 101.25$$

$$a = 114.75/6 = 19.125$$

and

$$Y^* = 19.125 + 5.625X$$

When $X = 5$, we would estimate Y as $Y^* = 19.125 + (5.625)(5) = 47.25$ papers. $Y^* = 47.25$ is also our estimate of the mean of the $X = 5$ subpopulation. The slope estimate, $b = 5.625$, implies that Y^* increases by 5.625 papers every time X increases by one unit (i.e., one store). The sample regression line and the residuals are shown in Figure 16–5.

FIGURE 16–5

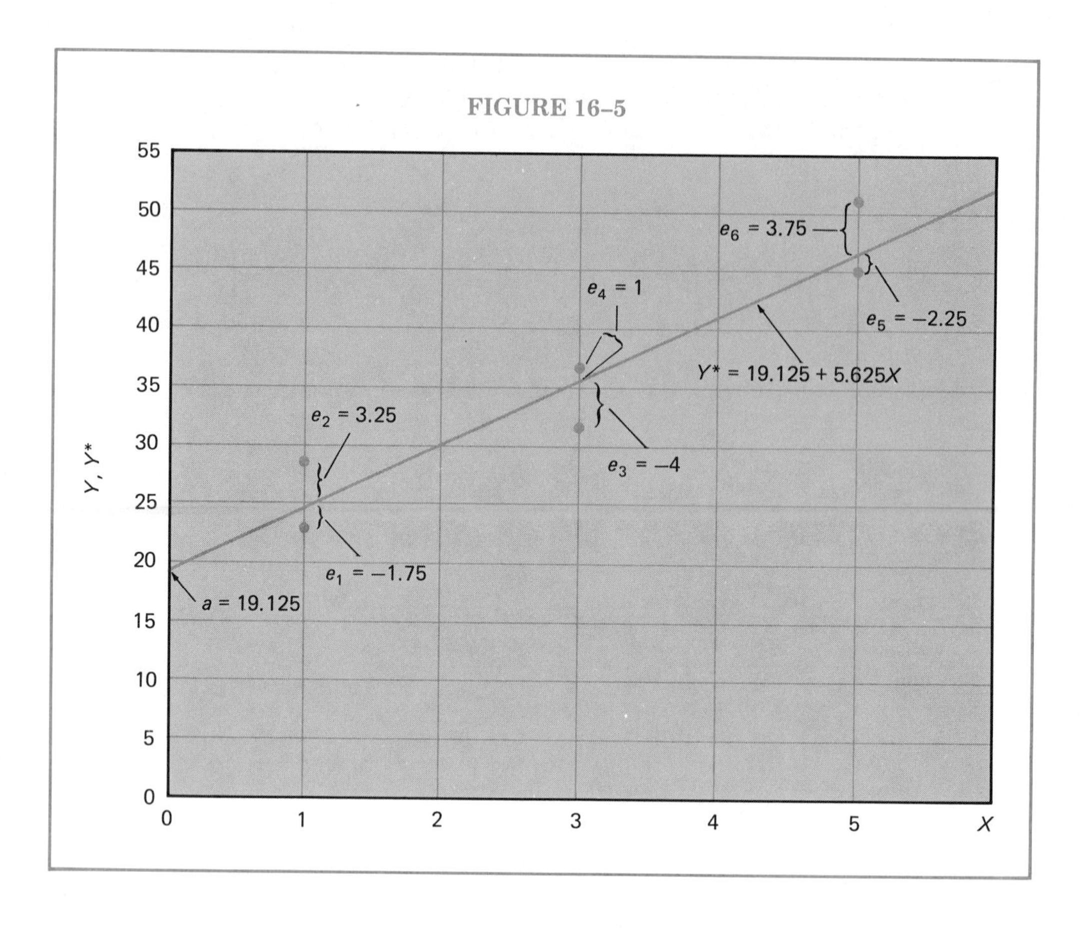

Two characteristics of regression follow from Equation 16–4:

1. Since $\Sigma e = \Sigma(Y - Y^*) = \Sigma(Y - a - bX) = \Sigma Y - an - b\Sigma X$, Σe is *always* zero.
2. Because $\Sigma Y^* = \Sigma(a + bX) = an + b\Sigma X$, ΣY^* is *always* equal to ΣY.

Instead of using the algebraic procedure indicated earlier, we can compute a and b by way of Equations 16–6 and 16–7 or 16–6 and 16–8:

$$a = \overline{Y} - b\overline{X} \tag{16–6}$$

$$b = \frac{\Sigma(X - \overline{X})(Y - \overline{Y})}{\Sigma(X - \overline{X})^2} \tag{16–7}$$

$$b = \frac{n\Sigma XY - \Sigma X \Sigma Y}{n\Sigma X^2 - (\Sigma X)^2} \tag{16–8}$$

Equality 16–8 is more efficient than 16–7, but 16–7 will play a role in our discussions in Chapter Seventeen. Table 16–7 contains the relevant totals. We are, therefore, able to calculate

$$b = 90/16 = 5.625$$

$$b = \frac{(6)(738) - (18)(216)}{(6)(70) - (18)^2} = 540/96 = 5.625$$

$$a = 36 - (5.625)(3) = 19.125$$

TABLE 16–7

Y	$\overline{Y}$	$Y - \overline{Y}$	X	X^2	XY	$\overline{X}$	$X - \overline{X}$	$(X - \overline{X})^2$	$(X - \overline{X})(Y - \overline{Y})$
23	36	−13	1	1	23	3	−2	4	26
28	36	−8	1	1	28	3	−2	4	16
32	36	−4	3	9	96	3	0	0	0
37	36	1	3	9	111	3	0	0	0
45	36	9	5	25	225	3	2	4	18
51	36	15	5	25	255	3	2	4	30
216			18	70	738			16	90

EVALUATING THE SAMPLE REGRESSION LINE

Although the regression technique produces a *minimum* sum of squared deviations, $\Sigma(Y - Y^*)^2$, such a sum, nonetheless, can be large. If this is so, (1) $Y^* = a + bX$ will not come close to the sample pairs or, equivalently, (2) $Y^* = a + bX$ will not describe the sample relation between Y and X very well. To determine how good our line is, we'll compute two summary measures involving $\Sigma(Y - Y^*)^2$.

We measured the variability of Y around $\mu_{y|x} = A + BX$ with $\sigma^2_{y|x}$. We could, therefore, measure the variability of Y around $Y^* = a + bX$ with $s^2_{y|x} = \Sigma(Y - Y^*)^2/n$. Since we want $s^2_{y|x}$ to be an unbiased estimator of $\sigma^2_{y|x}$, we will introduce the following equality:

$$s^2_{y|x} = \frac{\Sigma(Y - Y^*)^2}{n - 2} \qquad (16\text{–}9)$$

The right side of Equation 16–9 represents a sum of squares divided by its degrees of freedom. There are $n - 2$ degrees of freedom because only $n - 2$ of the $Y - Y^*$ residuals are "free." This fact becomes evident if we rewrite the normal equations as

$$(1) \quad \Sigma(Y - a - bX) = 0$$
$$\text{or } \Sigma(Y - Y^*) = 0$$
$$(2) \quad \Sigma X(Y - a - bX) = 0$$
$$\text{or } \Sigma X(Y - Y^*) = 0$$

Two restrictions, then, are placed on the residuals: (1) they must sum to zero, and (2) the sum of the products of X and a residual must be zero as well.

Computing $s^2_{y|x}$ is easier when we express Equation 16–9 as

$$s^2_{y|x} = \frac{\Sigma Y^2 - a\Sigma Y - b\Sigma XY}{n - 2} \qquad (16\text{–}10)$$

After calculating b from Equation 16–8, only ΣY^2 is still needed. Table 16–8 displays the appropriate summations. Thus,

$$s^2_{y|x} = \frac{49.75}{6 - 2} = 12.44$$

or

$$s^2_{y|x} = \frac{8{,}332 - (19.125)(216) - (5.625)(738)}{6 - 2} = 12.44$$

TABLE 16–8

Y	X	XY	Y^2	$Y^* = 19.125 + 5.625X$	$Y - Y^*$	$(Y - Y^*)^2$
23	1	23	529	24.75	−1.75	3.0625
28	1	28	784	24.75	3.25	10.5625
32	3	96	1,024	36.00	−4.00	16.0000
37	3	111	1,369	36.00	1.00	1.0000
45	5	225	2,025	47.25	−2.25	5.0625
51	5	255	2,601	47.25	3.75	14.0625
216		738	8,332	216.00		49.7500

When a researcher believes that 12.44 is "small," he correspondingly believes that $Y^* = 19.125 + 5.625X$ is "close" to the sample points. The sample *coefficient of determination,* r^2, is an alternative to $s^2_{y|x}$. We'll examine r^2 next.

We have indicated the point at $X = 5$, $Y = 51$ in Figure 16–6. Using $\overline{Y}$ to estimate Y, we see that $Y - \overline{Y} = 51 - 36 = 15$. Based on the regression line, however, we would estimate Y as $Y^* = 19.125 + (5.625)(5) = 47.25$. Since $Y - Y^* = 51 - 47.25 = 3.75$, part of the 15-paper difference between Y and $\overline{Y}$, that is, $(Y - \overline{Y}) - (Y - Y^*) = 15 - 3.75 = 11.25$, is "explained" by X (see Figure 16–7).

FIGURE 16–6

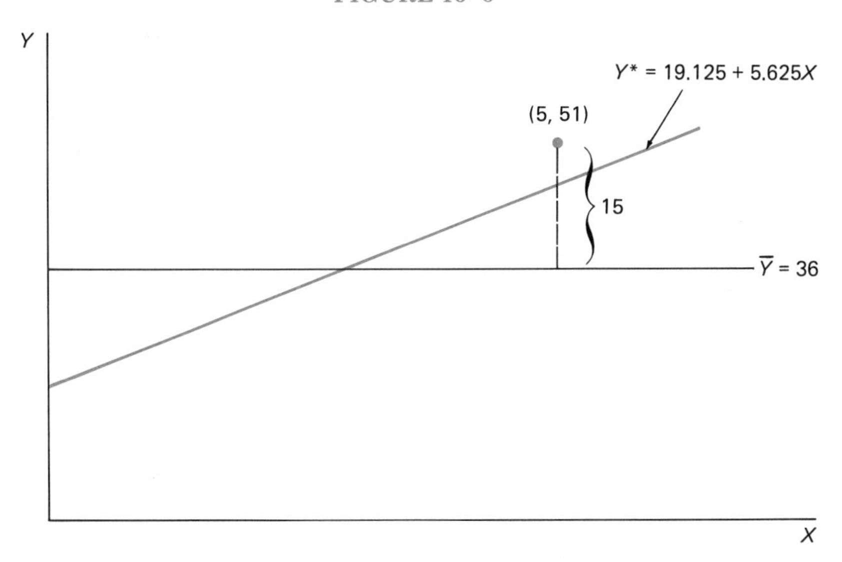

FIGURE 16–7

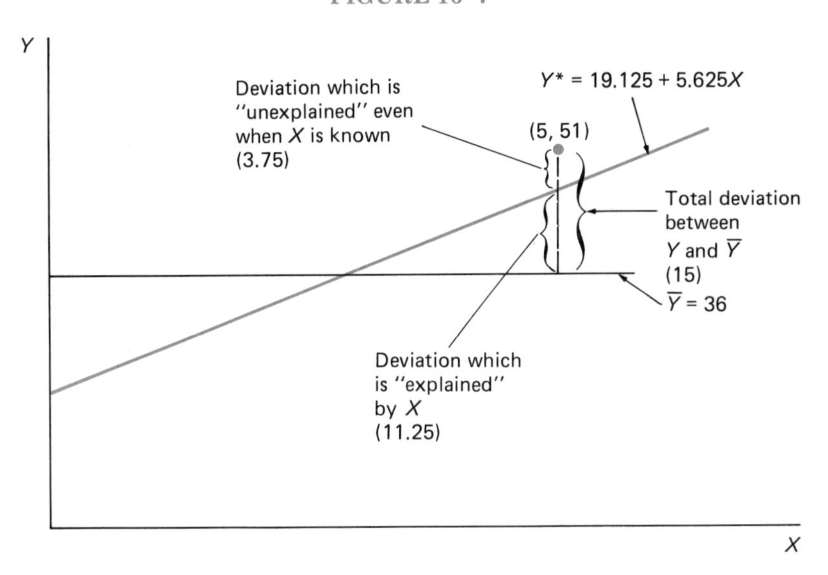

Note that

$$(Y - \overline{Y}) - (Y - Y^*) = Y - \overline{Y} - Y + Y^* = (Y^* - \overline{Y})$$

As such, we can write

$$\underbrace{(Y - \overline{Y})}_{\text{total deviation}} = \underbrace{(Y - Y^*)}_{\text{unexplained deviation}} + \underbrace{(Y^* - \overline{Y})}_{\text{explained deviation}} \qquad (16\text{–}11)$$

If the regression line is close to all of the sample pairs, the n unexplained deviations will be small. Because $\Sigma(Y - Y^*)$ is always zero, though, we can't use this summation to establish whether our line is good.

Suppose that we square both sides of Equation 16–11 and sum the squares for each Y:

$$\Sigma(Y - \overline{Y})^2 = \Sigma[(Y - Y^*) + (Y^* - \overline{Y})]^2$$

Whenever Y^* is determined by the regression technique, the last equality will simplify to

$$\underbrace{\Sigma(Y - \overline{Y})^2}_{\substack{\text{total sum}\\ \text{of squares}\\ (SS_T)}} = \underbrace{\Sigma(Y - Y^*)^2}_{\substack{\text{error sum}\\ \text{of squares}\\ (SS_E)}} + \underbrace{\Sigma(Y^* - \overline{Y})^2}_{\substack{\text{regression sum}\\ \text{of squares}\\ (SS_R)}} \qquad (16\text{–}12)$$

Equation 16–12 is important for two reasons:

1. The coefficient of determination is defined as

$$r^2 = 1 - \frac{\Sigma(Y - Y^*)^2}{\Sigma(Y - \overline{Y})^2} = 1 - \frac{SS_E}{SS_T} \qquad (16\text{–}13)$$

   If the regression line goes through every sample point, $\Sigma(Y - Y^*)^2$ will equal zero, and r^2 will equal one. Otherwise, r^2 will be less than one (i.e., $0 \leq r^2 \leq 1$).
2. Sums of squares are included in the F test constructed for regression inferences. This test will be studied in Chapter Seventeen.

The proof of Equation 16–12 relies on the normal equations. After squaring each side of Equation 16–11 and adding the squares for every Y, we have

$$\Sigma(Y - \overline{Y})^2 = \Sigma(Y - Y^*)^2 + 2\Sigma(Y - Y^*)(Y^* - \overline{Y}) + \Sigma(Y^* - \overline{Y})^2$$

We must show that $\Sigma(Y - Y^*)(Y^* - \overline{Y})$ is zero. Because

$$\Sigma(YY^* - Y\overline{Y} - Y^*Y^* + Y^*\overline{Y}) = \Sigma(YY^* - Y^{*2}) - \overline{Y}\Sigma(Y - Y^*),$$

and $\Sigma(Y - Y^*) = 0$, we have to confirm that $\Sigma(YY^* - Y^{*2}) = 0$.

We know that

$$\Sigma(YY^* - Y^{*2}) = \Sigma YY^* - \Sigma Y^{*2}$$

Moreover,

$$\Sigma YY^* = \Sigma Y(a + bX) = \Sigma aY + \Sigma bXY,$$

$$\Sigma Y^{*2} = \Sigma(a + bX)^2 = \Sigma(a^2 + 2abX + b^2X^2),$$

and

$$\begin{aligned}\Sigma YY^* - \Sigma Y^{*2} &= \Sigma aY + \Sigma bXY - \Sigma a^2 - \Sigma 2abX - \Sigma b^2X^2\\ &= a\Sigma Y + b\Sigma XY - a^2n - 2ab\Sigma X - b^2\Sigma X^2\\ &= a(\Sigma Y - an) + b(\Sigma XY - b\Sigma X^2) - 2ab\Sigma X\end{aligned}$$

If we substitute

$$b\Sigma X = \Sigma Y - an \quad \text{By virtue of Equation 16–4}$$

and

$$a\Sigma X = \Sigma XY - b\Sigma X^2 \quad \text{By virtue of Equation 16–5}$$

into the above, we find that

$$\begin{aligned}\Sigma YY^* - \Sigma Y^{*2} &= a(b\Sigma X) + b(a\Sigma X) - 2ab\Sigma X\\ &= ab\Sigma X + ab\Sigma X - 2ab\Sigma X = 0\end{aligned}$$

We can calculate r^2 using either Equation 16–13 or

$$r^2 = \frac{a\Sigma Y + b\Sigma XY - n\overline{Y}^2}{\Sigma Y^2 - n\overline{Y}^2} \tag{16–14}$$

All of the necessary summations are contained in Table 16–8. Due to the fact that

$$r^2 = \frac{(19.125)(216) + (5.625)(738) - (6)(36)^2}{8{,}332 - (6)(36)^2} = .91,$$

the regression line is said to "explain" 91 percent of Y's movement around $\overline{Y}$.

We might argue that X doesn't help us estimate Y if r^2 is low. Recall, however, that we are viewing the relationship between Y and X as linear. Such an assumption could be false. Consider the sample in Table 16–9. Table 16–10 verifies that the regression line for this sample is $Y^* = 35 + 0X = 35 = \overline{Y}$. Because Y^* and $\overline{Y}$ are the same, $\Sigma(Y - \overline{Y})^2 = \Sigma(Y - Y^*)^2$, and r^2 is 0.

TABLE 16–9

Y	X
40	0
35	1
32	2
31	3
32	4
35	5
40	6

TABLE 16–10

Y	$\overline{Y}$	$Y - \overline{Y}$	X	$\overline{X}$	$X - \overline{X}$	$(X - \overline{X})^2$	$(X - \overline{X})(Y - \overline{Y})$
40	35	5	0	3	−3	9	−15
35	35	0	1	3	−2	4	0
32	35	−3	2	3	−1	1	3
31	35	−4	3	3	0	0	0
32	35	−3	4	3	1	1	−3
35	35	0	5	3	2	4	0
40	35	5	6	3	3	9	15
245						28	0

$$b = 0/28 = 0$$

$$a = \overline{Y} - b\overline{X} = 35 - (0)(3) = 35$$

Every sample point, though, lies on the curve $Y^* = 40 - 6X + X^2$ (see Figure 16–8). Given this equation and a value of X, we can estimate Y *exactly*. We should conclude, therefore, that r^2 measures only the *linear association* between the sample Y and X values.

FIGURE 16–8

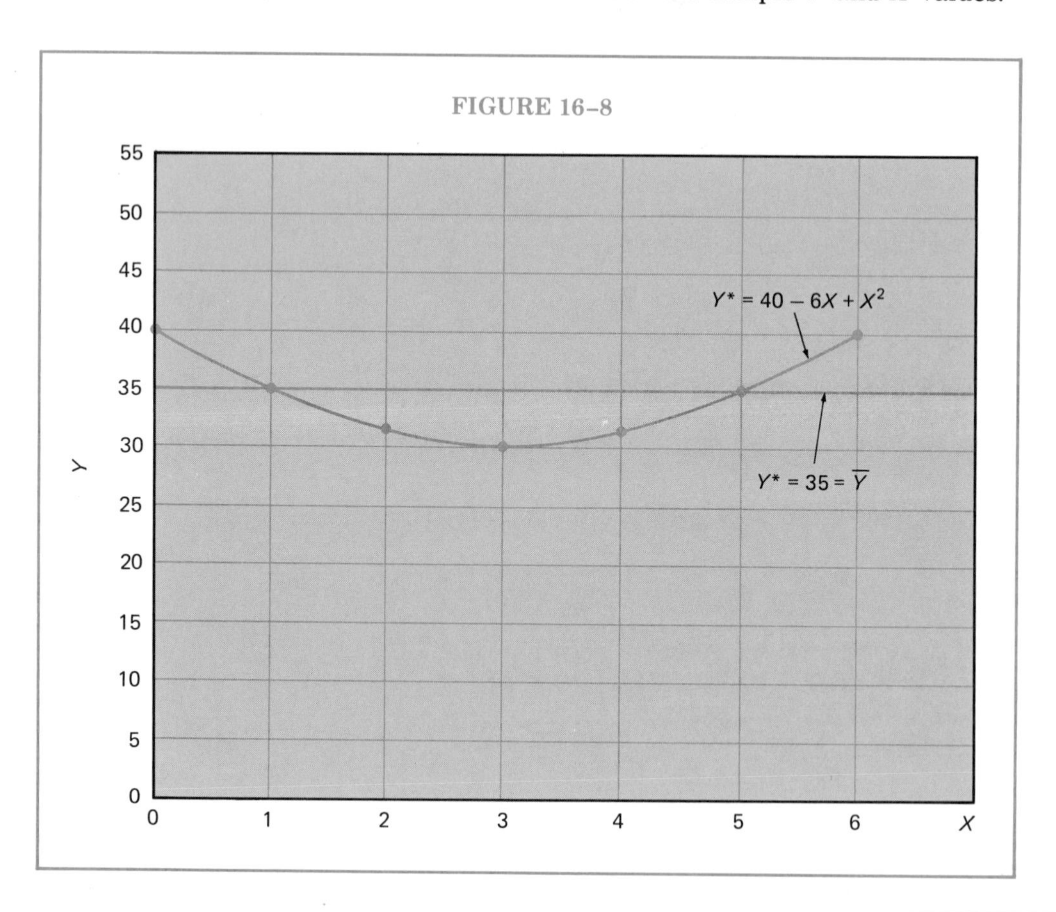

SUMMARY

Linear regression is used extensively in business and economics to describe how variables are related. This chapter has focused on the rationale and procedure for finding a least-squares line. Whereas researchers identify relationships quantitatively, the average person may be content to establish only that Y increases as X decreases (or some other consistent relationship). Thus, Ms. Jones might add more to her checking account in December, January, and February because she finds that

Decrease in X = temperature → Increase in heat for her home → Increase in energy consumption → Increase in Y = utilities bill

In the next chapter, we'll examine the way in which confidence intervals and hypothesis tests can be developed in connection with $Y^* = a + bX$.

EXERCISES

1. Show that Equation 16–6 derives from Equation 16–4.

2. Given Equations 16–3 and 16–4, show that the point $\overline{X},\overline{Y}$ always lies on the sample regression line.

3. Identify the sample regression line when the data are

Y	X
1.5	1.6
1.9	2.3
1.2	1.4
3.6	3.9
4.8	4.2
9.3	8.8
6.7	7.5
2.4	3.1

4. For the following sample, compute $Y^* = a + bX$.

Y	X
8	1
6	3
7	3
10	2
5	4
6	3
4	4
9	2

5. Based on the sample regression line in question 3, what is r^2?

6. Using the sample regression line in question 4, show that $SS_T = SS_E + SS_R$.

7. Suppose a researcher calculates a sample regression line when Y is measured in feet and X is measured in ounces.

a. What will happen to a and b if Y is converted to yards?
b. What will happen to a and b if Y is converted to inches?
c. Given that Y is in feet, what will happen to a and b if X is converted to pounds?

8. Company W intends to bid on a particular government contract. In preparing its bid, W must determine the relationship between Y = cost of production and X = number of units produced. Information from a sample of production runs is detailed below.

Production Run	Y	X
#2187	$223	30
#3043	147	18
#0765	482	60
#1184	302	43
#3240	368	49
#2379	291	41

a. Calculate the sample regression line.
b. What is the estimated average of the $X = 35$ subpopulation?
c. What is r^2?

9. An economist is studying the consumption habits of urban families having three members. She is interested in determining the relationship between Y = 1979 food expenditures (in $ thousands) and X = 1979 family income (in $ thousands). Based on the sample below,

Family	Y	X
A	2.2	6.5
B	5.1	18.2
C	9.3	29.5
D	7.5	22.0
E	4.2	12.8
F	3.9	9.2
G	8.6	20.7
H	6.2	14.6

a. What is the sample regression line?
b. What is the estimated average of the $X = 18.3$ subpopulation?
c. What is the estimated average of the $X = 14.9$ subpopulation?

10. Given the sample regression line in question 9, show that $\Sigma e = 0$.

11. Mr. Bailey uses a certain chemical mixture to improve the soil on his farm. He would like to estimate the relationship between Y = number of sacks of potatoes harvested on an acre and X = number of bags of mixture per acre. Calculate the sample regression line associated with the following data:

Y	X
15	2.0
21	3.0
9	1.0
17	2.5
11	1.5
27	4.0
23	3.5

12. Using the sample regression line in question 11, show that (a) $\Sigma Y = \Sigma Y^*$ and (b) $\Sigma(Xe) = 0$.

13. A financial analyst wants to identify the relationship between Y = retained earnings (in \$ billions) and X = after-tax profits (in \$ billions) of all manufacturing corporations in the United States. Based on the following data,

Year	Y	X
1971	16	31
1972	20	36
1973	30	48
1974	39	59
1975	29	49
1976	42	64

Source: *Statistical Abstract of the United States, 1977*, p. 797.

a. Calculate $Y^* = a + bX$.
b. What is the estimated level of retained earnings when $X = 40$?

14. The analyst in question 13 is also interested in determining, for all U.S. manufacturers, the relationship between Y = profits (in \$ billions) and X = sales (in \$ tens of billions). Based on the information below,

Year	Y	X
1971	55	75.1
1972	66	85.0
1973	86	101.7
1974	84	106.1
1975	77	106.5
1976	97	120.3

Source: *Statistical Abstract of the United States, 1977*, p. 797.

a. Graph the X,Y pairs.
b. Calculate $Y^* = a + bX$.
c. Plot $Y^* = a + bX$ on the graph derived in (a).
d. What is the estimated level of profits when $X = 100$?

15. Suppose that a researcher wishes to determine the relationship between Y = total department store sales (in $ billions) and X = total personal income (in $ tens of billions) in the United States. Given the following data,

Year	Y	X
1972	51.1	94.3
1973	52.3	105.2
1974	55.9	115.3
1975	60.7	125.0
1976	68.0	137.5

Source: *Statistical Abstract of the United States, 1977,* pp. 435 and 832.

a. Calculate the sample regression line.
b. What is the estimated level of sales when personal income is 1 trillion dollars?

Linear Regression Analysis: *Inferential Aspects*

CHARACTERISTICS OF THE SAMPLING DISTRIBUTIONS OF *a* AND *b*

We have previously indicated the relationship between Y and X as $Y_i = A + BX_i + \epsilon_i$, where X is known to the researcher before the experiment and ϵ is a random variable. When the following assumptions can be made, the least-squares estimators of A and B have a variety of important properties:

1. The expected value of ϵ in a particular subpopulation is zero [i.e., $E(\epsilon) = 0$]. Since, for any subpopulation, $A + BX$ is constant, this implies

$$\begin{aligned} E(Y|X) &= E(A + BX + \epsilon) \\ &= E(A + BX) + E(\epsilon) \\ &= E(A + BX) + 0 \\ &= A + BX + 0 \end{aligned}$$

or

$$\mu_{y|x} = A + BX$$

2. The variances, $\sigma^2_{y|x}$s, of the subpopulations are identical.
3. X is nonrandom and independent of ϵ.
4. Each ϵ_i is independent of the other ϵ_is. As such, each Y_i is independent of the other Y_is.
5. The subpopulations are normally distributed.

These assumptions can be tested to determine whether they appear reasonable in the context of a specific regression problem. Researchers often study the *residuals.* Consider assumption #2. Inasmuch as

$$\begin{aligned} V(Y|X) \text{ or } \sigma^2_{y|x} &= V(A + BX + \epsilon) \\ &= V(A + BX) + V(\epsilon) \\ &= 0 + V(\epsilon), \end{aligned}$$

the variance of a given subpopulation is equivalent to the variance of the error terms for that subpopulation. Hence, assumption #2 could be revised to: The error variances, $V(\epsilon)$s, of the subpopulations are identical.

The residuals produced by the regression line are estimates of the ϵ_is. If the homoscedasticity assumption is true, and we plot e_i with X_i, we would expect to see roughly the same distribution of e for each X, as in Figure 17–1. In Figure 17–2, however, the dispersion of the residuals increases as X increases. Under these circumstances, assumption #2 would seem doubtful.

FIGURE 17–1

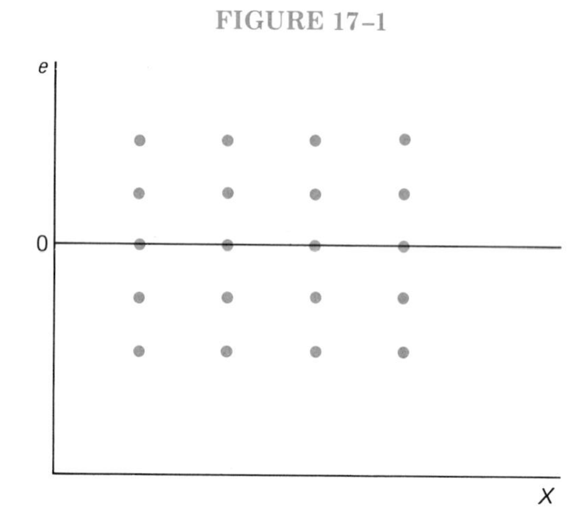

We shall now examine how the assumptions establish the characteristics of the least-squares estimators. To do this, we introduce the Gauss-Markov theorem:

> When assumptions #1, #2, #3, and #4 are true, a and b are *best linear unbiased estimators,* or BLUE, of A and B.

An estimator is BLUE if it meets the following criteria:

1. It is a linear combination of the sample observations.
2. It has an expected value equal to the population parameter θ.
3. Its variance is smaller than that of any other linear and unbiased estimator of θ.

FIGURE 17–2

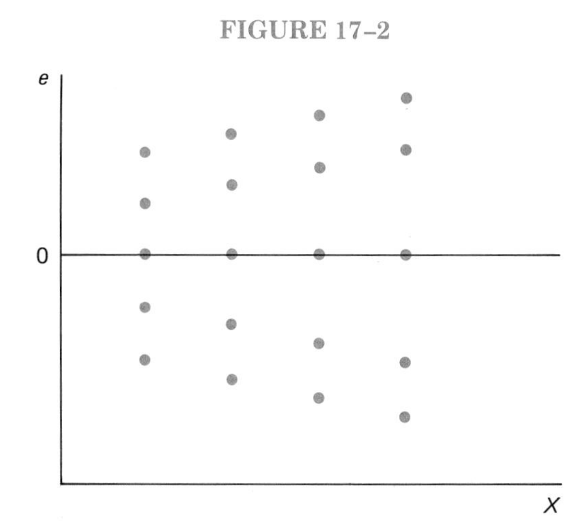

The sample average, for instance, is the best linear unbiased estimator of μ.

To derive the sampling distribution of b, we'll imagine a situation in which samples of X,Y pairs are repeatedly chosen from the same subpopulations. Only the Ys will differ from sample to sample. Since we wish to express b as a linear combination of Y, we have to rewrite Equation 16–7:

$$b = \frac{\Sigma(X - \overline{X})(Y - \overline{Y})}{\Sigma(X - \overline{X})^2} = \frac{\Sigma[(X - \overline{X})Y - (X - \overline{X})\overline{Y}]}{\Sigma(X - \overline{X})^2}$$

$$= \frac{\Sigma(X - \overline{X})Y - \Sigma(X - \overline{X})\overline{Y}}{\Sigma(X - \overline{X})^2} = \frac{\Sigma(X - \overline{X})Y - \overline{Y}\Sigma(X - \overline{X})}{\Sigma(X - \overline{X})^2}$$

Because $\Sigma(X - \overline{X}) = 0$, the last ratio becomes

$$b = \frac{\Sigma(X - \overline{X})Y}{\Sigma(X - \overline{X})^2},$$

that is,

$$b = \frac{(X_1 - \overline{X})Y_1}{\Sigma(X - \overline{X})^2} + \frac{(X_2 - \overline{X})Y_2}{\Sigma(X - \overline{X})^2} + \ldots + \frac{(X_n - \overline{X})Y_n}{\Sigma(X - \overline{X})^2}$$

Suppose that w_i is set equal to

$$\frac{(X_i - \overline{X})}{\Sigma(X - \overline{X})^2}$$

Then,

$$b = w_1Y_1 + w_2Y_2 + \ldots + w_nY_n \tag{17–1}$$

Due to the way every sample is selected, the Xs and therefore $\overline{X}$ are constant. [Note: With respect to the sampling procedure in Chapter Sixteen, we would always have $X = 1, 1, 3, 3, 5, 5$ and $\overline{X} = 3$. As such, $w_1 = (X_1 - \overline{X})/\Sigma(X - \overline{X})^2 = (1 - 3)/16 = -2/16$.] Since $w_1, w_2, \ldots, w_n$ are constants themselves, Equation 17–1 shows that b is a linear combination of Y.

The expected value of b is

$$\begin{aligned} E(b) &= E(w_1Y_1) + E(w_2Y_2) + \ldots + E(w_nY_n) \\ &= w_1E(Y_1) + w_2E(Y_2) + \ldots + w_nE(Y_n) \\ &= \Sigma w_iE(Y_i) \end{aligned}$$

From assumption #1, we know $E(Y|X) = A + BX$. Hence,

$$\begin{aligned} E(b) &= \Sigma w_i(A + BX_i) = \Sigma w_iA + \Sigma w_iBX_i \\ &= A\Sigma w_i + B\Sigma w_iX_i \end{aligned} \tag{17–2}$$

Note that

$$\Sigma w_i = \frac{\Sigma(X_i - \overline{X})}{\Sigma(X - \overline{X})^2} = \frac{0}{\Sigma(X - \overline{X})^2} = 0$$

and

$$\Sigma w_iX_i = \frac{\Sigma(X_i - \overline{X})X_i}{\Sigma(X - \overline{X})^2} = \frac{\Sigma(X_i - \overline{X})(X_i - \overline{X})}{\Sigma(X - \overline{X})^2} = 1$$

Equality 17–2, therefore, simplifies to

$$E(b) = A(0) + B(1) = B \tag{17–3}$$

Based on Equation 17–1,

$$V(b) = V(w_1Y_1 + w_2Y_2 + \ldots + w_nY_n)$$

If the Y observations are independent (assumption #4), the variance of the sum of the w_iY_i terms is equal to the sum of the variances of every w_iY_i:

$$V(b) = w_1^2V(Y_1) + w_2^2V(Y_2) + \ldots + w_n^2V(Y_n)$$

By virtue of Equation 5–17

Moreover, when the subpopulation variances, $V(Y|X)$ or $\sigma^2_{y|x}$, are identical (assumption #2),

$$\begin{aligned} V(b) &= w_1^2\sigma^2_{y|x} + w_2^2\sigma^2_{y|x} + \ldots + w_n^2\sigma^2_{y|x} \\ &= \Sigma w_i^2\sigma^2_{y|x} = \sigma^2_{y|x}\Sigma w_i^2 \end{aligned}$$

Inasmuch as

$$\Sigma w_i^2 = \frac{\Sigma(X - \overline{X})^2}{[\Sigma(X - \overline{X})^2]^2} = \frac{1}{\Sigma(X - \overline{X})^2},$$

we find that

$$V(b) = \frac{\sigma^2_{y|x}}{\Sigma(X - \overline{X})^2} \tag{17-4}$$

This variance is the smallest of the variances of the linear and unbiased estimators of B. When assumptions #2 and #4 are false, $V(b)$ will not be equal to the right side of Equation 17–4 and b will not possess the smallest variance.

The assumption of normal subpopulations is not needed to substantiate the BLUE properties of a and b. This assumption, however, is relevant for testing hypotheses about A and B. If the subpopulations are normal, as in Figure 17–3, a and b will be normal since they are linear combinations of normal Ys. The sampling distribution of b is presented in Figure 17–4.

FIGURE 17–3

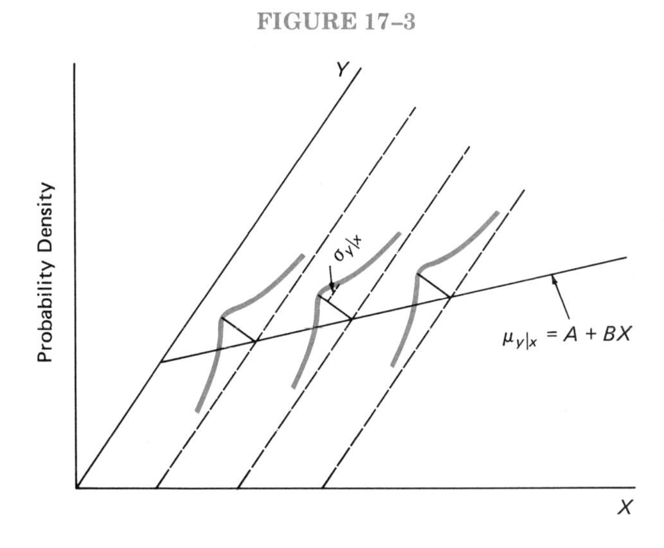

Since the parameters of the a distribution are used in other parts of the chapter, we shall mention that

$$E(a) = A \tag{17-5}$$

$$V(a) = \sigma^2_{y|x}\left[(1/n) + \frac{\overline{X}^2}{\Sigma(X - \overline{X})^2}\right] \tag{17-6}$$

FIGURE 17–4

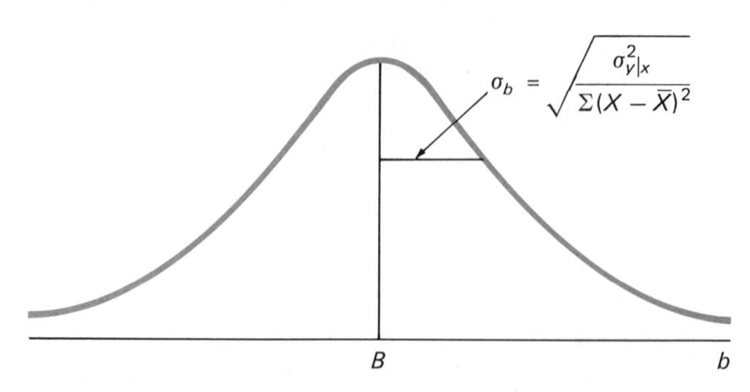

INFERENCES INVOLVING *b*

If the subpopulations are normal and we replace $\sigma^2_{y|x}$ with $s^2_{y|x}$, the $100(1 - \alpha)$ percent confidence interval estimator of B becomes

$$b - t^*_{(n-2)}s_b < B < b + t^*_{(n-2)}s_b \qquad (17\text{–}7)$$

where $t^*_{(n-2)}$ corresponds to an area of $(1 - \alpha)/2$ and

$$s_b = \sqrt{\frac{s^2_{y|x}}{\Sigma(X - \overline{X})^2}} \qquad (17\text{–}8)$$

The following example illustrates the way inferences are made in regression.

The S. L. Company specializes in constructing tennis courts, all of which are identical. However, the company's owner is often forced to change the size of her work crew. Because of vacations, illnesses, or obligations on other projects, she may assign only one or two laborers to a court. The largest crew consists of eight workers. For future scheduling purposes, the owner wants to determine how the size of a crew influences construction time. In one set of company records, past jobs are filed according to crew size (i.e., 1, 2, . . . , 8). She selects two jobs from each of the $X = 2$, 4, and 6 "subpopulations" and compiles the data in Table 17–1.

TABLE 17–1

Job	Y *Duration of Job (in hours)*	X *Number of Workers*
A	58	2
B	54	2
C	39	4
D	33	4
E	21	6
F	18	6

Using the regression technique, we get

$$Y^* = 73.667 - 9.125X$$

$$s^2_{y|x} = 8.65$$

$$s_b = \sqrt{8.65/16} = .735$$

The 95 percent confidence interval estimate of B is, then,

$$-9.125 - 2.776(.735) < B < -9.125 + 2.776(.735)$$

$$-11.2 < B < -7.1$$

The owner can be 95 percent confident that each additional worker will lower the average construction time by somewhere between 7 and 11 hours.

Since the length of the confidence interval is affected by the magnitude of s_b, researchers should plan to choose widely different Xs. This will increase $\Sigma(X - \overline{X})^2$ and reduce s_b, thereby increasing the precision of the estimate of B.

We can also use b to test hypotheses about B. The most popular null hypothesis is that $B = 0$. When H_0 is true, $\mu_{y|x} = A + (0)X = A$, and X won't "explain" Y. Given

$$H_0\colon B = 0$$

$$H_1\colon B \neq 0,$$

experimenters intend to investigate whether the value of b occurred because of chance (i.e., b is not significantly different from 0) or because Y and X are linearly related.

Since

$$t_{(n-2)} = \frac{b - B_0}{s_b}, \qquad (17\text{–}9)$$

that is, the ratio on the right has a t distribution with $n - 2$ degrees of freedom, a test of H_0 for the construction problem (given $\alpha = .05$) would involve the acceptance and rejection regions in Figure 17–5. Because

$$t = \frac{-9.125 - 0}{.735} = -12.4,$$

the owner of the S. L. Company would reject the null hypothesis and conclude that the duration of a job is related to the size of the crew.

Tests of B are frequently one-sided. The contractor, for instance, would certainly feel that increases in X should decrease Y or, equiv-

FIGURE 17–5

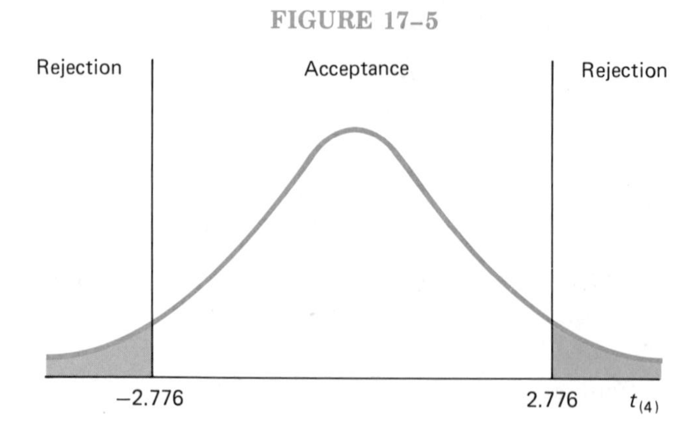

alently, that B is negative. Hence, instead of studying H_0: $B = 0$, she would form

$$H_0: B \geq 0$$

$$H_1: B < 0$$

The F test is an alternative to the two-tailed t test. Recall that

$$s^2_{y|x} = \frac{\Sigma(Y - Y^*)^2}{n - 2} = \frac{SS_E}{n - 2}$$

is an unbiased estimator of $\sigma^2_{y|x}$. This sample variance is unbiased even when $B = 0$. We will state that $\Sigma(Y^* - \overline{Y})^2$, or SS_R, is associated with 1 degree of freedom. If $B = 0$,

$$\frac{\Sigma(Y^* - \overline{Y})^2}{1} \text{ or } \frac{SS_R}{1}$$

is also an unbiased estimator of $\sigma^2_{y|x}$. Otherwise, the expected value of $SS_R/1$ is greater than $\sigma^2_{y|x}$. As long as H_0: $B = 0$ is true,

$$F_{(1,\, n\, -\, 2)} = \frac{\dfrac{SS_R}{1}}{\dfrac{SS_E}{n - 2}} \tag{17–10}$$

and we can refer to the F distribution having 1 and $n - 2$ degrees of freedom to determine whether the right side of Equation 17–10 is "very large." Under such circumstances, we would reject the null hypothesis.

Table 17–2 displays a typical computer print-out for a regression program, on which we have superimposed our symbols. (All print-outs in the text are based on the format used in the IBM 1130 Scientific Subroutine Package.) The information needed for the F

TABLE 17–2

	Variable No.	Mean	Standard Deviation	Correlation X vs Y (DISCUSSED IN CHAP. EIGHTEEN)	Regression Coefficient (b)	Std. Error of Reg. Coef. (s_b)	Computed t Value ($\frac{b-0}{s_b}$)
X	2	4.00000	1.78885	−0.98726	−9.12500	0.73509	−12.41344
	Dependent						
Y	1	37.16667	16.53380				

a	Intercept	73.66667
r^2	Coef. of Det.	0.97469
$s_{y\|x}$	Std. Error of Estimate	2.94037

Analysis of Variance for the Regression

Source of Variation	Degrees of Freedom	Sum of Squares		Mean Squares ($\frac{SS}{\text{DEG. OF FR.}}$)	F Value
Attributable to Regression	1	1332.25024	SS_R	1332.25024	154.09194
Deviation from Regression	4	34.58325	SS_E	8.64581	
Total	5	1366.83349	SS_T		

Table of Residuals

Case No. (OBSERVATIONS)	Y Value	Y Estimate (Y^*)	Residual (e)
1	58.00000	55.41666	2.58334
2	54.00000	55.41666	−1.41665
3	39.00000	37.16666	1.83334
4	33.00000	37.16666	−4.16665
5	21.00000	18.91666	2.08334
6	18.00000	18.91666	−0.91665

test is presented in the "analysis of variance for the regression" section. Letting $\alpha = .05$, the critical value of F is 7.71. Due to the fact that the computed F is 154.09, $B = 0$ seems doubtful. We will, therefore, reject H_0.

Hypothesis tests of A are based on

$$t_{(n-2)} = \frac{a - A_0}{s_a} \tag{17–11}$$

where s_a is obtained by substituting $s^2_{y|x}$ into Equation 17–6 and, thereafter, finding the square root. If H_0: $A = 0$ is true, the population regression line goes through the origin or (0,0) point on a graph; that is, $\mu_{y|x} = 0 + BX = BX$.

INFERENCES INVOLVING Y^*

We have emphasized that a and b will vary from sample to sample. Suppose that X_g is a particular value of X. Then, $Y^* = a + bX_g$ is the estimator of $\mu_{y|x_g}$. Figure 17–6 shows that different regression lines will produce different estimates of such a subpopulation average. We want to construct the sampling distribution of Y^* given X_g.

FIGURE 17–6

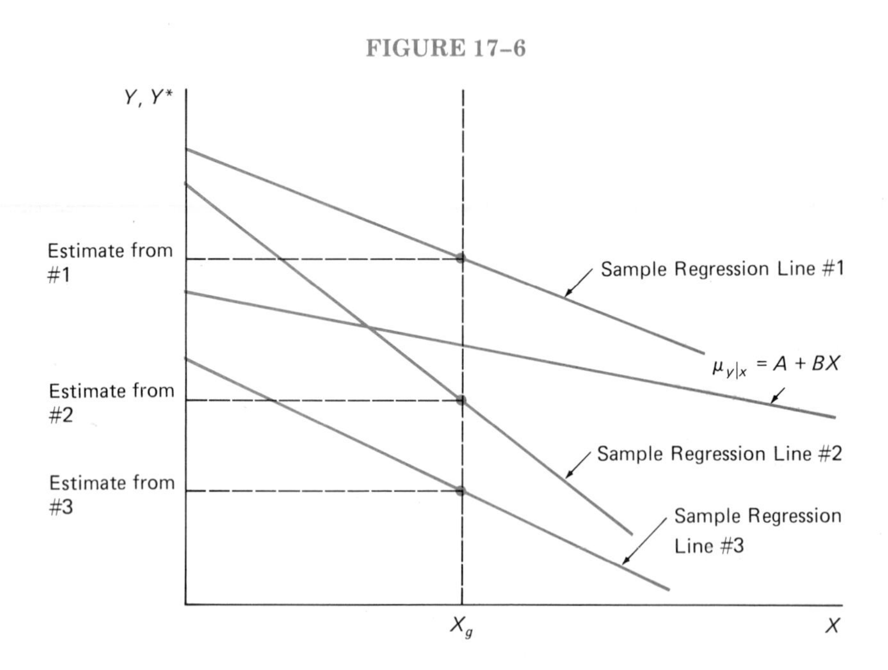

To begin with,

$$E(Y^*|X_g) = E(a + bX_g)$$
$$= E(a) + E(bX_g) \qquad \text{By virtue of Equation 5–14}$$
$$= E(a) + X_gE(b) \qquad \text{By virtue of Equation 5–7}$$
$$= A + BX_g \qquad \text{By virtue of Equations 17–3 and 17–5}$$

Y^* is, thus, an unbiased estimator of the mean of the X_g subpopulation.

We can derive $V(Y^*|X_g)$ if we substitute the right side of $a = \overline{Y} - b\overline{X}$ into $Y^* = a + bX_g$:

$$Y^* = (\overline{Y} - b\overline{X}) + bX_g$$
$$= \overline{Y} + b(X_g - \overline{X})$$

At this stage, we will state, but not prove, that $\overline{Y}$ and b are independent. As such,

$$\begin{aligned} V(Y^*|X_g) &= V[\overline{Y} + b(X_g - \overline{X})] \\ &= V(\overline{Y}) + V[b(X_g - \overline{X})] \\ &= V(\overline{Y}) + (X_g - \overline{X})^2 V(b) \quad \text{By virtue of Equation 5–17} \end{aligned}$$

Additionally, since the Ys are independent and the subpopulation variances are equal to one another,

$$\begin{aligned} V(\overline{Y}) &= V[(1/n)(Y_1 + Y_2 + \ldots + Y_n)] \\ &= (1/n)^2(\sigma^2_{y|x} + \sigma^2_{y|x} + \ldots + \sigma^2_{y|x}) \end{aligned}$$

By virtue of Equation 5–17

$$= (1/n)^2[(n)(\sigma^2_{y|x})] = \frac{\sigma^2_{y|x}}{n}$$

The variance of the estimates of $\mu_{y|x_g}$ may now be written as

$$V(Y^*|X_g) = \sigma^2_{y^*|x_g} = \sigma^2_{y|x}\left[(1/n) + \frac{(X_g - \overline{X})^2}{\Sigma(X - \overline{X})^2}\right] \quad (17–12)$$

When $s^2_{y|x}$ replaces $\sigma^2_{y|x}$, we will indicate Equation 17–12 by $s^2_{y^*|x_g}$.

We see that Y^* is a linear combination of a and b. It will, then, be normal as long as a and b are normal. The $100(1 - \alpha)$ percent confidence interval estimator of $\mu_{y|x_g}$ is

$$Y^* - t^*_{(n-2)} s_{y^*|x_g} < \mu_{y|x_g} < Y^* + t^*_{(n-2)} s_{y^*|x_g} \quad (17–13)$$

Consider the following example.

The owner of the S. L. Company wishes to construct a 95 percent confidence interval estimate of the average time two workers take to complete a tennis court. This estimate will be useful to her when she prepares bids for future projects. Since $Y^* = 73.667 - 9.125X$, the point estimate of $\mu_{y|x=2}$ is $73.667 - 9.125(2) = 55.4$ hours. Inasmuch as

$$\begin{aligned} s_{y^*|x_g} &= \sqrt{(8.65)[(1/6) + (2 - 4)^2/16]} \\ &= \sqrt{3.61} = 1.9 \text{ hours,} \end{aligned}$$

the owner can be 95 percent confident that the average time lies between 50.1 and 60.7 hours; that is,

$$55.4 - 2.776(1.9) < \mu_{y|x=2} < 55.4 + 2.776(1.9)$$

$$50.1 < \mu_{y|x=2} < 60.7$$

Equality 17–12 suggests two ways in which researchers can increase the precision of their estimate of $\mu_{y|x_g}$ without selecting a larger sample. Besides choosing very different Xs, they should plan their choices so that X_g is the same as $\overline{X}$. Both of these actions will reduce the second term in the brackets.

Y^* is not only an estimator of $\mu_{y|x_g}$; it also enables a researcher to predict the value of a new Y drawn from the X_g subpopulation. We should expect a $100(1 - \alpha)$ percent confidence interval estimate of $\mu_{y|x_g}$ to be smaller than a $100(1 - \alpha)$ percent confidence interval estimate of a particular Y since the latter has to incorporate the variation of Y^* around $\mu_{y|x_g}$ as well as the variation of Y around $\mu_{y|x_g}$.

The variance of the estimates of individual or *future* Ys is

$$\sigma^2_{y^*_f|x_g} = \sigma^2_{y^*|x_g} + \sigma^2_{y|x}$$

$$= \sigma^2_{y|x}\left[1 + (1/n) + \frac{(X_g - \overline{X})^2}{\Sigma(X - \overline{X})^2}\right] \qquad (17\text{–}14)$$

We'll substitute $s^2_{y|x}$ for $\sigma^2_{y|x}$. Statisticians refer to confidence intervals developed for Y as *prediction intervals*. A $100(1 - \alpha)$ percent prediction interval is constructed from

$$Y^* - t^*_{(n-2)} s_{y^*_f|x_g} < Y < Y^* + t^*_{(n-2)} s_{y^*_f|x_g} \qquad (17\text{–}15)$$

The following example concerns such intervals.

The S. L. Company is scheduled to build a court on Mr. Smith's property. Smith wants to leave town while the work is progressing and has asked the company to estimate the length of the job. S. L. is planning to assign a crew of six laborers.

When $X = 6$, $Y^* = 73.667 - 9.125(6) = 18.92$. Additionally,

$$s_{y^*_f|x_g} = \sqrt{(8.65)[1 + (1/6) + (6 - 4)^2/16]}$$

$$= \sqrt{12.26} = 3.5 \text{ hours}$$

The 95 percent prediction interval becomes

$$18.92 - 2.776(3.5) < Y < 18.92 + 2.776(3.5)$$

$$9.2 \text{ hours} < Y < 28.6 \text{ hours}$$

Since the crew will put in an 8-hour day, the company advises Smith to go on a trip of two to four days' duration.

Whether a researcher is interested in estimating $\mu_{y|x_g}$ or Y, he should be hesitant to extend his estimates to subpopulations beyond the range of Xs covered by his sample. The relationship between Y and X may be linear only within this range. For instance, theoretically, Y = time in our example might actually increase whenever the size of a crew passes a certain level. Such a situation illustrates *diminishing returns* (see Figure 17–7).

FIGURE 17–7

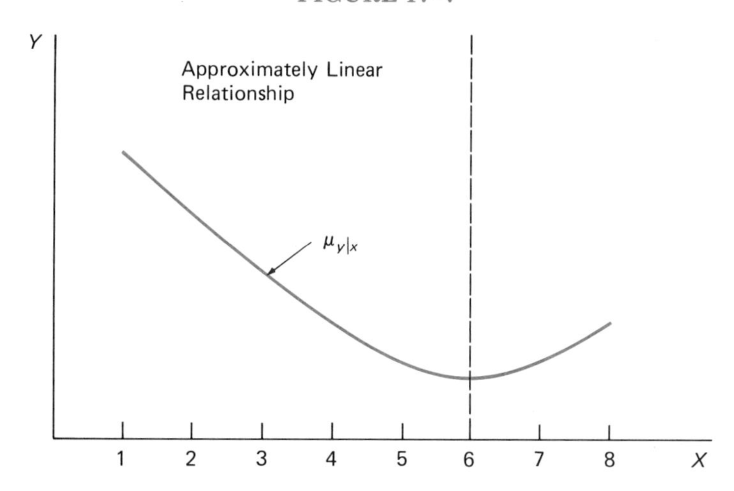

SUMMARY

Chapter Seventeen has explored the types of inferences that can be derived from a regression line. Because the properties of a and b are a consequence of some assumptions, researchers should test these assumptions before establishing hypotheses about A, B, $\mu_{y|x_g}$, and Y. We turn next to correlation and multiple regression, which will continue our study of the relationship between Y and other variables.

EXERCISES

1. Consider the following data:

Y	X
3	2
10	6
7	4
4	2
9	7
12	10
16	11

a. Using Equations 16–6 and 16–7 or Equations 16–6 and 16–8, determine the sample regression line.
b. Derive the value of b in light of Equation 17–1.

2. Suppose all subpopulation variances equal 5. A researcher selects four observations at random from the $X = 3$ subpopulation, four from the $X = 6$ subpopulation, and four from the $X = 9$ subpopulation.

a. What is $V(b)$?
b. If eight observations instead of four are chosen from each of the above subpopulations, what is $V(b)$?

3. Suppose all subpopulation variances equal 6. A researcher selects three observations at random from the $X = 2$ subpopulation, three from the $X = 4$ subpopulation, and three from the $X = 6$ subpopulation.

a. What is $V(a)$?
b. If five observations instead of three are chosen from each of the above subpopulations, what is $V(a)$?

4. Rework questions 2 and 3 under the condition that the researcher draws his observations from the $X = 3$, $X = 4$, and $X = 5$ subpopulations.

5. Part of the computer print-out for a regression problem is reproduced below ($n = 6$).

Variable No.	*Mean*	*Standard Deviation*	*Correlation X vs Y*	*Regression Coefficient*	*Std. Error of Reg. Coef.*	*Computed t Value*
2	3.50000	1.87082	0.78980	1.11428	0.43267	
Dependent						
1	6.83333	2.63944				

Intercept	2.93333
Coef. of Det.	0.62379
Std. Error of Estimate	1.81002

a. What is the equation of the sample regression line?
b. What is the computed t value?
c. Is b significantly different from 0 at the .01 level of significance?
d. What is $s_{y|x}^2$?

6. Part of the computer print-out for a regression problem is reproduced below ($F^*_{(1, 34)} \approx 4.17$).

Analysis of Variance for the Regression

Source of Variation	*Degrees of Freedom*	*Sum of Squares*	*Mean Squares*	*F Value*
Attributable to Regression	1	4010.90137	4010.90137	
Deviation from Regression	34		22.02075	
Total	35	4759.60743		

a. What is SS_E?
b. What is $s^2_{y|x}$?
c. What is the F value?
d. Is b significantly different from 0 at the .05 level of significance?
e. What is r^2?
f. How many observations were included in this regression problem?

7. Part of the computer print-out for a regression problem is reproduced below.

Table of Residuals

Case No.	*Y Value*	*Y Estimate*	*Residual*
1	2.00000	3.51950	−1.51950
2		3.79374	1.20625
3	7.00000	6.27142	0.72857
4	8.00000	8.74909	−0.74909
5	10.00000	10.72241	
6	11.00000	9.94381	1.05618

a. What is the second observation?
b. What is the fifth residual?
c. What is $s^2_{y|x}$?
d. What is SS_R?
e. What is r^2?

8. Using the information in question 5 and the fact that $n = 6$, construct (a) a 95 percent confidence interval estimate of B and (b) a 99 percent confidence interval estimate of B.

9. For scheduling purposes, a plant manager wishes to determine the relationship between Y = a worker's speed in assembling part W (in minutes) and X = a worker's length of employment with Company R (in months). Given the following data,

Employee	*Y*	*X*
A	70	6
B	54	8
C	49	10
D	45	11
E	37	12
F	30	13
G	36	14
H	27	15

a. Construct the sample regression line.
b. Construct a 95 percent confidence interval estimate of B.
c. Construct a 99 percent confidence interval estimate of B.

10. Based on the information in question 9 and the computed value of t, is H_0: $B \geq 0$ acceptable at the .05 level of significance?

11. Based on the information in question 9,

a. What is the estimated average assembly time for all workers having 12 months employment at Company R?
b. What is the estimated average assembly time for all workers having 9 months employment at Company R?
c. Construct a 95 percent confidence interval estimate of the $X = 11$ subpopulation average assembly time.

12. Suppose employee L has just been asked to assemble a part. Based on the information in question 9 and the fact that L has been employed at Company R for 13 months,

a. What is his predicted assembly time?
b. Construct a 95 percent prediction interval for his assembly time.

13. After looking over the balance sheets and income statements of many companies in industry Q, a financial analyst became interested in identifying the relationship between Y = a company's 1979 level of sales (in \$ millions) and X = a company's assets (in \$ millions). Given the data below, (a) construct the sample regression line and (b) compute r^2.

Company	Y	X
A	2.6	0.4
B	3.7	1.2
C	3.1	0.7
D	4.0	1.3
E	5.3	1.5
F	2.9	0.6

14. Using the data in question 13,

a. Construct a 90 percent confidence interval estimate of B.
b. What is the estimated average level of sales for all companies having \$1.1 million in assets?
c. Construct a 95 percent confidence interval estimate of the average level of sales for all companies having \$1,350,000 in assets.

15. Based on the data in question 13, (a) construct an analysis of variance for the regression table and (b) test

$$H_0: B = 0$$

$$H_1: B \neq 0$$

at the .05 level of significance.

Correlation and Multiple Regression Analysis

THE COEFFICIENT OF CORRELATION

Interpretation

To derive $Y^* = a + bX$, we identified a dependent and an independent variable and picked Ys randomly from particular subpopulations. In *correlation analysis,* researchers are interested in the *strength* of the linear relation between X and Y (e.g., Are "high" values of X usually paired with "high" values of Y?) rather than in the *specific way* in which X and Y are associated (i.e., What are the values of a and b?). The sampling plan for correlation makes *both* X and Y random variables. A variable, moreover, isn't categorized as dependent or independent. (Note: If correlation analysis were introduced into the contractor example in Chapter Seventeen, the experimenter would have to choose n jobs at random and *then* record the values of Y = duration of the job and X = number of workers.)

We'll establish the basis for correlation by exploring a population of $N = 10$ objects where each object possesses an X value (0, 1, or 2) and a Y value (3, 5, or 7). The sample space for "select one object at random" appears in Figure 18–1. When we graph the joint events, they all lie on a positively sloped line (see Figure 18–2). We now wish to consider the covariance between X and Y, or

$$\text{Cov}(X,Y) = \Sigma\Sigma(X - \mu_x)(Y - \mu_y)P(X,Y) \qquad (18\text{–}1)$$

Cov(X,Y) was initially discussed in Chapter Five.

By virtue of the joint probability distribution of X and Y (see Table 18–1), we find that

$$\mu_x = \Sigma XP(X) = (0)(.4) + (1)(.3) + (2)(.3) = .9$$

$$\mu_y = \Sigma YP(Y) = (3)(.4) + (5)(.3) + (7)(.3) = 4.8$$

and

$$\text{Cov}(X,Y) = (0 - .9)(3 - 4.8)(.4) + (1 - .9)(5 - 4.8)(.3) + (2 - .9)(7 - 4.8)(.3) = 1.38$$

The sign of the covariance arises from the fact that above-average Ys are paired with above-average Xs, while below-average Ys are paired with below-average Xs. Hence, $(X - \mu_x)(Y - \mu_y)$ is either $(+)(+) = +$ or $(-)(-) = +$.

FIGURE 18–1

· A $(X = 0, Y = 3)$	· B $(X = 1, Y = 5)$
· C $(X = 1, Y = 5)$	· D $(X = 1, Y = 5)$
· E $(X = 0, Y = 3)$	· F $(X = 2, Y = 7)$
· G $(X = 0, Y = 3)$	· H $(X = 2, Y = 7)$
· I $(X = 0, Y = 3)$	· J $(X = 2, Y = 7)$

FIGURE 18–2

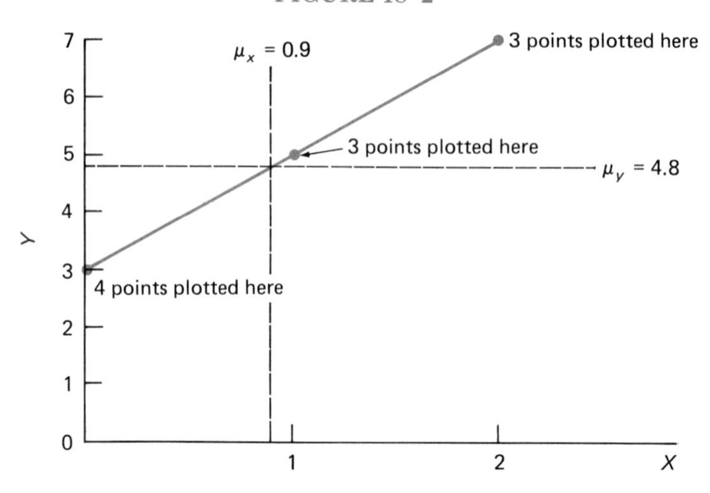

TABLE 18–1

		X			Marginal Probability of Y
		0	1	2	
	3	4/10 = .4			.4
Y	5		3/10 = .3		.3
	7			3/10 = .3	.3
Marginal Probability of X		.4	.3	.3	

By contrast, the joint events in Figure 18–3 lie on a negatively sloped line (see Figure 18–4) and have the joint probability distribution in Table 18–2. Therefore

$$\mu_x = (0)(.1) + (1)(.5) + (2)(.4) = 1.3$$

$$\mu_y = (0)(.4) + (2)(.5) + (4)(.1) = 1.4$$

and

$$\text{Cov}(X,Y) = (0 - 1.3)(4 - 1.4)(.1) + (1 - 1.3)(2 - 1.4)(.5) + (2 - 1.3)(0 - 1.4)(.4) = -.82$$

FIGURE 18–3

· A ($X = 2, Y = 0$)	· B ($X = 1, Y = 2$)
· C ($X = 1, Y = 2$)	· D ($X = 1, Y = 2$)
· E ($X = 2, Y = 0$)	· F ($X = 2, Y = 0$)
· G ($X = 0, Y = 4$)	· H ($X = 1, Y = 2$)
· I ($X = 2, Y = 0$)	· J ($X = 1, Y = 2$)

FIGURE 18–4

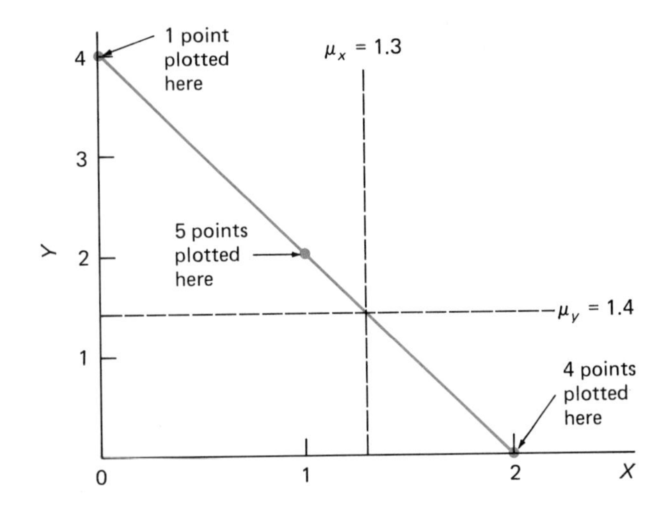

TABLE 18–2

		X			Marginal Probability of Y
		0	1	2	
	0			4/10 = .4	.4
Y	2		5/10 = .5		.5
	4	1/10 = .1			.1
Marginal Probability of X		.1	.5	.4	

The graph shows that above-average Ys are always associated with below-average Xs. Additionally, below-average Ys are paired with above-average Xs. Thus, $(X - \mu_x)(Y - \mu_y)$ is negative in every instance; that is, $(-)(+) = -$ and $(+)(-) = -$.

When X and Y are independent, $\text{Cov}(X,Y) = 0$. Since the covariance was 1.38 and $-.82$ in the previous instances, X and Y were dependent. However, to justify calling 1.38 and $-.82$ measures of *linear* dependence, we must show how a *nonlinear* association can affect the value of the covariance.

Imagine that Figure 18–5 is a sample space. Ys are paired with Xs according to the "rule" $Y = X^2 - 4X + 5$ (see Figure 18–6). Given the joint probability distribution in Table 18–3,

$$\mu_x = (1)(.3) + (2)(.4) + (3)(.3) = 2.0$$

$$\mu_y = (1)(.4) + (2)(.6) = 1.6$$

The covariance is, therefore,

$$\text{Cov}(X,Y) = (1 - 2)(2 - 1.6)(.3) + (2 - 2)(1 - 1.6)(.4) + (3 - 2)(2 - 1.6)(.3) = 0$$

FIGURE 18–5

· A ($X = 2, Y = 1$)	· B ($X = 1, Y = 2$)
· C ($X = 1, Y = 2$)	· D ($X = 3, Y = 2$)
· E ($X = 2, Y = 1$)	· F ($X = 1, Y = 2$)
· G ($X = 2, Y = 1$)	· H ($X = 3, Y = 2$)
· I ($X = 2, Y = 1$)	· J ($X = 3, Y = 2$)

FIGURE 18–6

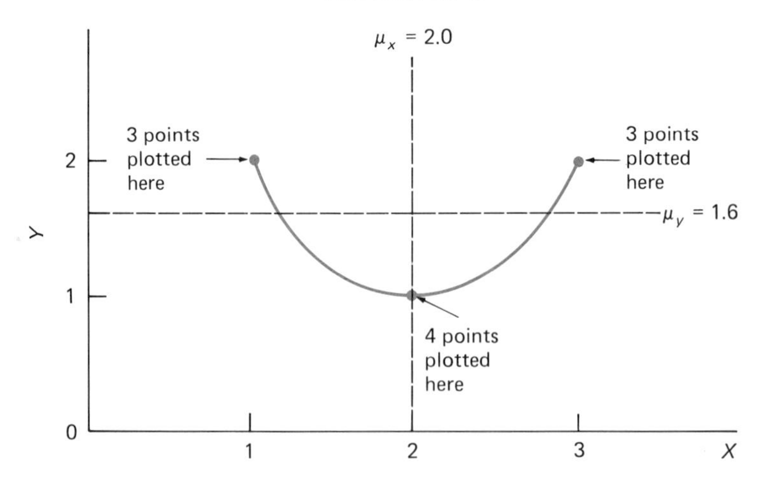

TABLE 18–3

		X			*Marginal Probability of Y*
		1	2	3	
Y	1		4/10 = .4		.4
	2	3/10 = .3		3/10 = .3	.6
Marginal Probability of X		.3	4	.3	

Despite the fact that Cov(X,Y) is zero, X and Y are, nonetheless, related to each other. Put another way, a nonlinear association can result in Cov(X,Y) = 0.

Our first two illustrations represented *perfect positive linear* dependence (i.e., all joint events were on a positively sloped line) and *perfect negative linear* dependence (i.e., all joint events were on a negatively sloped line). The sign of Cov(X,Y) indicates the direction of the line, but the absolute value of the covariance (i.e., 1.38 and .82) doesn't indicate whether every joint event lies on a line or whether a relationship is only approximately linear, as in Figure 18–7.

FIGURE 18–7

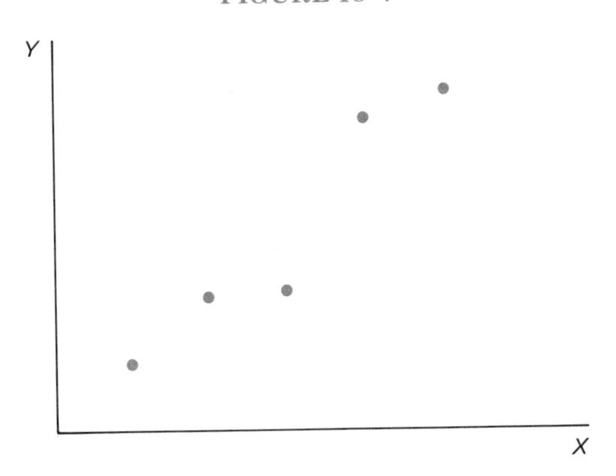

We can state that $|\text{Cov}(X,Y)| = (\sigma_x)(\sigma_y)$ when each X,Y pair is on a line. Otherwise, $|\text{Cov}(X,Y)| < (\sigma_x)(\sigma_y)$. Because of this, $\text{Cov}(X,Y)/(\sigma_x)(\sigma_y)$ will (1) equal 1.0 if X and Y have a perfect positive association, (2) equal -1.0 if X and Y have a perfect negative association, and (3) range between -1.0 and 1.0. These comments are valid irrespective of the X units or the Y units.

The above ratio is the *population coefficient of correlation* and is symbolized by the Greek letter *rho*, ρ:

$$\rho = \frac{\text{Cov}(X,Y)}{(\sigma_x)(\sigma_y)} \tag{18–2}$$

Using the joint distributions in Tables 18–1 and 18–2, we can confirm that ρ is 1.0 or -1.0 when a line passes through all joint events. Given the Table 18–1 distribution,

$$\sigma_x = \sqrt{\Sigma(X - \mu_x)^2 P(X)}$$
$$= \sqrt{(0 - .9)^2(.4) + (1 - .9)^2(.3) + (2 - .9)^2(.3)} = .831$$

$$\sigma_y = \sqrt{\Sigma(Y - \mu_y)^2 P(Y)}$$
$$= \sqrt{(3 - 4.8)^2(.4) + (5 - 4.8)^2(.3) + (7 - 4.8)^2(.3)} = 1.661$$

and

$$\rho = \frac{1.38}{(.831)(1.661)} = 1.0$$

Given the Table 18–2 distribution,

$$\sigma_x = \sqrt{(0 - 1.3)^2(.1) + (1 - 1.3)^2(.5) + (2 - 1.3)^2(.4)} = .64$$

$$\sigma_y = \sqrt{(0 - 1.4)^2(.4) + (2 - 1.4)^2(.5) + (4 - 1.4)^2(.1)} = 1.28$$

and

$$\rho = \frac{-.82}{(.64)(1.28)} = -1.0$$

Hypothesis Testing

We will assume that inferences about ρ are derived from a sample of n X,Y pairs picked from a population that has a *bivariate normal distribution*. With such a distribution, (1) every conditional distribution of Y given X or X given Y is normal, and (2) the marginal distribution of Y and the marginal distribution of X are also normal. The *sample coefficient of correlation*, or r, is

$$r = \frac{\dfrac{\Sigma(X - \overline{X})(Y - \overline{Y})}{n - 1}}{\sqrt{\dfrac{\Sigma(X - \overline{X})^2}{n - 1}}\sqrt{\dfrac{\Sigma(Y - \overline{Y})^2}{n - 1}}}$$

(numerator: sample covariance; first root: s_x; second root: s_y)

$$= \frac{\Sigma(X - \overline{X})(Y - \overline{Y})}{\sqrt{\Sigma(X - \overline{X})^2}\sqrt{\Sigma(Y - \overline{Y})^2}} \tag{18–3}$$

Both the value of ρ and the sample size influence the shape of the sampling distribution of r. If X and Y are independent, $\rho = 0$ since $\text{Cov}(X,Y) = 0$. Under these circumstances, r is distributed symmetrically around $\rho = 0$ (see Figure 18–8). Moreover,

$$t_{(n-2)} = \frac{r}{\sqrt{(1 - r^2)/(n - 2)}}, \qquad (18\text{–}4)$$

that is, the ratio on the right has a t distribution with $n - 2$ degrees of freedom. When the null hypothesis includes a ρ which isn't zero, the t distribution is not applicable. The following example provides a test of ρ.

FIGURE 18–8

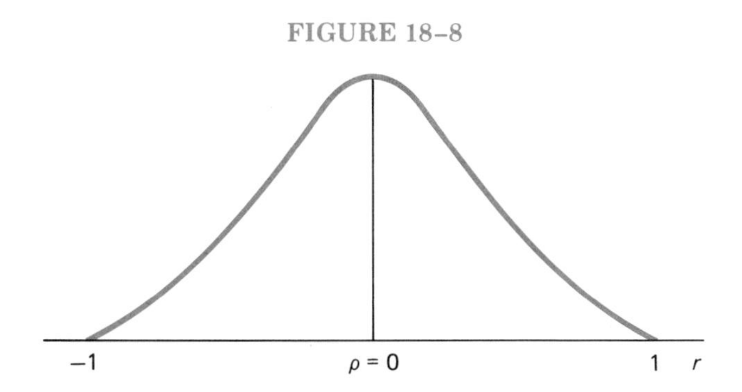

An industrial psychologist is investigating the characteristics of senior corporate executives. She believes that such people thrive on challenging managerial problems. Additionally, she feels that intelligent executives marry intelligent spouses. Ten executives are randomly selected. They and their mates are subsequently asked to complete an I.Q. test. The outcomes are presented in Table 18–4.

TABLE 18–4

Executive	X *Executive's I.Q.*	Y *Spouse's I.Q.*
A	112	116
B	123	119
C	142	136
D	125	131
E	122	117
F	157	148
G	127	136
H	118	117
I	134	130
J	120	110

When the level of significance is .05, the acceptance and rejection regions for testing

H_0: $\rho \leq 0$ (*X* and *Y* are not related or are not related positively.)

H_1: $\rho > 0$ (*X* and *Y* are related positively.)

are as shown in Figure 18–9. The relevant totals are provided in Table 18–5. Thus,

$$r = \frac{1{,}244}{\sqrt{1{,}564}\sqrt{1{,}292}} = .88$$

and

$$t = \frac{.88}{\sqrt{(1 - .88^2)/(10 - 2)}} = 5.24$$

FIGURE 18–9

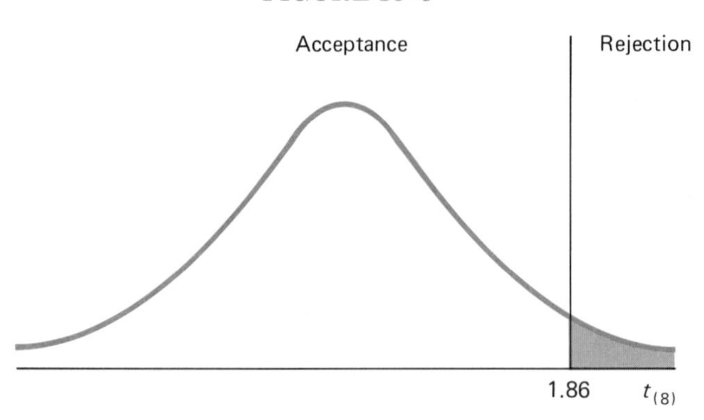

TABLE 18–5

X	$\overline{X}$	$X - \overline{X}$	$(X - \overline{X})^2$	Y	$\overline{Y}$	$Y - \overline{Y}$	$(Y - \overline{Y})^2$	$(X - \overline{X})(Y - \overline{Y})$
112	128	−16	256	116	126	−10	100	160
123	128	−5	25	119	126	−7	49	35
142	128	14	196	136	126	10	100	140
125	128	−3	9	131	126	5	25	−15
122	128	−6	36	117	126	−9	81	54
157	128	29	841	148	126	22	484	638
127	128	−1	1	136	126	10	100	−10
118	128	−10	100	117	126	−9	81	90
134	128	6	36	130	126	4	16	24
120	128	−8	64	110	126	−16	256	128
			1,564				1,292	1,244

Since 5.24 is greater than 1.86, the researcher rejects H_0 and concludes that there is a positive correlation between the I.Q.s of executives and those of their spouses.

Inferring statistical dependence implies only that X and Y are *associated.* Whether, in fact, *X causes Y* or *Y causes X* is something that experimenters have to justify in terms of theory or practical experience. Dependence frequently arises, for instance, because both X and Y are simultaneously affected by another variable.

MULTIPLE REGRESSION

Since business and economic data are generally produced by the interactions of a host of factors, researchers may want to "explain" the movements in Y by identifying more than one independent variable. If X_1 and X_2 are independent variables, we might write

$$Y_i = A + B_1X_{i1} + B_2X_{i2} + \epsilon_i \tag{18-5}$$

A particular combination of X_1 and X_2 values will, then, indicate a particular subpopulation of Ys; for example, $\mu_{y|x_1 = 3,\, x_2 = 4}$ is the average of the conditional distribution of Y given $X_1 = 3$ and $X_2 = 4$. When the expected value of the error term is zero,

$$\mu_{y|x_1, x_2} = A + B_1X_1 + B_2X_2 \tag{18-6}$$

The graph of Equation 18–6 is three dimensional and appears as a *plane.* We would like to select a sample of n Y,X_1,X_2 collections of observations and estimate Equation 18–6 with

$$Y^* = a + b_1X_1 + b_2X_2 \tag{18-7}$$

The least-squares criterion [i.e., make $\Sigma(Y - Y^*)^2$ or Σe^2 or $\Sigma(Y - a - b_1X_1 - b_2X_2)^2$ as small as possible] will be used since a, b_1, and b_2 are BLUE as long as the following conditions are met:

1. $E(\epsilon) = 0$.
2. The subpopulation variances are identical.
3. The ϵs are independent.
4. X_1 and X_2 are nonrandom or they are independent of ϵ.

We will further assume:

5. The subpopulations are normal.
6. X_1 and X_2 are not linear combinations of each other (e.g., $X_1 \neq 3X_2$).

The estimation technique is referred to as *multiple regression.*

In the light of calculus, the least-squares plane is determined by the solution of the following *normal* equations:

$$\Sigma Y = an + b_1\Sigma X_1 + b_2\Sigma X_2 \tag{18–8}$$

$$\Sigma X_1Y = a\Sigma X_1 + b_1\Sigma X_1^2 + b_2\Sigma X_1X_2 \tag{18–9}$$

$$\Sigma X_2Y = a\Sigma X_2 + b_1\Sigma X_1X_2 + b_2\Sigma X_2^2 \tag{18–10}$$

We won't find a, b_1, and b_2 manually because the arithmetic is quite lengthy. We will discuss the following example within the context of a typical computer print-out for a multiple regression program.

A marketing analyst intends to investigate the relationship between company sales (Y) and the use of TV advertising. He wishes to "explain" Y in terms of X_1 = the number of minutes a company's commercials are aired per week in "prime time" and X_2 = the number of different types of commercials aired by a company per week in "prime time." The researcher believes that both X_1 and X_2 increase sales. This implies that b_1 and b_2 should be positive. Six companies are picked. The data are displayed in Table 18–6 and generated the print-out in Table 18–7.

TABLE 18–6

Company	Y *(in \$10 millions)*	X_1	X_2
A	2	5	1
B	5	6	5
C	7	10	3
D	8	14	4
E	10	17	7
F	11	15	3

We see that $Y^* = 0.133 + 0.617X_1 + 0.038X_2$. Given all companies advertising for ten minutes and having three different commercials, the estimated sales average would be

$$Y^* = \text{estimated } \mu_{y|x_1 = 10,\, x_2 = 3} = 0.133 + 0.617(10) + 0.038(3)$$

$$= 6.417, \text{ or } \$64{,}170{,}000$$

Both b_1 and b_2 are called *net regression coefficients*. The coefficient b_1 represents the impact of a one-unit change in X_1, with X_2 *held constant,* while b_2 represents the impact of a one-unit change in X_2, with X_1 *held constant.* To clarify the distinction between b_1 and the b in a *simple,* or two-variable, regression where X_1 is the only independent variable, consider $Y^* = a + bX_1$. If X_1 and X_2 are correlated and X_2 influences Y, b will incorporate both the *direct* and the *indirect* effect of X_1 on Y (see Figure 18–10). Multiple

TABLE 18–7

	Variable No.	*Mean*	*Standard Deviation*	*Correlation X vs Y*	*Regression Coefficient*	*Std. Error of Reg. Coef.*	*Computed t Value*
X_1	2	11.16666	4.95647	0.93620	0.61676	0.16296	3.78473
X_2	3	3.83333	2.04124	0.53749	0.03813	0.39570	0.09636
					b_1 b_2	s_{b_2} s_{b_1}	
	Dependent						
Y	1	7.16666	3.31159				

a	Intercept	0.13328
R^2	Coef. of Mult. Det.	0.87686
$s_{y\|x_1, x_2}$	Std. Error of Estimate	1.50027

Analysis of Variance for the Regression

Source of Variation	*Degrees of Freedom*	*Sum of Squares*	*Mean Squares*	*F Value*
Attributable to Regression	2	48.08082 (SS_R)	24.04041	10.68069
Deviation from Regression	3	6.75250 (SS_E)	2.25083	
Total	5	54.83332 (SS_T)		

Table of Residuals

Case No.	*Y Value*	*Y Estimate*	*Residual*
1	2.00000	3.25523	−1.25523
2	5.00000	4.02454	0.97545
3	7.00000	6.41533	0.58467
4	8.00000	8.92052	−0.92052
5	10.00000	10.88521	−0.88521
6	11.00000	9.49915	1.50085

FIGURE 18–10

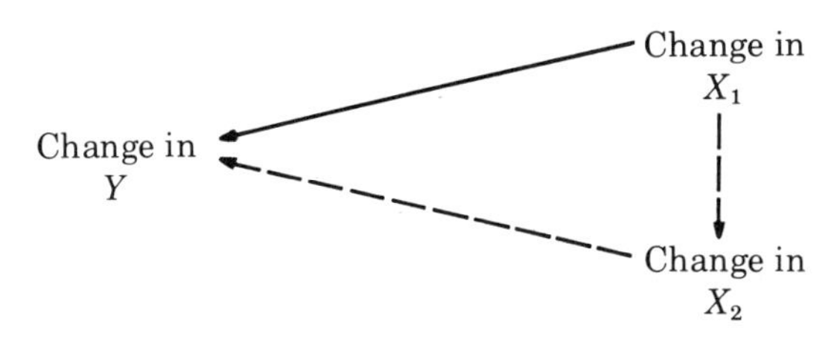

regression is, therefore, a procedure for "netting out" the indirect effect.

Suppose X_3 is another independent variable. The coefficients of X_1 in $Y^* = a + b_1X_1 + b_2X_2$ and $Y^* = a + b_1X_1 + b_3X_3$ will differ because X_2 isn't held constant in the latter regression. For the TV example, we would estimate that each one-unit increase in X_1, holding X_2 constant, increases the sales average by .617, or $6,170,000. The estimated average of the $X_1 = 6$, $X_2 = 3$ subpopulation will be .617 higher than the estimate of $\mu_{y|x_1 = 5,\, x_2 = 3}$.

We are able to measure the *fit* of the regression plane through the sample data in several ways. For instance, a researcher may be interested in the variance of Y around Y^*. For our TV example,

$$s^2_{y|x_1,x_2} = \frac{\Sigma(Y - Y^*)^2}{n - 3} = \frac{SS_E}{n - 3} \qquad (18\text{–}11)$$

$$= \frac{6.75250}{6 - 3} = 2.25$$

SS_E is associated with $n - 3$ degrees of freedom since three restrictions are now placed on the residuals. We can present these restrictions by rewriting Equations 18–8, 18–9, and 18–10 as

$$\Sigma(Y - Y^*) = 0$$

$$\Sigma X_1(Y - Y^*) = 0$$

$$\Sigma X_2(Y - Y^*) = 0$$

Like r^2, the *coefficient of multiple determination,* or R^2, is the ratio of SS_R to SS_T:

$$R^2 = \frac{SS_R}{SS_T} = 1 - \frac{SS_E}{SS_T} \qquad (18\text{–}12)$$

$$= 1 - \frac{6.75250}{54.83332} = .87686$$

X_1 and X_2 *together* "explain" almost 88 percent of the movements in Y in our example.

Researchers use the F test to explore

H_0: $B_1 = B_2 = 0$

H_1: At least one B_j ($j = 1, \ldots, m$) isn't 0

When the null hypothesis is true or, equivalently, when neither X_1 nor X_2 "explains" any part of Y,

$$F_{(k - 1,\, n - k)} = \frac{\dfrac{SS_R}{k - 1}}{\dfrac{SS_E}{n - k}} \qquad (18\text{–}13)$$

where k = the number of independent variables + 1. With a level of significance of .05, the critical value of $F_{(2,3)}$ is 9.55. Inasmuch as the computed F in our example is

$$F = \frac{\frac{48.08082}{2}}{\frac{6.75250}{3}} = \frac{24.04}{2.25} = 10.68,$$

the marketing analyst would reject H_0.

We can also test the b_js individually by introducing

$$t_{(n-k)} = \frac{b_j - B_{j0}}{s_{b_j}} \tag{18–14}$$

Although we may reject H_0: $B_1 = B_2 = 0$ because of an F test, b_1 and b_2 may not be statistically significant from zero when they are tested separately. Such a situation can arise if X_1 and X_2 are highly correlated. Under these circumstances, the s_{b_j}s are generally very large. High correlation between the independent variables will bother a researcher seeking to study the individual effect of X_1 or X_2 because large s_{b_j}s imply imprecise estimates of B_1 and B_2. This so-called *multicollinearity*, however, won't be troublesome whenever experimenters simply wish to estimate a subpopulation average and Y.

Returning to our example, if

$$H_0\colon B_1 = 0$$

$$H_1\colon B_1 \neq 0$$

is tested at the .05 level of significance, the critical values are $t_{(3)} = -3.182$ and 3.182. Since the calculated t is

$$t = \frac{.61676 - 0}{.16296} = 3.78,$$

H_0 will be rejected. By contrast, if

$$H_0\colon B_2 = 0$$

$$H_1\colon B_2 \neq 0$$

is tested, the researcher will accept the null hypothesis because the calculated t is only $(.03813 - 0)/.39570 = .096$. The analyst might decide to drop X_2 from his study and re-estimate Y using the two-variable regression $Y^* = a + bX_1$.

SUMMARY

Correlation and multiple regression analysis involve studying the relationship between or among variables. In contrast to the way in which the sample regression line is determined, the sample coefficient of correlation is derived from a sampling plan in which both X and Y are random variables. This statistic is used to test hypotheses about ρ. Multiple regression is an extension of the least-squares technique introduced in Chapter Sixteen. Although we explored a situation in which Y was "explained" by two independent variables, more than two independent variables can be included in a multiple regression problem. We next wish to investigate observations that are associated with consecutive periods of time.

EXERCISES

1. For the joint probability distribution of X and Y shown below, calculate ρ.

		X		
		0	1	2
Y	0	.1	.1	.2
	1	.2	.1	.1
	2	.1	0	.1

2. For the joint probability distribution of X and Y shown below, calculate ρ.

		X		
		3	5	7
Y	1	0	.1	.1
	3	.1	.3	.1
	5	.1	.1	.1

3. A researcher wants to select one of ten objects (i.e., A, B, etc.) at random. The sample space for this experiment appears below.

· A ($X = 1, Y = 2$)	· B ($X = 3, Y = 1$)
· C ($X = 0, Y = 3$)	· D ($X = 1, Y = 2$)
· E ($X = 1, Y = 2$)	· F ($X = 0, Y = 3$)
· G ($X = 2, Y = 2$)	· H ($X = 3, Y = 1$)
· I ($X = 2, Y = 1$)	· J ($X = 0, Y = 3$)

a. Construct the joint probability distribution of X and Y.
b. Calculate ρ.

4. An investor has constructed the following joint probability distribution of X and Y, where X = return on Project A (in $ millions) and Y = return on Project B (in $ millions). Calculate ρ.

		X		
		1	2	3
	1	.2	.1	.1
Y	2	.1	.1	.2
	3	.1	.1	0

5. The sample coefficient of correlation isn't affected by the units in which X and Y are measured. Therefore the correlation between X = weight (in pounds) and Y = height (in inches) is the same as the correlation between X = weight (in ounces) and Y = height (in feet). Using Equation 18–3, show that multiplying each X by b and each Y by c will not change the right side of the equation.

6. Based on the pattern that appears in Equations 18–8, 18–9, and 18–10, what are the normal equations when three independent variables are included in a multiple regression problem?

7. A researcher chooses a random sample of eight objects from a bivariate normal population. Given the data below and $\alpha = .05$, determine whether X and Y are independent random variables.

Object	X	Y
A	5	7
B	4	10
C	8	3
D	7	12
E	2	9
F	1	6
G	4	14
H	6	8

8. A financial analyst is studying the correlation between two measures of corporate profitability: X = profit margin (i.e., after-tax profits divided by sales) and Y = return on stockholders' equity (i.e., after-tax profits divided by stockholders' equity). Based on the following information, compute r.

Company	X	Y
A	3.8%	8.7%
B	4.2	9.3
C	5.6	10.8
D	3.2	7.6
E	6.1	11.2
F	4.0	8.9

9. For the companies in question 8, the analyst also collected data on X = assets (in $ hundreds of thousands) and Y = before-tax profits (in $ tens of thousands). Using the information below, compute r.

Company	X	Y
A	5.1	3.6
B	8.6	4.3
C	7.4	4.1
D	12.3	5.7
E	15.2	6.8
F	10.7	6.2

10. An economist wants to develop a multiple regression equation to "explain" Y = the personal consumption component of Gross National Product. Her Ys will involve the years 1970 to 1979.

a. What are some of the independent variables that might be relevant?
b. What are the expected signs of the coefficients of these variables?

11. Using the same time period as in question 10, the economist would also like to "explain" Y = the business fixed investment component of Gross National Product.

a. What are some of the independent variables that might be relevant?
b. What are the expected signs of the coefficients of these variables?

12. Part of the computer print-out for a multiple regression problem is reproduced below (n = 6).

Variable No.	*Mean*	*Standard Deviation*	*Correlation X vs Y*	*Regression Coefficient*	*Std. Error of Reg. Coef.*	*Computed t Value*
2	11.16666	4.95647	.93620	.44227	.11898	
3	4.66666	1.50554	.85577	.85472	.39170	
Dependent						
1	7.16666	3.31159				
Intercept		−1.76073				

a. What is the multiple regression equation?
b. What is the estimated mean of the $X_1 = 10$, $X_2 = 6$ subpopulation?
c. What is the estimated mean of the $X_1 = 5$, $X_2 = 2$ subpopulation?
d. Given b_1, what is the computed t value?
e. Given b_2, what is the computed t value?
f. Is b_1 significantly different from 0 at the .05 level of significance?

13. The following table is associated with question 12. Using $\alpha =$ .05, test

$$H_0: B_1 = B_2 = 0$$

$$H_1: \text{At least one } B_j \text{ isn't } 0$$

Analysis of Variance for the Regression

Source of Variation	*Degrees of Freedom*	*Sum of Squares*	*Mean Squares*	*F Value*
Attributable to Regression	2	52.21524	26.10762	29.91626
Deviation from Regression	3	2.61808	.87269	
Total	5	54.83332		

14. The following table is associated with question 12.

Table of Residuals

Case No.	*Y Value*	*Y Estimate*	*Residual*
1	2.00000	2.16006	−.16006
2	5.00000	4.31178	.68821
3	7.00000	7.79031	−.79031
4	8.00000	8.70467	−.70467
5	10.00000	10.03149	−.03148
6	11.00000	10.00167	.99832

Given

X_1	X_2
5	2
6	4
10	6
14	5
17	5
15	6

show that

a. $\Sigma e = 0$
b. $\Sigma(X_1 e) = 0$
c. $\Sigma(X_2 e) = 0$

15. Based on the following table, what is R^2?

Analysis of Variance for the Regression

Source of Variation	*Degrees of Freedom*	*Sum of Squares*	*Mean Squares*	*F Value*
Attributable to Regression	2		26.10762	29.91626
Deviation from Regression	3		.87269	
Total	5			

Time Series Analysis

THE COMPONENTS OF A TIME SERIES

Since long-range planning is crucial to the survival of any company, management must speculate about the future. Before building a new plant, for instance, U.S. Steel would consider such things as:

1. the projected level of steel demand,
2. possible revisions in E.P.A. pollution standards,
3. potential actions by Congress to restrict steel imports, and
4. the projected direction of interest rates.

The following example illustrates how *forecasting* can lead to a particular action.

Because of the post–World War II "baby boom," many colleges and universities increased their faculties, as well as classroom and dormitory space, during the 1960s. Inasmuch as substantial declines in the traditional college-age population (i.e., 18 to 22) are forecasted for the coming years, further construction is unwise. Quite a few administrators, however, predict that adult, or "continuing," education courses will become attractive to more and more "older" students. Hence, if schools aggressively promote these courses, income from tuition can be boosted.

Chapter Nineteen introduces several ways in which management can use data to make forecasts. In this regard, we are going to investigate a *time series*. For some variable, Y, such a series is a *chronological* listing of n observations. Table 19–1 presents a time series of Y = W. M. Clothing Company's quarterly sales.

We will not try to *explain* Y. Instead, we want to *describe* its behavior. Our approach involves looking for patterns in the data. Some patterns will appear after Y is graphed against *time* (see Figure 19–1). Despite the ups and downs in sales, we can see (1) that Y rises over time and (2) that Y is always greater in the second and fourth quarters than in the first and third. In the following sections, we will express these patterns mathematically.

TABLE 19-1
W. M. Clothing Company Sales
(in $ thousands)

	Quarter			
Year	I	II	III	IV
1970	49.5	56.6	54.2	59.2
1971	48.8	54.8	54.4	60.9
1972	52.2	59.8	58.0	65.6
1973	55.1	62.6	60.9	68.2
1974	58.5	66.0	65.3	72.0
1975	62.1	71.2	70.0	80.8
1976	68.1	76.3	75.7	83.9
1977	70.0	79.3	78.5	86.0
1978	75.3	85.8	84.9	93.6

FIGURE 19-1
W. M. Clothing Company Sales (in $ thousands)

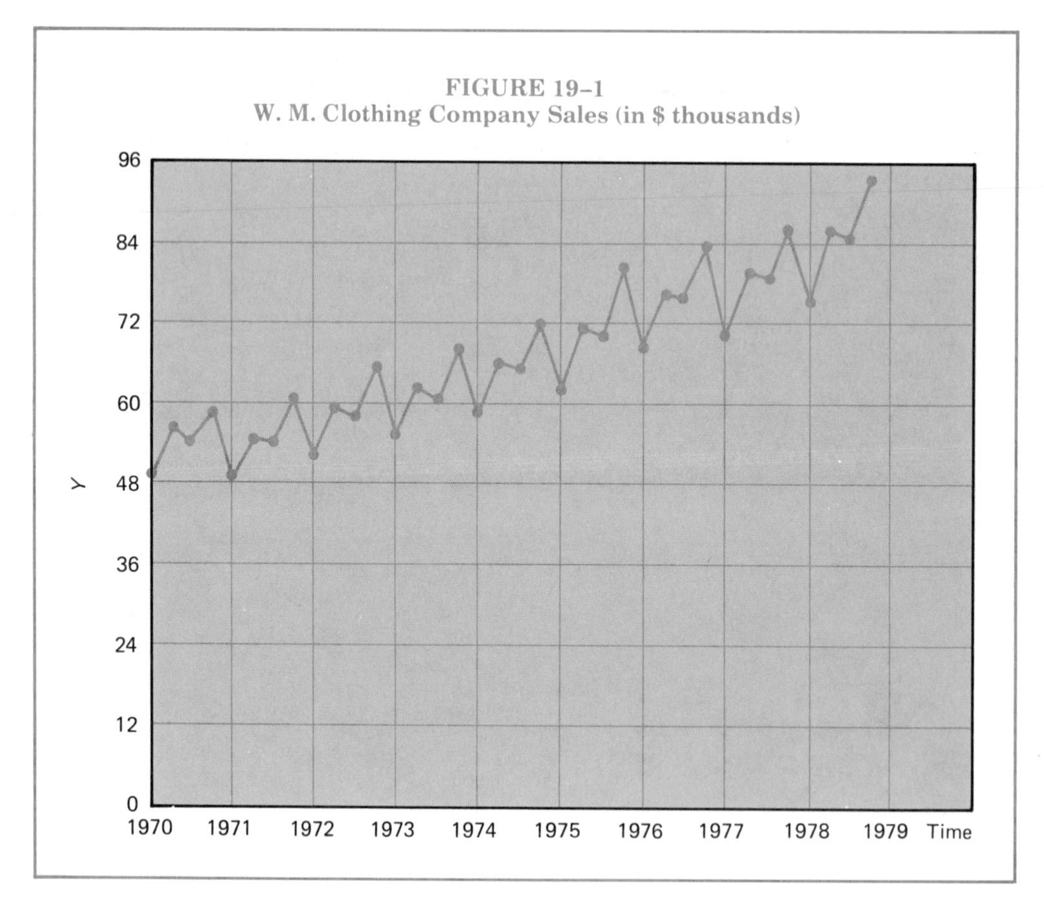

We can view Y at time t (i.e., Y_t) as the product of four influences:

1. *Trend:* A smooth, progressive movement due to such things as the steady growth in both population and personal income (see Figure 19–2).

FIGURE 19–2

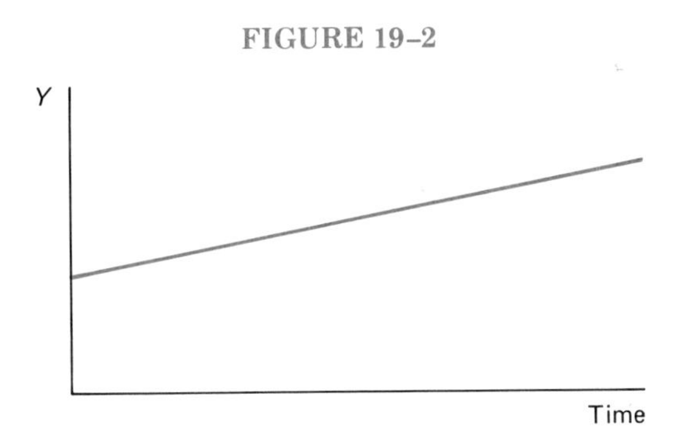

2. *Seasonal variation:* Fluctuations that are repeated every year at the same time because of the weather or the placement of holidays (see Figure 19–3).

FIGURE 19–3

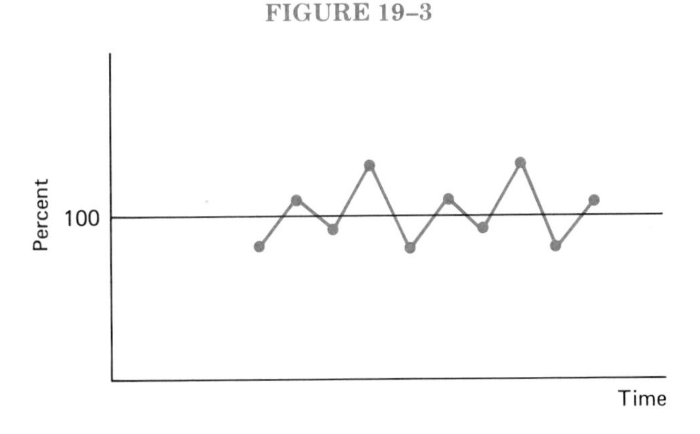

3. *Cyclical variation:* Fluctuations that are repeated at intervals longer than a year as a result, perhaps, of fiscal and monetary policy. Unlike seasonal variation, cycles have differing lengths, as well as differing "peaks" and "valleys" (see Figure 19–4). Hence, they are difficult to forecast. The National Bureau of Economic Research, however, has tried to determine when past cycles for a variety of time series have occurred.
4. *Irregular variation:* Fluctuations that can't be assigned to any of the above (see Figure 19–5). Irregular variations may be caused by nonrecurring identifiable events like floods or by unknown chance happenings.

FIGURE 19–4

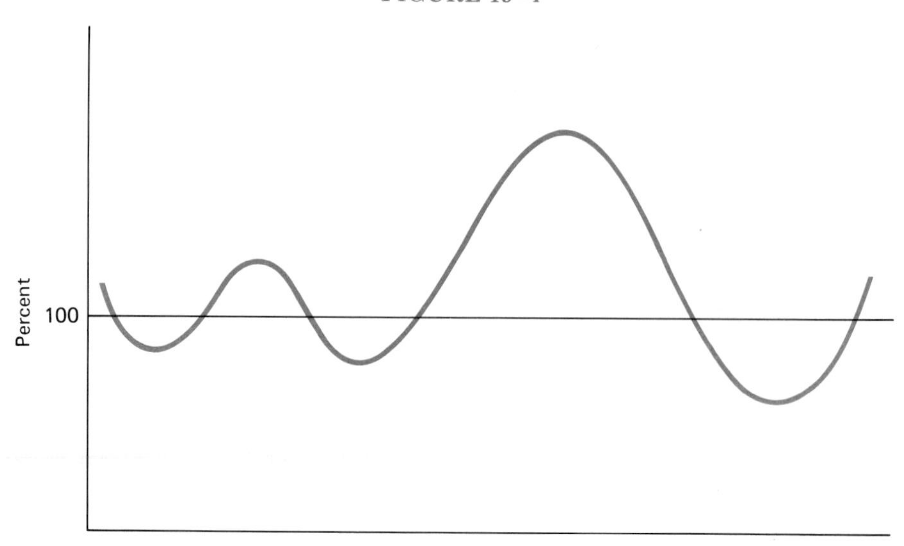

FIGURE 19–5

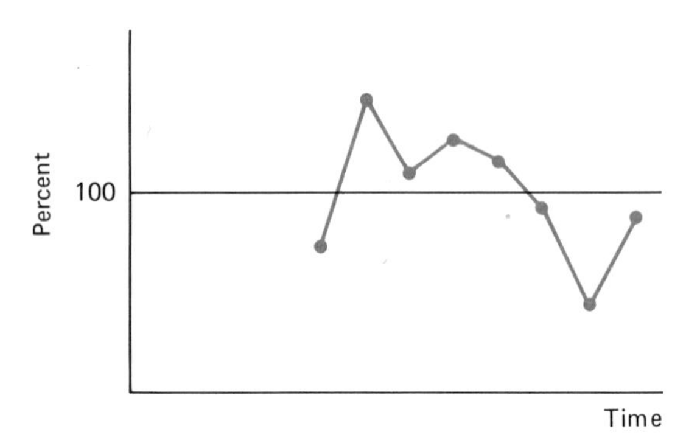

If we represent these influences as T, S, C, and I, respectively,

$$Y_t = (T_t)(S_t)(C_t)(I_t) \quad (19\text{–}1)$$

The latter three factors pull Y away from its trend value. We intend to isolate T and S and predict Y in the light of these components. Bear in mind that T, S, C, and I are only *symptoms* of the variables (e.g., income) that affect Y. To forecast confidently, we have to feel that the patterns we discover will persist.

THE TREND LINE

The general behavior of the W. M. Clothing Company's sales can be indicated by a line. To establish this line, that is,

$$Y^* = T = a + bX, \quad (X = \text{time})$$

objectively, we'll apply the least-squares technique. "Time" is recorded in terms of the following code:

$$\begin{aligned} 1970\text{ I} &\rightarrow X = 0 \\ 1970\text{ II} &\rightarrow X = 1 \\ &\vdots \\ 1978\text{ IV} &\rightarrow X = 35 \end{aligned}$$

Table 19–2 displays the summations associated with

$$\Sigma Y = an + b\Sigma X$$

$$\Sigma XY = a\Sigma X + b\Sigma X^2$$

for the data in Table 19–1.

TABLE 19–2

Quarter		Y	X	XY	X^2
1970	I	49.5	0	0.0	0
	II	56.6	1	56.6	1
	III	54.2	2	108.4	4
	IV	59.2	3	177.6	9
1971	I	48.8	4	195.2	16
	II	54.8	5	274.0	25
	III	54.4	6	326.4	36
	IV	60.9	7	426.3	49
1972	I	52.2	8	417.6	64
	II	59.8	9	538.2	81
	III	58.0	10	580.0	100
	IV	65.6	11	721.6	121
1973	I	55.1	12	661.2	144
	II	62.6	13	813.8	169
	III	60.9	14	852.6	196
	IV	68.2	15	1,023.0	225

Table 19–2 continued

Quarter	Y	X	XY	X^2
1974 I	58.5	16	936.0	256
II	66.0	17	1,122.0	289
III	65.3	18	1,175.4	324
IV	72.0	19	1,368.0	361
1975 I	62.1	20	1,242.0	400
II	71.2	21	1,495.2	441
III	70.0	22	1,540.0	484
IV	80.8	23	1,858.4	529
1976 I	68.1	24	1,634.4	576
II	76.3	25	1,907.5	625
III	75.7	26	1,968.2	676
IV	83.9	27	2,265.3	729
1977 I	70.0	28	1,960.0	784
II	79.3	29	2,299.7	841
III	78.5	30	2,355.0	900
IV	86.0	31	2,666.0	961
1978 I	75.3	32	2,409.6	1,024
II	85.8	33	2,831.4	1,089
III	84.9	34	2,886.6	1,156
IV	93.6	35	3,276.0	1,225
	2,424.1	630	46,369.2	14,910

From these totals, we find that

$$2{,}424.1 = 36a + 630b$$

$$46{,}369.2 = 630a + 14{,}910b$$

and

$$T = 49.556 + 1.016X$$

(1970 I is $X = 0$; one quarter = one unit of X; T is in \$ thousands)

Sales typically increase \$1,016 per quarter. Figure 19–6 shows the relationship between the trend line and the actual time series.

Let's assume that the president of the W. M. Clothing Company has created an incentive plan for employees. If sales in any quarter are higher than the trend value, each worker gets a \$100 bonus. The president is interested in the fourth quarter of 1980. She wants to identify the sales level on which the bonus will be based.

We know that 1978 IV corresponds to $X = 35$. Since 1980 IV is eight quarters, or eight units, beyond $X = 35$,

$$T_{1980\ \text{IV}} = 49.556 + 1.016(43) = 93.244$$

Therefore, a bonus will be granted only if $Y_{1980\ \text{IV}}$ is above \$93,244.

FIGURE 19–6
W. M. Clothing Company Sales (in $ thousands)

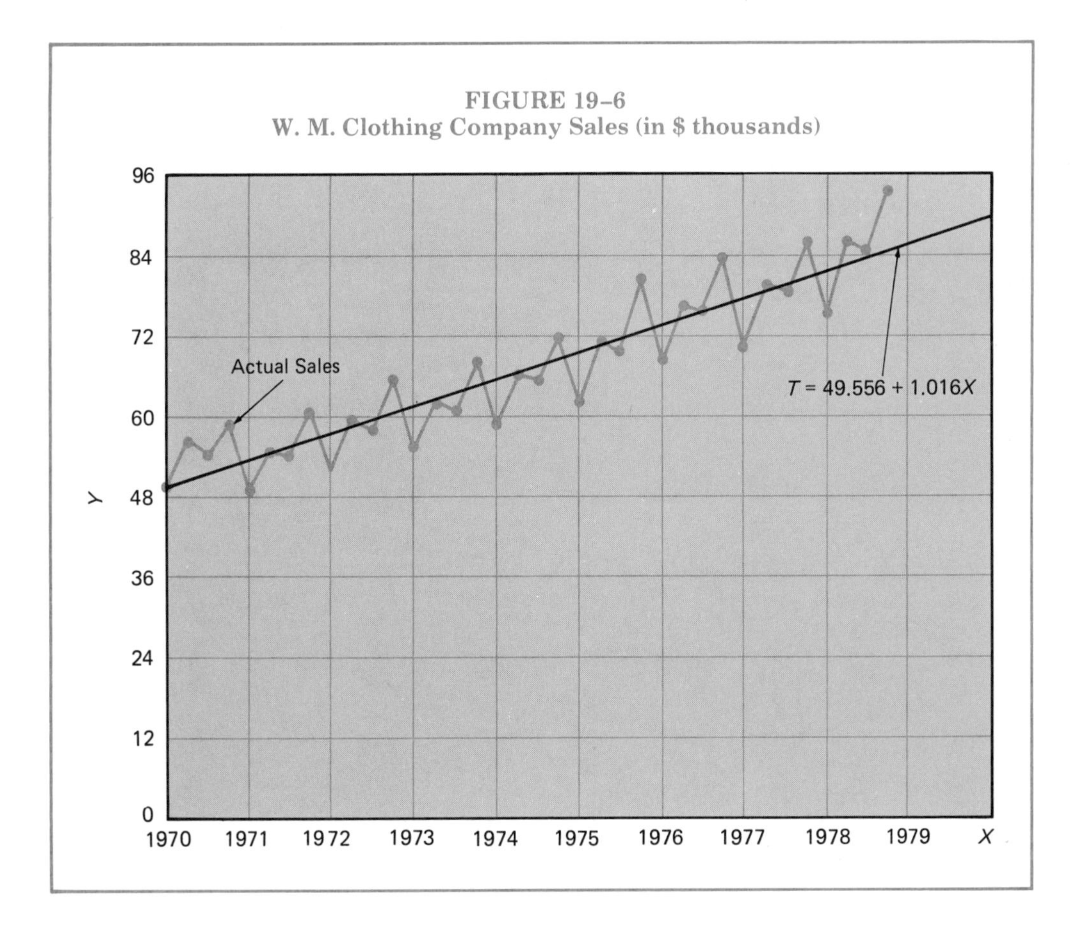

THE SEASONAL INDEX

We now wish to determine whether W. M.'s sales in some quarter (i.e., I, II, III, or IV) are usually above or below the quarterly average for the year. To obtain a *seasonal index,* or S, we have to investigate the *ratio-to-moving-average* procedure.

A *moving average* is a sequence of averages where:

1. The first member of the sequence is the average of the first m Ys:

$$\frac{Y_1 + Y_2 + \ldots + Y_m}{m}$$

2. The second member equals

$$\frac{Y_2 + Y_3 + \ldots + Y_{m+1}}{m}$$

3. Other members are derived by deleting one Y, adding another, and dividing by m.

TABLE 19–3
W. M. Clothing Company Sales (in $ thousands)

(1) *Quarter*	(2) *Y*	(3) *Four-Quarter Moving Average*	(4) *Average of Two Adjacent Moving Averages*	(5) *Y Divided by the Corresponding Average in Column 4 (in %)*
1970 I	49.5			
II	56.6			
		54.9		
III	54.2		54.8	98.9
		54.7		
IV	59.2		54.5	108.6
		54.3		
1971 I	48.8		54.3	89.9
		54.3		
II	54.8		54.5	100.6
		54.7		
III	54.4		55.2	98.6
		55.6		
IV	60.9		56.2	108.4
		56.8		
1972 I	52.2		57.3	91.1
		57.7		
II	59.8		58.3	102.6
		58.9		
III	58.0		59.3	97.8
		59.6		
IV	65.6		60.0	109.3
		60.3		
1973 I	55.1		60.7	90.8
		61.1		
II	62.6		61.4	102.0
		61.7		
III	60.9		62.1	98.1
		62.6		
IV	68.2		63.0	108.3
		63.4		
1974 I	58.5		64.0	91.4
		64.5		
II	66.0		65.0	101.5
		65.5		
III	65.3		65.9	99.1
		66.4		
IV	72.0		67.0	107.5
		67.7		

Table 19–3 continued

(1) Quarter	(2) Y	(3) Four-Quarter Moving Average	(4) Average of Two Adjacent Moving Averages	(5) Y Divided by the Corresponding Average in Column 4 (in %)
1975 I	62.1		68.2	91.1
		68.8		
II	71.2		69.9	101.9
		71.0		
III	70.0		71.8	97.5
		72.5		
IV	80.8		73.2	110.4
		73.8		
1976 I	68.1		74.5	91.4
		75.2		
II	76.3		75.6	100.9
		76.0		
III	75.7		76.2	99.3
		76.5		
IV	83.9		76.9	109.1
		77.2		
1977 I	70.0		77.6	90.2
		77.9		
II	79.3		78.2	101.4
		78.5		
III	78.5		79.1	99.2
		79.8		
IV	86.0		80.6	106.7
		81.4		
1978 I	75.3		82.2	91.6
		83.0		
II	85.8		84.0	102.1
		84.9		
III	84.9			
IV	93.6			

Each average is placed in the center of the time interval covered by the m Ys. The calculations in Table 19–3 are based on $m = 4$.

The rationale for Table 19–3 can be stated as follows:

1. Column 3: Seasonal fluctuations cause Ys to deviate from their average quarterly values. Since these fluctuations complete their pattern in four quarters, they will "cancel out" when we average four quarters.

Because moving averages are always centered, the first moving average, $(49.5 + 56.6 + 54.2 + 59.2)/4 = 54.9$, appears *between* 1970 II and 1970 III. The second moving average, $(56.6 + 54.2 + 59.2 + 48.8)/4 = 54.7$, is recorded between 1970 III and 1970 IV.

2. Column 4: To pair a moving average with an actual Y, we average two adjacent moving averages, for example, $(54.9 + 54.7)/2 = 54.8$. When these new averages are centered, they are placed *at* a quarter.

   Due to the fact that irregular fluctuations are temporary and random, creating moving averages probably eliminates I in addition to S. Figure 19–7 compares the moving average series in column 4 with the Ys.

FIGURE 19–7
W. M. Clothing Company Sales (in $ thousands)

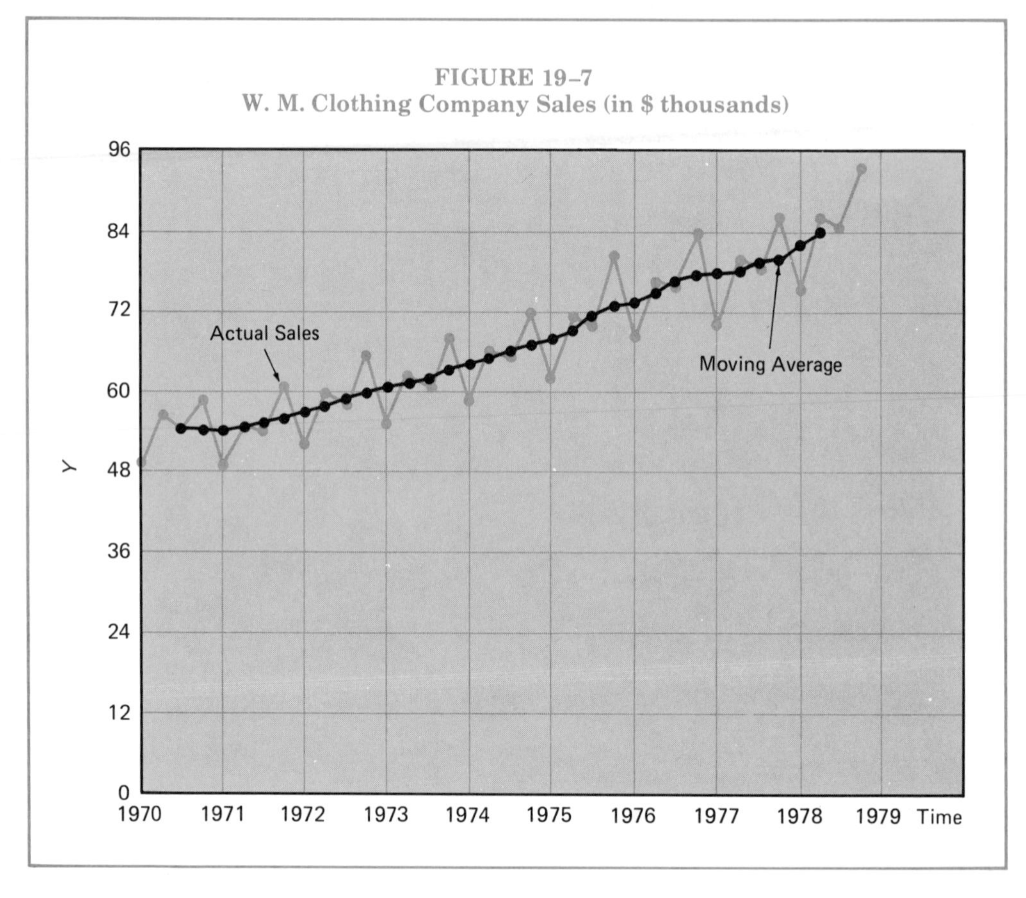

3. Column 5: Since $Y = (T)(S)(C)(I)$ and column 4 contains a series possessing only T and C, we can isolate S and I by dividing the moving average into Y:

$$\frac{Y}{(T)(C)} = \frac{(T)(S)(C)(I)}{(T)(C)} = (S)(I)$$

moving average

We see, for instance, that $Y_{1970\ \mathrm{III}}$ was 98.9 percent of the typical quarter.

The differences in the percentages for all of the first quarters, all of the second quarters, and so on, are attributed to irregular forces. We remove I by *averaging* (see Table 19–4).

A "typical" quarter should have an index of 100. For consistency, therefore, the column averages in Table 19–4 should total 400. We note, however, that $90.9 + 101.6 + 98.6 + 108.5 = 399.6$. We will adjust each average by a factor of $400/399.6 = 1.001$ (see Table 19–5).

TABLE 19–4
Y Divided by the Moving Average (in %)

	Quarter			
Year	I	II	III	IV
1970			98.9	108.6
1971	89.9	100.6	98.6	108.4
1972	91.1	102.6	97.8	109.3
1973	90.8	102.0	98.1	108.3
1974	91.4	101.5	99.1	107.5
1975	91.1	101.9	97.5	110.4
1976	91.4	100.9	99.3	109.1
1977	90.2	101.4	99.2	106.7
1978	91.6	102.1		
Column Average	90.9	101.6	98.6	108.5

TABLE 19–5

Quarter	*Seasonal Index* *(S)*
I	91.0
II	101.7
III	98.7
IV	108.6

The trend and seasonal components are often used to develop short-range forecasts. To predict sales in 1980 IV, we would compute $(T)(S) = (93.244)(1.086) = 101.263$. Inasmuch as sales are generally above average in the fourth quarter, the forecast has been raised from $93,244 (trend only) to $101,263.

Let's look at another example. Suppose the W. M. Clothing Company relied on "word-of-mouth" advertising before trying a "media blitz" in the fourth quarter of 1979. Since $Y_{1979\ \mathrm{III}} = \$89,000$, while $Y_{1979\ \mathrm{IV}} = \$99,000$, it might seem that the new strategy had a big impact on sales.

Based on the seasonal indexes, however, we would expect fourth-quarter sales to be higher than third-quarter in any case. To compare the sales levels, we should *deseasonalize,* or *seasonally adjust,* the data; that is,

$$\text{Seasonally adjusted } Y_t = \frac{Y_t}{S_t}(100) \qquad (19\text{–}2)$$

Given Equation 19–2,

$$\text{Seasonally adjusted } Y_{1979\ \text{III}} = \frac{\$89{,}000}{98.7}(100) = \$90{,}172$$

$$\text{Seasonally adjusted } Y_{1979\ \text{IV}} = \frac{\$99{,}000}{108.6}(100) = \$91{,}160$$

With the seasonal influences removed, the company's performance appears roughly the same for the third and fourth quarters of 1979. Hence, the advertising policy was not particularly effective.

PRICE INDEXES

Inflation has been a major problem in the United States for many years. If the general level of prices rises by 10 percent, workers receiving only 5 percent wage hikes will suffer from a reduction in the *purchasing power* of their earnings. In this section, we will see how to summarize price information through a *price index.* Viewed in practical terms, these indexes are extremely important. Increases in the *consumer price index* (CPI), for instance, (1) trigger higher wages for employees having "escalator clauses" and (2) generate larger Social Security payments. (Note: The importance of the CPI is reflected in the fact that the Bureau of Labor Statistics recently spent about $50 million to revise the way it is constructed. [*Business Week,* March 13, 1978, p. 33.])

A price index expresses the relationship between prices at time t and time 0, or the *base period.* Consider the following example.

The S. L. Restaurant is widely known for its steak sandwiches. Because of increased operating costs, S. L. has been forced to raise its prices frequently, as shown in Table 19–6. Using 1977 as the base, we can compare 1978 and 1977 prices:

$$\frac{P_{1978}}{P_{1977}}(100) = \frac{\$4.20}{\$3.85}(100) = 109.1\%$$

TABLE 19–6

Year	*Price of a Steak Sandwich (P)*
1976	$3.70
1977	3.85
1978	4.20
1979	4.45

The price of a sandwich was 9.1 percent higher in 1978. Notice that the ratio of prices is 100 percent when t and 0 are the same, for example,

$$\frac{P_{1977}}{P_{1977}}(100) = \frac{\$3.85}{\$3.85}(100) = 100\%$$

Index numbers are always interpreted relative to the base. We will delete the "%" sign from future calculations. Table 19–7 indicates the usual format for presenting a series of indexes.

TABLE 19–7
Index of Steak Sandwich Prices
(1977 = 100)

1976	96.1
1977	100.0
1978	109.1
1979	115.6

We now want to investigate price indexes that incorporate the prices of k commodities. Assume that commodities A, B, C, and D represent a substantial portion of the Smith family's diet. Data on these foods are reproduced in Table 19–8.

TABLE 19–8

	1978		1979	
Commodity	P_0 Price per Unit	Q_0 Quantity Purchased	P_t Price per Unit	Q_t Quantity Purchased
A	$2.00/lb.	100 pounds	$2.50/lb.	90 pounds
B	5.00/qt.	300 quarts	5.50/qt.	285 quarts
C	4.00/oz.	400 ounces	4.75/oz.	370 ounces
D	6.00/gal.	500 gallons	7.00/gal.	480 gallons

For the *simple aggregate index,*

$$\text{INDEX}_t = \frac{\sum^{k} P_{it}}{\sum^{k} P_{i0}}(100) \qquad (19\text{–}3)$$

P_{it} is the price of the ith commodity ($i = 1, \ldots, k$) at time t. Both the numerator and denominator identify the cost of a market basket consisting of one unit of each commodity. Since

$$\frac{2.50 + 5.50 + 4.75 + 7.00}{2.00 + 5.00 + 4.00 + 6.00}(100) = \frac{19.75}{17.00}(100) = 116.2,$$

such a collection cost 16.2 percent more in 1979.

Two deficiencies occur in connection with the simple aggregate index:

1. The selection of units is arbitrary. If D were reported as $1.50 per quart in 1978 and as $1.75 per quart in 1979, the index would fall from 116.2 to 116.0
2. The prices are not "weighted." A 5¢ price increase for a frequently consumed item is treated the same as a 5¢ increase for an item that is rarely purchased.

We can remedy the first problem by using the *simple average of price relatives index*:

$$\text{INDEX}_t = \frac{\sum^{k} \frac{P_{it}}{P_{i0}}}{k}(100) \tag{19–4}$$

The choice of units will not affect this index. We can verify this with D:

$$\frac{P_t}{P_0} = \frac{7.00}{6.00} = 1.17 = \frac{1.75}{1.50}$$

Returning to Table 19–8, we get

$$\frac{\frac{2.50}{2.00} + \frac{5.50}{5.00} + \frac{4.75}{4.00} + \frac{7.00}{6.00}}{4}(100)$$

$$= \frac{1.25 + 1.10 + 1.19 + 1.17}{4}(100) = 117.8$$

Unlike Equation 19–4, the *Laspeyres index* "weights" the price of the ith commodity by the quantity that is consumed:

$$\text{INDEX}_t = \frac{\Sigma P_{it}Q_{i0}}{\Sigma P_{i0}Q_{i0}}(100) \tag{19–5}$$

Because the Qs refer to the base period, the Laspeyres index shows how the cost of *all* of the quantities in the base-period market basket changes. Inasmuch as

$$\frac{2.50(100) + 5.50(300) + 4.75(400) + 7.00(500)}{2.00(100) + 5.00(300) + 4.00(400) + 6.00(500)}(100)$$

$$= \frac{7{,}300}{6{,}300}(100) = 115.9,$$

we find that this market basket is 15.9 percent more expensive in 1979.

The same quantity weights will be used for 1980 or for any other time periods that we want to evaluate in relation to 1978. As such, the 1980 index can be compared with the 1979 index (e.g., if the 1980 index is 120.4, we not only can say that the market basket costs more in 1980 than it did in 1978 but also that it costs more in 1980 than it did in 1979).

When the price of a commodity increases, we would expect a family to reduce its consumption of that item. Because the Laspeyres index assumes that the quantities in the market basket will not change as prices rise at different rates, the index exaggerates the general price increase.

In contrast, the *Paasche index* uses quantity weights from the current period:

$$\text{INDEX}_t = \frac{\Sigma P_{it} Q_{it}}{\Sigma P_{i0} Q_{it}}(100) \qquad (19\text{–}6)$$

We would calculate

$$\frac{2.50(90) + 5.50(285) + 4.75(370) + 7.00(480)}{2.00(90) + 5.00(285) + 4.00(370) + 6.00(480)}(100)$$

$$= \frac{6{,}910}{5{,}965}(100) = 115.8$$

The Paasche index is not popular because the weights have to be revised every year.

Price indexes are used to *deflate* time series. This process yields *real,* or *constant-dollar,* values:

$$\begin{array}{l}\text{Constant-dollar}\\ \text{value of } Y_t\end{array} = \frac{Y_t}{\text{INDEX}_t}(100) \qquad (19\text{–}7)$$

The following example illustrates this operation.

Corporation H has just concluded its most profitable year. Because of this, union leaders feel that H will be receptive to many of labor's demands. To reinforce the justification for a large wage hike, the union's negotiating team is prepared to argue that previous contracts have resulted in a decline in the purchasing power of H's workers. The union will introduce the information displayed in Table 19–9.

TABLE 19–9

Year	*Hourly Wages of Corporation H Workers*	*Consumer Price Index* (1967 = 100)*
1974	$4.00	147.7
1975	4.30	161.2
1976	4.45	170.5
1977	4.73	181.5

* Source: *Economic Indicators,* August 1979, p. 23.

With the CPI as an indicator of the general level of prices, the negotiating team can convert hourly wages into *real hourly wages* (see Table 19–10). In 1974 $4.00 could purchase the goods and services that $2.71 bought in 1967, while $4.73 in 1977 could purchase only what $2.60 did in 1967. Although the actual wages of the workers have increased every year, their real wages have consistently fallen. In other words, the small size of the wage increases has prevented these workers from exchanging their paychecks for a larger volume of commodities over the past four years.

TABLE 19–10

Year	*Wages in 1967 Dollars (i.e., real wages)*
1974	$\frac{\$4.00}{147.7}(100) = \$2.71$
1975	$\frac{\$4.30}{161.2}(100) = \$2.67$
1976	$\frac{\$4.45}{170.5}(100) = \$2.61$
1977	$\frac{\$4.73}{181.5}(100) = \$2.60$

SUMMARY

Time series analysis is another statistical procedure that is relevant for business and economic decision making. Although forecasting with trend and seasonal components ignores the *causes* of Y, $(T)(S)$ forecasts may, nonetheless, be very accurate. As inflation continues in the United States, more and more of the public will become aware of price indexes such as the CPI. In the final chapter of the text, we'll study hypothesis tests which necessitate few assumptions about the population.

EXERCISES

1. Orders for the output of industry H are strongly influenced by the demand for automobiles. As part of his study of industry H, an economist would like to forecast Y = retail sales of new passenger cars (in millions of cars). Given the data below,

Year	Y
1971	8.7
1972	9.3
1973	9.7
1974	7.5
1975	7.1
1976	8.6
1977	9.1
1978	9.3

Source: *Economic Indicators*, August 1979, p. 4.

a. Calculate $T = a + bX$.
b. What is T_{1981}?

2. The Evelyn Miller Company, a retailer of expensive clothes, has a market share of 40 percent, where the "market" corresponds to City A. Based on the following information, (a) calculate the trend equation and (b) forecast the Evelyn Miller Company's level of sales in 1981.

Year	*Sales of Expensive Clothes in City A (in \$ millions)*
1973	1.2
1974	1.5
1975	1.6
1976	1.9
1977	2.3
1978	2.5
1979	2.9

3. Given the GNP time series (in $ trillions) below,

Year	*GNP in Current Dollars*
1971	1.06
1972	1.17
1973	1.31
1974	1.41
1975	1.53
1976	1.70
1977	1.90
1978	2.13

Source: *Economic Indicators*, August 1979, p. 1.

a. Calculate $T = a + bX$.
b. Plot the actual data and the trend equation on the same graph.
c. What is T_{1981}?

4. The "implicit GNP price deflator" is the index involved in the relation

$$\frac{\text{GNP in current dollars}}{\text{INDEX}}(100) = \begin{array}{c}\text{GNP in 1972 dollars (or}\\ \text{constant dollars)}\end{array}$$

Using the following information, deflate the time series in question 3.

Year	*Implicit GNP Price Deflator (1972 = 100)*
1971	96.02
1972	100.00
1973	105.80
1974	116.02
1975	127.15
1976	133.71
1977	141.70
1978	152.05

Source: *Economic Indicators*, August 1979, p. 2.

5. Information on Y = average weekly earnings of workers in manufacturing and on the consumer price index is displayed below. Calculate the "real" earnings of these workers.

Year	*Y*	*CPI (1967 = 100)*
1975	$190.79	161.2
1976	209.32	170.5
1977	228.90	181.5
1978	249.27	195.4

Source: *Economic Indicators*, August 1979, pp. 15 and 23.

6. People often comment on the fact that the dollar is "shrinking." Based on the index numbers in question 5, determine how many cents in 1967 were used to purchase what $1 could buy in (a) 1975, (b) 1976, (c) 1977, and (d) 1978.

7. A company calculated the following seasonal indexes for its sales:

Quarter	*S*
I	85.6
II	104.7
III	93.2
IV	?

If sales in the fourth quarter of 1979 equaled $2,500,000, what was the seasonally adjusted value?

8. Given the following time series of Y = corporate profits before taxes (in $ hundreds of billions),

Year	*Y*
1971	0.82
1972	0.96
1973	1.16
1974	1.27
1975	1.20
1976	1.56
1977	1.77
1978	2.06

Source: *Economic Indicators*, August 1979, p. 4.

a. Calculate $T = a + bX$.
b. Plot the actual data and the trend equation on the same graph.
c. What is T_{1981}?

9. Average retail prices for one dozen Grade A large eggs are presented in the table at the top of p. 380.

Year	*Price*
1972	52.4¢
1973	78.1
1974	78.3
1975	77.0
1976	84.1

Source: *Statistical Abstract of the United States, 1977*, p. 486.

a. Restate the prices as index numbers where the base period is 1973.
b. Restate the prices as index numbers where the base period is 1975.

10. Average retail prices for one pound of hamburger meat are given below.

Year	*Price*
1972	74.4¢
1973	95.7
1974	97.2
1975	87.8
1976	87.6

Source: *Statistical Abstract of the United States, 1977*, p. 486.

a. Restate the prices as index numbers where the base period is 1972.
b. Restate the prices as index numbers where the base period is 1974.

11. Consider the "market baskets" of food shown below.

	1971		*1976*	
Commodity	Quantity (in pounds)	Average Retail Price (per pound)	Quantity (in pounds)	Average Retail Price (per pound)
Rib roast	10	$1.18	9	$1.77
Sliced bacon	3	.80	2	1.71
Turkey	6	.55	8	.74
Frying chickens	7	.41	11	.60

Source: *Statistical Abstract of the United States, 1977*, p. 486.

Using 1971 as the base period, calculate the value of the following price indexes for 1976:

a. Simple aggregate
b. Simple average of price relatives
c. Laspeyres
d. Paasche

12. Suppose that the quantities in question 11 are 7, 5, 8, and 6, respectively, for 1971, and 8, 4, 10, and 12, respectively, for 1976. Recompute (a) the Laspeyres index and (b) the Paasche index.

13. Information on department store sales (in $ billions) appears below.

	1972	*1973*	*1974*	*1975*	*1976*
January	2.7	3.1	3.3	3.3	3.7
February	2.7	2.9	3.1	3.2	3.6
March	3.4	3.8	4.0	4.2	4.5
April	3.4	4.0	4.4	4.2	5.0
May	3.7	4.2	4.5	4.9	4.8
June	3.8	4.3	4.4	4.6	5.0
July	3.5	3.8	4.2	4.4	4.8
August	3.8	4.2	4.6	4.8	5.0
September	3.9	4.1	4.2	4.6	5.0
October	4.0	4.3	4.6	4.8	5.4
November	4.6	5.2	5.2	5.6	6.2
December	7.1	7.6	7.7	8.8	9.8

Source: U.S. Department of Commerce, *Business Statistics*, 1975, p. 60, and *Business Statistics*, 1977, p. 61.

a. Calculate monthly seasonal indexes. (Note: The seasonal pattern is repeated every 12 months. Hence, a moving average with $m = 12$ will eliminate seasonal influences.)
b. Graph the seasonal indexes.

14. Based on the data in question 13,

a. Develop a time series of quarterly observations.
b. Calculate quarterly seasonal indexes.
c. Graph the seasonal indexes.

15. Information concerning unemployment in the United States is presented below.

Month	*Number of People Unemployed (in thousands)*	*Seasonally Adjusted Number of People Unemployed (in thousands)*
August, 1978	5,931	5,940
September	5,797	5,964
October	5,460	5,836
November	5,629	5,877
December	5,725	6,012
January, 1979	6,431	5,883
February	6,484	5,881
March	6,165	5,871
April	5,561	5,937
May	5,253	5,929
June	6,235	5,774
July	6,104	5,848

Source: *Economic Indicators*, August 1979, p. 11.

a. Compute the monthly seasonal indexes.
b. Based on the indexes, which months typically have "high" levels of unemployment (i.e., the corresponding index is above 100)?
c. Suppose the June 1980 level of unemployment is 5.8 million. What is the seasonally adjusted level?

16. Unemployment as a percentage of the labor force (i.e., the employed plus people actively seeking work) is equivalent to the "unemployment rate." Using the information in question 15 and the following time series, calculate the monthly unemployment rates (seasonally adjusted).

Month	*Seasonally Adjusted Civilian Employment (in thousands)*
August, 1978	94,723
September	95,010
October	95,241
November	95,751
December	95,855
January, 1979	96,300
February	96,647
March	96,842
April	96,174
May	96,318
June	96,754
July	97,210

Source: *Economic Indicators*, August 1979, p. 11.

20

Nonparametric Hypothesis Tests

CHARACTERISTICS OF NONPARAMETRIC TESTS

In Chapters Nine, Ten, Twelve, and Fifteen, we constructed hypotheses concerning μ. The tests of these hypotheses are said to be *parametric* because parameters appear in H_0 and H_1. We have also seen that the parametric t and ANOVA tests force researchers to assume that samples are selected from *normal* populations.

Suppose $\alpha = .05$, $n = 22$, and we wish to test

$$H_0: \mu = \mu_0$$

$$H_1: \mu \neq \mu_0$$

We would reject the null hypothesis if the computed t fell outside the interval $-2.08 < t < 2.08$. The rationale derives from the fact that $t < -2.08$ or $t > 2.08$ is "possible but improbable" when a random sample of 22 observations is drawn from a normal population having a mean of μ_0.

If the population is not normal, however,

$$P(-2.08 < t_{(21)} < 2.08) \neq .95$$

or

$$P(\text{Type I error}) \neq .05$$

Hence, the actual level of significance will differ from .05, and the researcher's calculation of β will also be incorrect. Table 20–1 indicates several other parametric tests and the population assumptions associated with them.

During the past thirty years, statisticians have developed a collection of inferential techniques involving very few assumptions about the population. Because of this, the techniques are called

TABLE 20–1

Null Hypothesis	*Assumption*	*Test Quantity*
H_0: $\mu_1 - \mu_2 = 0$	Population 1 and Population 2 are normal and $\sigma_1^2 = \sigma_2^2$.	$t = \dfrac{(\overline{X}_1 - \overline{X}_2) - (\mu_1 - \mu_2)}{s_{\overline{x}_1 - \overline{x}_2}}$
H_0: $\sigma_1^2 = \sigma_2^2$	Population 1 and Population 2 are normal.	$F = \dfrac{s_1^2}{s_2^2}$
H_0: $\rho = 0$	X and Y are bivariate normal.	$t = \dfrac{r}{\sqrt{(1 - r^2)/(n - 2)}}$

distribution free. Many distribution-free procedures can be used merely if the population is continuous. By contrast, assuming normality implies that the population is (1) continuous, (2) symmetrical, and (3) generated by the density function in Equation 6–1.

Hypotheses connected with distribution-free tests may not identify parameter values, for example,

H_0: Population 1 and Population 2 are the same.

H_1: Population 1 and Population 2 are not the same.

Testing such hypotheses isn't equivalent to testing

$$H_0: \mu_1 = \mu_2$$
$$H_1: \mu_1 \neq \mu_2$$

When we accept "The populations are the same," we can also conclude that $\mu_1 = \mu_2$. On the other hand, when we reject "The populations are the same," we can't conclude $\mu_1 \neq \mu_2$ since two populations can be different and yet have equal means, as in Figure 20–1. Even though there are both parametric and nonparametric distribution-free tests, "distribution-free" and "nonparametric" are often treated as synonyms.

FIGURE 20–1

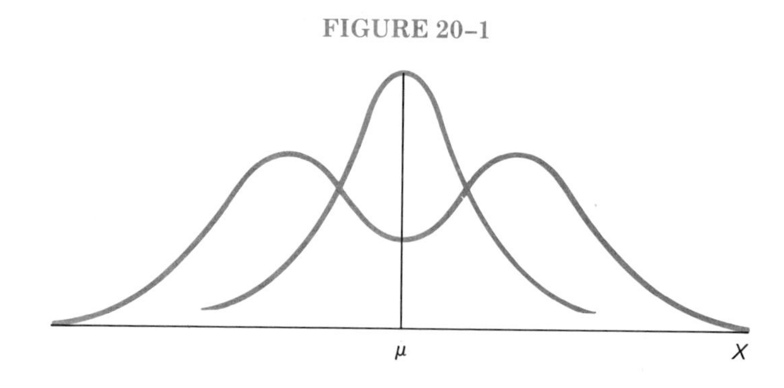

Like the t and F tests, nonparametric procedures arise from sampling distributions constructed when observations are chosen at random. Because the characteristics of these sampling distributions don't depend on whether the population is, in fact, normal, experimenters can generally attain the actual level of significance that they want.

Imagine that the assumptions of a parametric test are fulfilled and that a nonparametric alternative is available. If each test is conducted with the same α and n, the parametric test will be more *powerful* (i.e., it will produce lower βs). The nonparametric *sign test* illustrates a reason for this. As we will shortly discuss, the sign test is relevant where the hypotheses are

$$H_0: \mu = \mu_0$$

$$H_1: \mu \neq \mu_0$$

Observations are replaced by a "+" when $X_i > \mu_0$ and by a "−" when $X_i < \mu_0$. The number of "+"s determines whether H_0 is accepted or rejected.

Given either of the samples in Table 20–2, we would accept H_0 and conclude $\mu = 20$ if we viewed seven "+"s as "possible and probable." Inasmuch as the sign test "ignores" some information (e.g., the size of the difference between X_i and μ_0), the probability of accepting a false H_0 is high.

TABLE 20–2

X	*Sign*	*X*	*Sign*
19.14	−	19.14	−
19.26	−	19.26	−
19.35	−	19.35	−
19.49	−	19.49	−
19.58	−	19.58	−
20.01	+	200.01	+
20.03	+	200.03	+
20.15	+	200.15	+
20.32	+	200.32	+
20.46	+	200.46	+
20.59	+	2,000.59	+
20.63	+	2,000.63	+

Nonparametric tests are frequently applied to (1) data that occur originally as *ranks* (i.e., the observations are measured ordinally) and (2) data that have been converted from interval and ratio scales to ordinal scales. We will now explore five of the most common nonparametric approaches to hypothesis testing.

THE SIGN TEST

Whenever a population is both continuous and symmetrical, as in Figure 20–2, two conditions apply:

1. $P(X > \text{median}) = .5$, $P(X < \text{median}) = .5$, and $P(X = \text{median}) = 0$.
2. The median equals the mean.

FIGURE 20–2

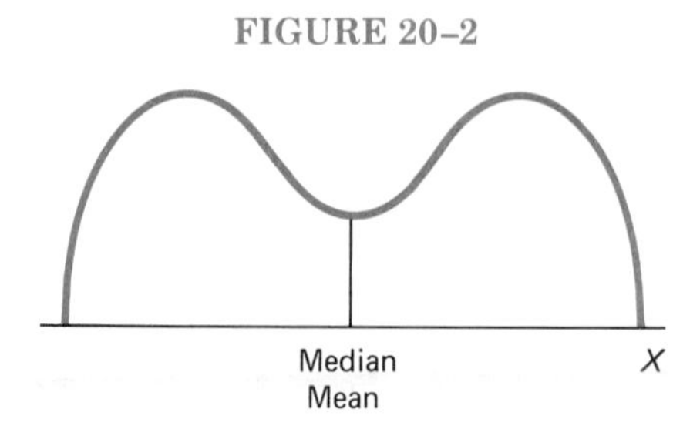

Consequently, if we randomly select an observation from such a population and indicate the sign of $(X - \text{mean})$, we can write $P(+) = .5$ and $P(-) = .5$.

If a random sample of n observations is chosen, we can determine n "+" and "−" signs. The "+"s represent r "successes" in n independent trials of "select an item." The probability of getting a particular r, therefore, is given by the binomial probability function

$$P(r) = \frac{n!}{r!(n-r)!}(.50)^r(.50)^{n-r} \qquad (20\text{–}1)$$

The *sign test* is used to investigate

$$H_0: \mu = \mu_0$$

$$H_1: \mu \neq \mu_0$$

Unlike with the t test, with the sign test we don't have to work with a normal population. Consider the following example.

The federal income tax structure is the target of much debate. Middle-class Americans, for instance, object to the fact that rich people can drastically cut their taxes by taking advantage of loopholes. Congress can change the tax laws, moreover, so as to influence the economy (e.g., stimulate it).

A congressional subcommittee is currently studying how a presidential tax-reform plan will affect families having incomes between \$15,000 and \$20,000. The committee members believe that this income group paid an average of \$3,000 in taxes in 1978. To test H_0: $\mu = \$3{,}000$, they ask the Internal Revenue Service to randomly pick twelve of the tax returns submitted by families in the \$15,000–\$20,000 bracket. The results are displayed in Table 20–3.

We'll delete Taxpayer I from the sample since we are assuming

TABLE 20–3

Taxpayer	X 1978 Taxes	Sign of $(X - \$3{,}000)$
A	\$3,200	+
B	\$2,800	−
C	\$3,500	+
D	\$1,200	−
E	\$2,600	−
F	\$3,900	+
G	\$4,100	+
H	\$1,500	−
I	\$3,000	
J	\$2,700	−
K	\$3,300	+
L	\$4,200	+

$P(X = \mu_0) = 0$. Six "+" signs are present in the table. If H_0 is true, the probability of drawing a "+" return is .5. We will be suspicious of $\mu = \$3{,}000$ as long as r = the number of "+"s is "very large" or "very small." Instead of relating r to the binomial distribution, we're going to introduce the normal curve approximation. Then the basis for evaluating the "evidence" will be

$$Z = \frac{r - n\pi}{\sqrt{n\pi(1 - \pi)}} \qquad (20\text{–}2)$$

Letting $\alpha = .05$, the acceptance and rejection regions for

$$H_0\colon \mu = \$3{,}000 \text{ (or } \pi = .5)$$

$$H_1\colon \mu \neq \$3{,}000 \text{ (or } \pi \neq .5)$$

are as shown in Figure 20–3. Since the computed Z is

$$Z = \frac{6 - (11)(.5)}{\sqrt{(11)(.5)(.5)}} = .30,$$

FIGURE 20–3

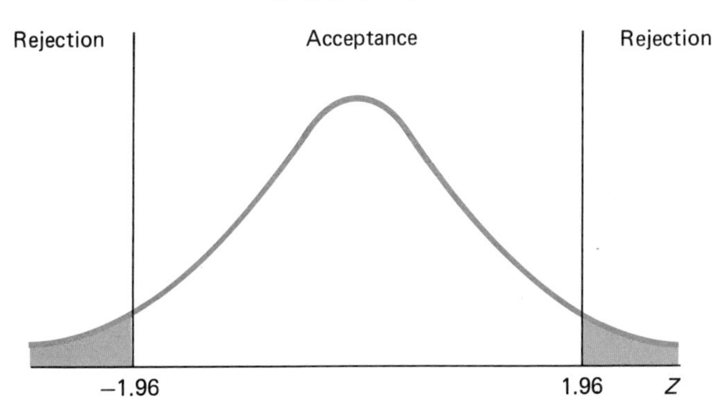

the committee accepts the null hypothesis; the members feel that finding six "+" signs isn't "improbable" when $\mu = \$3{,}000$. By accepting H_0, however, the congressmen could be making a Type II error. Now let's see how the sign test can be applied to matched-pairs sampling.

The tax-reform plan being considered will lower tax rates, eliminate certain exemptions and tax shelters, and establish several new write-offs. The congressional committee wants to assess how the plan would affect the distribution of taxes paid by families in the \$15,000–\$20,000 bracket. (The elimination of some exemptions might offset the lower tax rates.) Therefore, the committee recalculates, based on the proposed changes, the taxes of all twelve families in the IRS sample. Table 20–4 gives the results.

TABLE 20–4

Taxpayer	X_1 *Actual 1978 Taxes*	X_2 *1978 Taxes if the Reform Plan Had Been in Operation*	*Sign of* $(X_1 - X_2)$
A	\$3,200	\$2,900	+
B	\$2,800	\$2,300	+
C	\$3,500	\$3,400	+
D	\$1,200	\$1,400	−
E	\$2,600	\$2,100	+
F	\$3,900	\$3,400	+
G	\$4,100	\$3,600	+
H	\$1,500	\$1,800	−
I	\$3,000	\$2,500	+
J	\$2,700	\$2,400	+
K	\$3,300	\$2,800	+
L	\$4,200	\$3,700	+

As long as the "actual" and "recalculated" tax distributions are the same, the probability of a "+" difference and the probability of a "−" difference are both .5. If $\alpha = .05$, the critical values for testing

H_0: The tax populations are identical (or $\pi = .5$)

H_1: The tax populations are not identical (or $\pi \neq .5$)

are $Z = -1.96$ and $Z = 1.96$ (see Figure 20–4). Given ten "+"s,

$$Z = \frac{10 - 6}{\sqrt{(12)(.5)(.5)}} = 2.31$$

FIGURE 20–4

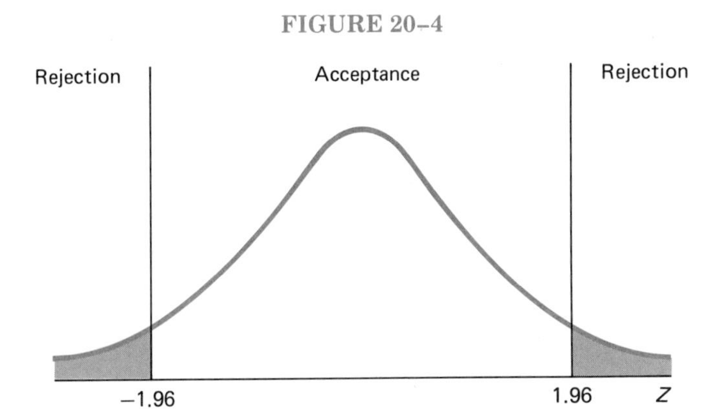

Since 2.31 lies in the rejection region, the committee concludes that the tax package alters the distribution of taxes.

THE MANN-WHITNEY TEST

The *Mann-Whitney test* is used to compare two populations. The populations are assumed to be continuous, and samples are selected independently. We will now see how the sampling distribution for this test is constructed.

The hypotheses are as follows:

H_0: Population 1 and Population 2 are identical.

H_1: Population 1 and Population 2 are not identical.

Suppose H_0 is true. Let's pick n_1 observations from Population 1 and n_2 from Population 2. Under these circumstances, there are

$$\frac{(n_1 + n_2)!}{n_1![(n_1 + n_2) - n_1]!}$$

equally likely ways to rank the n_1 items. We are interested in the sampling distribution of

$$U = n_1 n_2 + \frac{n_1(n_1 + 1)}{2} - R_1 \qquad (20\text{–}3)$$

where R_1 = the sum of the ranks for the observations in Sample 1.

We'll imagine that $n_1 = 2$ and $n_2 = 3$. If all five observations are listed in order (i.e., the smallest observation occupies rank #1, etc.), ten different Sample 1 rankings are possible (see Table 20–5). We may, then, create the probability distribution of U, shown in Table 20–6.

TABLE 20–5

Rankings of the Observations in Sample 1	R_1	U
1,2	3	6
1,3	4	5
1,4	5	4
1,5	6	3
2,3	5	4
2,4	6	3
2,5	7	2
3,4	7	2
3,5	8	1
4,5	9	0

TABLE 20–6

U	$P(U)$
0	1/10
1	1/10
2	2/10
3	2/10
4	2/10
5	1/10
6	1/10

For this probability distribution,

$$E(U) = \Sigma UP(U) = 3.0$$

$$V(U) = \Sigma[U - E(U)]^2P(U) = 3.0$$

The parameters of the U distribution can also be found via the following:

$$E(U) = \frac{n_1n_2}{2} \qquad (20\text{–}4)$$

$$V(U) = \frac{n_1n_2(n_1 + n_2 + 1)}{12} \qquad (20\text{–}5)$$

When n_1 and n_2 are both at least ten, we can approximate the sampling distribution with a normal curve. Let's look at an example.

A researcher feels that male and female students will perform the same on a current affairs quiz. She has collected pictures concerning 100 "front-page" stories and independently draws a random sample of eleven male undergraduates and a random sample of twelve female undergraduates. The students are told to write one paragraph about each picture. These responses are graded for accuracy and depth of understanding. Table 20–7 shows the scores.

We wish to test the following:

H_0: The population of male scores is the same as the population of female scores.

H_1: The populations are different.

TABLE 20–7

Males ($n_1 = 11$)	*Females* ($n_2 = 12$)
18	42
31	16
40	10
7	38
52	43
36	54
24	14
15	26
37	29
23	49
29	32
	17

To apply the Mann-Whitney test, we must first combine the samples and rank the scores. This is accomplished in Table 20–8. Two of the observations (i.e., 29 and 29) "tie" for ranks #11 and #12. We have recorded the average of these ranks. Given Table 20–8,

$$R_1 = 1 + 4 + 7 + 8 + 9 + 11.5 + 13 + 15 + 16 + 18 + 22$$
$$= 124.5$$

and

$$U = (11)(12) + \frac{(11)(11 + 1)}{2} - 124.5 = 73.5$$

Provided the male and female distributions are identical, the n_1 items would rarely cluster at the lower or upper rankings. Because of this, "small" and "large" values of R_1 or "large" and "small" values of U are unlikely.

When the level of significance is .05, the acceptance and rejection regions are as shown in Figure 20–5. Since

$$Z = \frac{73.5 - E(U)}{\sqrt{V(U)}} = \frac{73.5 - \frac{(11)(12)}{2}}{\sqrt{\frac{(11)(12)(11 + 12 + 1)}{12}}} = 0.46,$$

TABLE 20–8

Observation	*Rank*	*Sample Containing the Observation*
7	1	#1
10	2	2
14	3	2
15	4	1
16	5	2
17	6	2
18	7	1
23	8	1
24	9	1
26	10	2
29	11.5	1
29	11.5	2
31	13	1
32	14	2
36	15	1
37	16	1
38	17	2
40	18	1
42	19	2
43	20	2
49	21	2
52	22	1
54	23	2

FIGURE 20–5

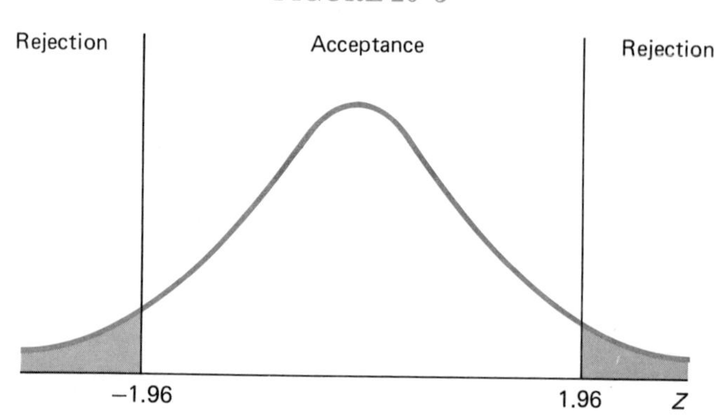

the researcher accepts H_0; she concludes that the population of male scores is the same as the population of female scores.

If the researcher is willing to assume at the beginning that the populations have the same *shape,* as in Figure 20–6, for example, she can test hypotheses concerning the relationship between μ_1 and μ_2. Such an assumption implies $\sigma_1^2 = \sigma_2^2$. Suppose the researcher believes that females have more knowledge of current affairs than males do. She is going to investigate

H_0: $\mu_1 \geq \mu_2$ (i.e., Population 1 is identical to or "above" Population 2)

H_1: $\mu_1 < \mu_2$ (i.e., Population 1 is "below" Population 2)

FIGURE 20–6

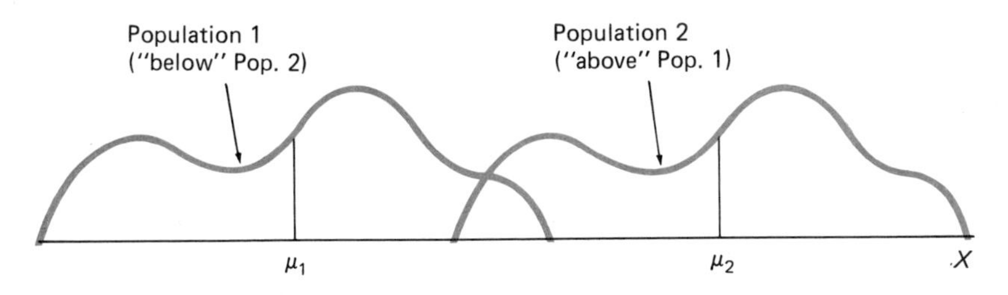

When H_0 is true, a small R_1 or a large U is "improbable." The rejection region, therefore, falls in the right tail. Using $\alpha = .01$, we get the acceptance and rejection regions in Figure 20–7. Because the computed Z is .46, the researcher accepts the null hypothesis. Her contention that females out-perform (i.e., have a higher average) than males is not supported by the data.

FIGURE 20–7

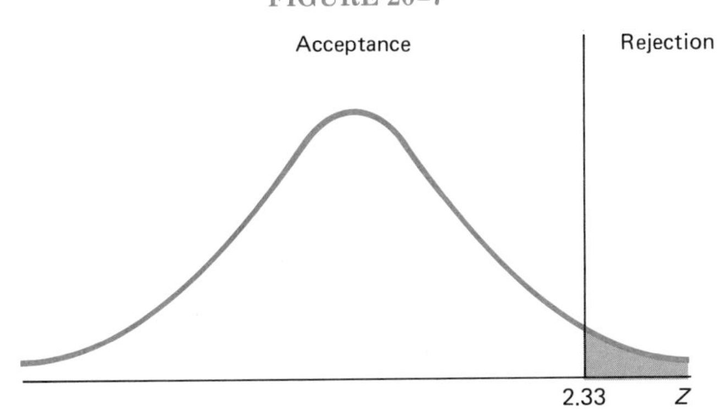

THE WILCOXON TEST

We have emphasized that the sign test "ignores" the absolute value of the difference between the members of a matched pair. Differences of −.001 and −2,500,000 are only recorded as "−"s. The *Wilcoxon test* uses more information than the sign test does and is, therefore, more powerful (i.e., the probability of rejecting a false null hypothesis is greater).

Suppose we compute $n = 3$ differences, rank the absolute values from smallest to largest, and attach the sign of the differences to the ranks. If Population 1 and Population 2 are identical, every possible arrangement of "signed" ranks is equally likely (see Table 20–9). Under these circumstances, we can determine the sampling distribution of R_+ = the sum of the positive ranks.

TABLE 20–9

Arrangements		
+,	+,	+
−,	−,	−
+,	+,	−
−,	+,	+
+,	−,	+
−,	−,	+
+,	−,	−
−,	+,	−

Table 20–10 presents the value of R_+ for the eight arrangements of three differences. Hence, the probability distribution of R_+ would be as shown in Table 20–11.

TABLE 20–10

Arrangements	R_+
+, +, +	6
−, −, −	0
+, +, −	3
−, +, +	5
+, −, +	4
−, −, +	3
+, −, −	1
−, +, −	2

TABLE 20–11

R_+	$P(R_+)$
0	1/8
1	1/8
2	1/8
3	2/8
4	1/8
5	1/8
6	1/8

We can derive the parameters of R_+ through either

$$E(R_+) = \Sigma R_+ P(R_+) = 3.0$$

$$V(R_+) = \Sigma[R_+ - E(R_+)]^2 P(R_+) = 3.5$$

or

$$E(R_+) = \frac{n(n+1)}{4} \qquad (20\text{–}6)$$

$$V(R_+) = \frac{n(n+1)(2n+1)}{24} \qquad (20\text{–}7)$$

If n is 12 or more, the sampling distribution of R_+ is approximately normal. The following example illustrates the Wilcoxon test.

An economist would like to investigate the behavior of *Fortune*-500 companies in the period following the 1974–1975 recession. Since the United States recovered very slowly from that recession, the economist feels that the 1975 and 1976 distributions of X = earnings per share of common stock were the same. The data from a random sample of fourteen companies are displayed in Table 20–12.

TABLE 20–12

Company	X_1 *1975 Earnings per Share*	X_2 *1976 Earnings per Share*	$X_1 - X_2$
Twentieth Century–Fox	\$3.00	\$1.41	\$1.59
Stokely–Van Camp	3.09	2.54	.55
Pitney-Bowes	1.95	1.55	.40
Hershey Foods	2.53	3.32	− .79
Revlon	2.18	2.68	− .50
Polaroid	1.91	2.43	− .52
Time, Inc.	2.26	3.32	−1.06
Pillsbury	2.75	2.73	.02
Kellogg	1.40	1.71	− .31
Texas Instruments	2.71	4.25	−1.54
Johnson & Johnson	3.18	3.53	− .35
Xerox	3.07	4.51	−1.44
Aluminum Co. of America	1.85	4.14	−2.29
Greyhound	1.87	1.76	.11

Source: *Fortune,* May 1977, pp. 366–381.

Once the differences between X_{i1} and X_{i2} have been established, we must rank their absolute values and indicate the sign of the ranks (see Table 20–13). We find that $R_+ = 1 + 2 + 5 + 8 + 13 = 29$. If a difference were zero, we would delete the corresponding matched pair from the sample. Ties are handled as they are in the Mann-Whitney test.

TABLE 20–13

$\lvert X_1 - X_2 \rvert$	*Rank*	$X_1 - X_2$	*Sign of the Rank*
.02	1	.02	+
.11	2	.11	+
.31	3	−.31	−
.35	4	−.35	−
.40	5	.40	+
.50	6	−.50	−
.52	7	−.52	−
.55	8	.55	+
.79	9	−.79	−
1.06	10	−1.06	−
1.44	11	−1.44	−
1.54	12	−1.54	−
1.59	13	1.59	+
2.29	14	−2.29	−

The economist will test the following:

H_0: The distribution of X_1 is identical to the distribution of X_2.

H_1: The distributions are not identical.

The acceptance and rejection regions for a level of significance of .01 are shown in Figure 20–8. Based on $R_+ = 29$ and $n = 14$, the researcher calculates

$$Z = \frac{29 - E(R_+)}{\sqrt{V(R_+)}}$$

$$= \frac{29 - \dfrac{(14)(14 + 1)}{4}}{\sqrt{\dfrac{(14)(14 + 1)[(2)(14) + 1]}{24}}} = -1.48$$

$Z = -1.48$ lies in the acceptance region. As such, $R_+ = 29$ is "possible and probable" when the null hypothesis is true, and the researcher concludes that the distribution of earnings per share was the same in 1975 and 1976.

(Note: If the sign test were applied to this example, the researcher would compute

$$Z = \frac{r - n\pi}{\sqrt{n\pi(1 - \pi)}} = \frac{5 - (14)(.5)}{\sqrt{(14)(.5)(.5)}} = -1.07$$

In this instance, each test leads to accepting H_0.)

FIGURE 20–8

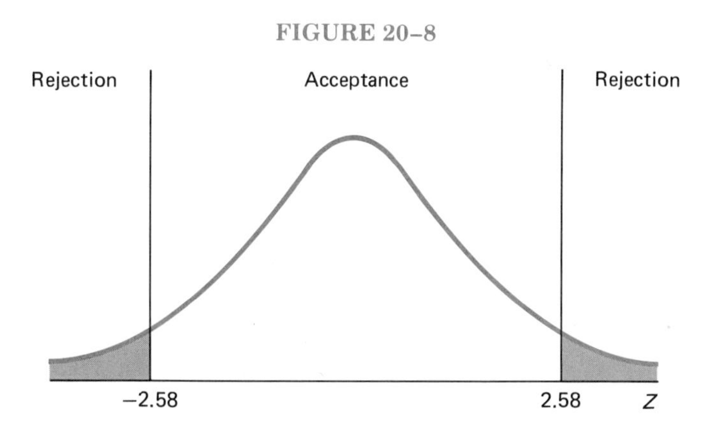

THE RUNS TEST

The *runs test* is used to investigate whether a sample has been chosen at random. We'll list the observations in the order of their selection, and let A = an observation above the median and B = an observation below the median. Given n_1 As and n_2 Bs, the number of permutations of the $n_1 + n_2$ items is

$$\frac{(n_1 + n_2)!}{n_1![(n_1 + n_2) - n_1]!}$$

When observations are randomly drawn, each permutation is equally likely.

Suppose $n_1 = 2$ and $n_2 = 2$. Table 20–14 lists the possible permutations. Each uninterrupted sequence of As or sequence of Bs is referred to as a "run." A run can also consist of a single A or B; for example, A B A represents three runs. Table 20–15 shows the values of T = the number of runs when $n_1 = 2$ and $n_2 = 2$.

TABLE 20–14

Permutations
A, *A*, *B*, *B*
A, *B*, *A*, *B*
B, *A*, *A*, *B*
B, *B*, *A*, *A*
B, *A*, *B*, *A*
A, *B*, *B*, *A*

TABLE 20–15

Permutations	T
$A\ A\ B\ B$	2
$A\ B\ A\ B$	4
$B\ A\ A\ B$	3
$B\ B\ A\ A$	2
$B\ A\ B\ A$	4
$A\ B\ B\ A$	3

The runs test develops from the sampling distribution of T when the null hypothesis

$$H_0\text{: The data are randomly picked}$$

is true. We have derived this distribution for $n_1 = 2$ and $n_2 = 2$ in Table 20–16. We find that

$$E(T) = \Sigma TP(T) = 3.0$$

$$V(T) = \Sigma[T - E(T)]^2 P(T) = 0.67$$

TABLE 20–16

T	$P(T)$
2	2/6
3	2/6
4	2/6

The following equations are an alternative approach to determining $E(T)$ and $V(T)$:

$$E(T) = \frac{2n_1n_2}{n_1 + n_2} + 1 \qquad (20\text{–}8)$$

$$V(T) = \frac{2n_1n_2(2n_1n_2 - n_1 - n_2)}{(n_1 + n_2)^2(n_1 + n_2 - 1)} \qquad (20\text{–}9)$$

As long as either n_1 or n_2 is greater than 20, the sampling distribution of T is approximately normal.

We will be suspicious of randomness if T is "small" or "large." Such outcomes imply "patterns" in the data, for example,

$$A\ B\ A\ B\ A\ B\ A\ B\ A\ B\ A\ B\ A \qquad (T = 13)$$

$$A\ A\ A\ A\ A\ A\ A\ B\ B\ B\ B\ B\ B \qquad (T = 2)$$

In both of these instances, the researcher would feel that something other than chance was responsible for the items actually selected. Consider the following example.

President Nixon tried to attack inflation by using wage and price controls. Although inflation declined while the controls were in effect, prices shot up substantially when the controls were removed. Many economists believe that this experience confirms that controls are not a long-range answer to the inflation problem. A researcher has obtained the membership roster of a local chamber of commerce. He wants to pick a random sample of 49 members and estimate π = the proportion of members favoring wage and price controls. A number is assigned to each member, and 49 numbered slips are chosen. Table 20–17 presents the numbers and the order of their selection.

TABLE 20–17

1.	20	18.	760	34.	13
2.	510	19.	959	35.	544
3.	783	20.	18	36.	758
4.	243	21.	92	37.	57
5.	1,100	22.	701	38.	96
6.	87	23.	22	39.	185
7.	637	24.	633	40.	95
8.	462	25.	1,005	41.	932
9.	573	26.	774	42.	481
10.	891	27.	400	43.	93
11.	37	28.	89	44.	7
12.	68	29.	922	45.	842
13.	915	30.	56	46.	73
14.	1,023	31.	781	47.	39
15.	46	32.	406	48.	808
16.	102	33.	62	49.	523
17.	358				

The median is 406. We'll eliminate it (i.e., observation #32), and label the other items A or B (see Table 20–18). We then have $n_1 = 24$, $n_2 = 24$, and $T = 26$.

TABLE 20–18

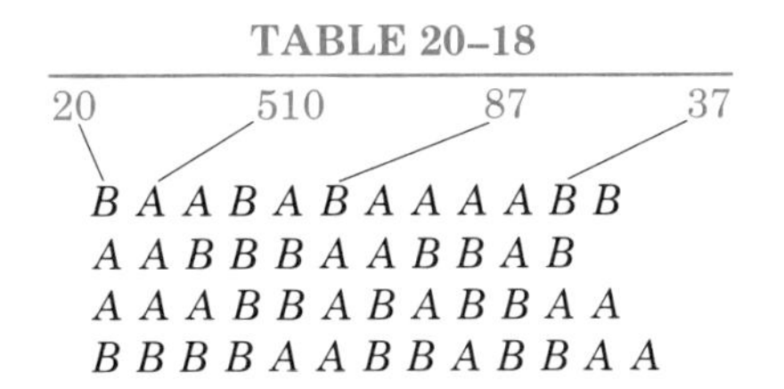

The researcher will test the following:

H_0: The data are randomly selected.

H_1: The data are not randomly selected.

When $\alpha = .05$, the acceptance and rejection regions are as shown in Figure 20–9. Inasmuch as the computed Z is

$$Z = \frac{26 - E(T)}{\sqrt{V(T)}}$$

$$= \frac{26 - \left[\dfrac{(2)(24)(24)}{(24 + 24)} + 1\right]}{\sqrt{\dfrac{(2)(24)(24)[(2)(24)(24) - 24 - 24]}{(24 + 24)^2(24 + 24 - 1)}}} = 0.29,$$

the researcher accepts his null hypothesis; $T = 26$ is not "improbable." He is now ready to contact the members of his sample and estimate π.

FIGURE 20–9

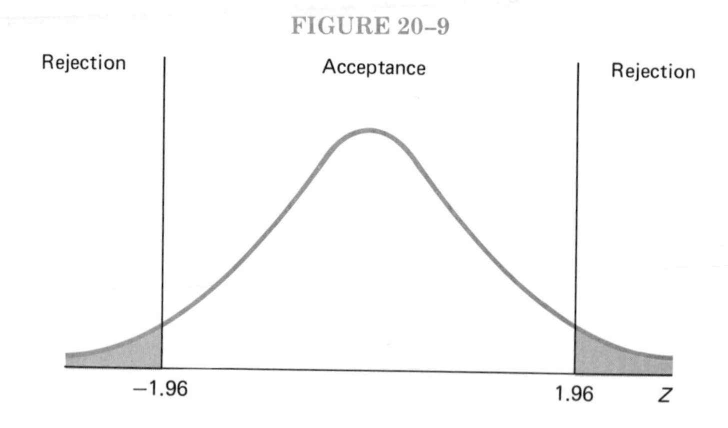

THE SPEARMAN RANK CORRELATION COEFFICIENT

Without assuming that X and Y are bivariate normal, we can test their independence by ranking these random variables. Formally, the hypotheses are:

H_0: X and Y are independent.

H_1: X and Y are not independent.

For each of $n = 3$ randomly selected observations, let's identify an X and a Y value, as in Table 20–19, and then rank the values in the X and Y columns (see Table 20–20). If X and Y aren't correlated, $n!$ equally likely sets of rankings are possible.

TABLE 20–19

Observation	X	Y
#1	16	58
#2	31	63
#3	42	29

TABLE 20–20

Observation	*X Rank*	*Y Rank*
#1	1	2
#2	2	3
#3	3	1

To evaluate the relation between the rankings, we might compute

$$d_i = X_i \text{ rank} - Y_i \text{ rank}$$

and total the deviations. However,

$$\Sigma d = \Sigma(X \text{ rank} - Y \text{ rank}) = \Sigma(X \text{ rank}) - \Sigma(Y \text{ rank})$$

$$\Sigma(X \text{ rank}) = \Sigma(Y \text{ rank})$$

$$\Sigma d = 0$$

We will, therefore, calculate Σd^2. Our attention will be directed at the *Spearman rank correlation coefficient:*

$$r_S = 1 - \frac{6\Sigma d^2}{n(n^2 - 1)} \tag{20–10}$$

All possible sets of rankings, given $n = 3$, as well as the r_Ss produced by them, are indicated in Table 20–21. When H_0 is true, and n is at least ten,

$$t_{(n-2)} = \frac{r_S}{\sqrt{(1 - r_S^2)/(n - 2)}} \tag{20–11}$$

that is, the ratio on the right possesses approximately a t distribution with $n - 2$ degrees of freedom. Let's look at an example.

A group of researchers is interested in determining whether there is any connection between the writing and speaking skills of corporate executives. The researchers have developed a "scoring system" to analyze these qualities. Fourteen executives are picked at random and read a report on how Company G convinced Com-

TABLE 20–21

X Rank	*Y Rank*	*X Rank*	*Y Rank*	*X Rank*	*Y Rank*
1	1	1	3	1	1
2	2	2	2	2	3
3	3	3	1	3	2
$r_S = 1.0$		$r_S = -1.0$		$r_S = 0.5$	

X Rank	*Y Rank*	*X Rank*	*Y Rank*	*X Rank*	*Y Rank*
1	3	1	2	1	2
2	1	2	1	2	3
3	2	3	3	3	1
$r_S = -0.5$		$r_S = 0.5$		$r_S = -0.5$	

pany H to merge with it. The executives are given thirty minutes to write a summary of the report. The papers are collected, and then the executives tape-record their version of the merger. Table 20–22 details the outcomes of this experiment.

The researchers will test the following:

H_0: X and Y are not correlated.

H_1: X and Y are correlated.

TABLE 20–22

Executive	X *Writing Score*	Y *Speaking Score*
A	30	42
B	36	28
C	41	57
D	49	63
E	52	41
F	54	78
G	57	85
H	63	38
I	63	74
J	71	68
K	75	80
L	78	83
M	81	66
N	93	89

Imagine that $\alpha = .05$. The acceptance and rejection regions for 14 − 2 = 12 degrees of freedom are shown in Figure 20–10. The derivation of Σd^2 is provided in Table 20–23.

FIGURE 20–10

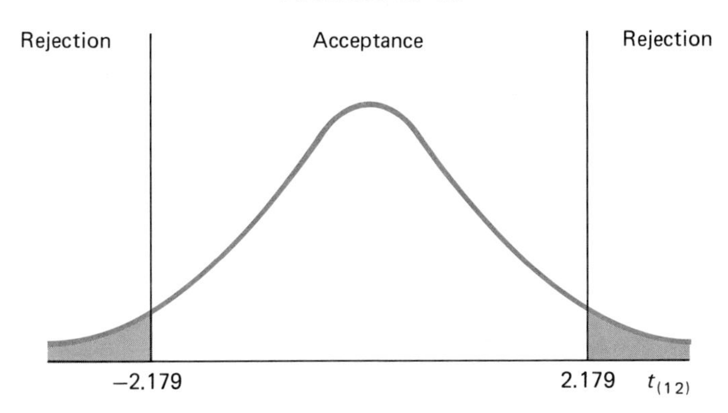

TABLE 20–23

Executive	*X Rank*	*Y Rank*	*d*	d^2
A	1	4	−3	9
B	2	1	1	1
C	3	5	−2	4
D	4	6	−2	4
E	5	3	2	4
F	6	10	−4	16
G	7	13	−6	36
H	8.5	2	6.5	42.25
I	8.5	9	−0.5	0.25
J	10	8	2	4
K	11	11	0	0
L	12	12	0	0
M	13	7	6	36
N	14	14	0	0
				156.5

Since

$$r_S = 1 - \frac{(6)(156.5)}{(14)(14^2 - 1)} = .66$$

and

$$t = \frac{.66}{\sqrt{(1 - .66^2)/12}} = 3.04,$$

the researchers reject H_0 and conclude that writing and speaking skills are related. An r_S of .66 is "possible but improbable" when X and Y are uncorrelated.

SUMMARY

We have said that the t and F tests depend upon the assumption of normality. If a t test is used when a population is not normal, the researcher will identify incorrect values for α and β. Although many of the sampling distributions constructed for nonparametric hypothesis tests approach the normal distribution as the sample size increases, the sampling distributions themselves are derived from very few assumptions regarding the population.

EXERCISES

1. A researcher wishes to test H_0: $\mu = \$7,000$. He selects a random sample of 20 observations from a nonnormal population and determines that 14 of them are greater than \$7,000. Is his null hypothesis acceptable at the .05 level of significance?

2. In how many different ways can $n_1 = 8$ items be ranked with $n_2 = 10$ items if there are no ties?

3. Suppose $n_1 = 3$ and $n_2 = 3$. Construct the sampling distribution of U under the assumption that Population 1 and Population 2 are identical.

4. An investigator believes that two particular populations have the same distribution. She independently selects random samples of $n_1 = 16$ and $n_2 = 14$ observations and records the following data:

Sample #1	Sample #2
2	4
12	21
17	13
6	11
14	15
11	6
10	3
5	17
20	25
4	8
7	19
18	16
16	23
9	30
3	
27	

Test her belief at the .05 level of significance.

5. Given $n = 4$, construct the sampling distribution of R_+ under the assumption that Population 1 and Population 2 are identical.

6. In preparing for a Wilcoxon test, a researcher calculated the following differences:

$X_1 - X_2$
2.03
−0.44
−1.36
−2.48
3.57
4.21
−0.86
2.75
3.14
4.08
1.13
−4.26
6.73
3.55
0.06

Using $\alpha = .05$, is H_0: The distribution of X_1 is the same as the distribution of X_2 acceptable?

7. Imagine that a researcher draws a sample. Given $n_1 = 40$, $n_2 = 40$, and $T = 19$, and using $\alpha = .05$, test H_0: The data are randomly selected.

8. An investigator chooses a random sample of 15 objects and identifies the values of X and Y corresponding to these objects. This information is displayed below.

Object	X	Y
A	15	3
B	19	27
C	42	57
D	76	61
E	51	89
F	82	90
G	33	98
H	41	25
I	87	14
J	65	36
K	54	31
L	16	44
M	73	20
N	92	7
O	48	95

a. Calculate r_S.
b. Using $\alpha = .05$, test H_0: X and Y are not correlated.

9. A financial analyst is studying the extent to which firms in industry G and industry H rely on borrowing to finance their operations. The "debt-to-equity ratio" is a measure of such reliance. The analyst feels that firms in both industries have the same distribution of debt-to-equity ratios. He independently selects random samples of 12 companies from G and 13 companies from H. Based on the following data and $\alpha = .05$, test his belief.

Debt-to-Equity Ratios in Industry G Sample	*Debt-to-Equity Ratios in Industry H Sample*
1.10	.85
.72	1.19
.31	1.01
1.54	1.63
1.73	1.44
.95	1.17
1.46	1.58
1.20	1.67
1.68	1.81
1.07	1.45
1.92	1.11
.84	1.02
	1.60

10. The analyst in question 9 also believes that debt-to-equity ratios differ between recession and non-recession years. The Sample G observations reported previously are associated with a non-recession year. The financial analyst compares these to data obtained for a recession year, as shown below. Using the sign test and $\alpha = .05$, will he be able to reject H_0: The distributions of debt-to-equity ratios are identical?

Company in Industry G	*Debt-to-Equity Ratios for a Non-Recession Year*	*Debt-to-Equity Ratios for a Recession Year*
#1	1.10	1.21
2	0.72	0.85
3	0.31	0.26
4	1.54	1.83
5	1.73	1.64
6	0.95	0.86
7	1.46	1.50
8	1.20	1.37
9	1.68	1.72
10	1.07	1.36
11	1.92	1.42
12	0.84	1.03

11. A researcher wants to test

$$H_0: \mu = 54$$
$$H_1: \mu \neq 54$$

He selects a random sample including the following observations:

X
39
21
36
43
62
71
25
27
60
18
24
65

a. Suppose the population is normal. Use t and $\alpha = .05$ to test H_0.
b. Suppose the population is nonnormal. Use the sign test and $\alpha = .05$ to test H_0.

12. The Phillips Curve indicates the historical relation between inflation and the unemployment rate. It associates "high" unemployment rates with "mild" inflation and "low" unemployment rates with "severe" inflation. During the 1970s the United States went through periods in which high unemployment rates and severe inflation occurred simultaneously. This condition has forced some economists to challenge the validity of the Phillips Curve. An investigator wishes to determine how members of the American Economic Association feel on this subject. A number is assigned to each member and 53 slips are chosen. The numbers and the order of their selection are as follows:

1.	804	15.	2,001	28.	1,751	41.	303
2.	546	16.	005	29.	888	42.	310
3.	075	17.	983	30.	238	43.	772
4.	769	18.	127	31.	390	44.	649
5.	1,067	19.	112	32.	508	45.	1,192
6.	2,087	20.	2,994	33.	193	46.	2,335
7.	073	21.	808	34.	1,645	47.	2,909
8.	2,908	22.	189	35.	1,689	48.	078
9.	441	23.	500	36.	2,631	49.	060
10.	592	24.	676	37.	009	50.	122
11.	618	25.	844	38.	1,894	51.	408
12.	033	26.	1,090	39.	2,111	52.	820
13.	488	27.	2,064	40.	668	53.	377
14.	319						

Given $\alpha = .05$, test H_0: The data are randomly selected.

13. Suppose a researcher draws a random sample of 14 objects and identifies the values of X and Y corresponding to these objects. Based on the following data,

Object	X	Y
A	3	72
B	9	80
C	6	14
D	15	55
E	12	63
F	14	7
G	23	16
H	18	20
I	40	32
J	38	59
K	43	17
L	11	26
M	52	67
N	61	49

a. Calculate r_S.
b. Using the .05 level of significance, can the researcher accept H_0: X and Y are not correlated?

14. In how many different ways can $n_1 = 11$ items be ranked with $n_2 = 15$ items if there are no ties?

15. A researcher is studying the behavior of Corporation D's common stock. Using a = an increase in price relative to the day before, b = a decrease in price relative to the day before, and c = no change in price relative to the day before, she recorded the following information concerning a particular 60-day period. (Note: The data are in chronological order.)

1. a	16. c	31. b	46. a
2. a	17. b	32. a	47. b
3. b	18. b	33. a	48. a
4. c	19. b	34. b	49. a
5. b	20. a	35. a	50. b
6. b	21. a	36. b	51. b
7. a	22. a	37. b	52. b
8. a	23. b	38. a	53. b
9. a	24. a	39. b	54. c
10. a	25. b	40. c	55. b
11. a	26. a	41. c	56. b
12. b	27. a	42. b	57. a
13. b	28. c	43. a	58. a
14. a	29. b	44. a	59. a
15. b	30. b	45. a	60. a

Eliminate the cs and let n_1 = the number of as and n_2 = the number of bs. If α = .05, can the researcher accept H_0: Changes in stock prices occur randomly?

THE STANDARD NORMAL DISTRIBUTION

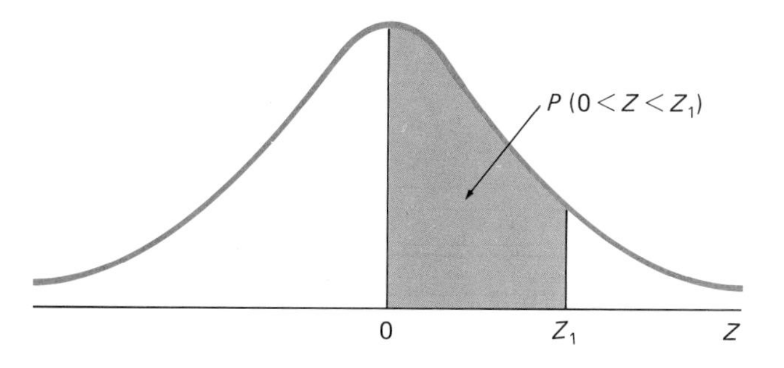

Areas between $Z = 0$ and another Z

z	*.00*	*.01*	*.02*	*.03*	*.04*	*.05*	*.06*	*.07*	*.08*	*.09*
0.0	.0000	.0040	.0080	.0120	.0160	.0199	.0239	.0279	.0319	.0359
0.1	.0398	.0438	.0478	.0517	.0557	.0596	.0636	.0675	.0714	.0753
0.2	.0793	.0832	.0871	.0910	.0948	.0987	.1026	.1064	.1103	.1141
0.3	.1179	.1217	.1255	.1293	.1331	.1368	.1406	.1443	.1480	.1517
0.4	.1554	.1591	.1628	.1664	.1700	.1736	.1772	.1808	.1844	.1879
0.5	.1915	.1950	.1985	.2019	.2054	.2088	.2123	.2157	.2190	.2224
0.6	.2257	.2291	.2324	.2357	.2389	.2422	.2454	.2486	.2517	.2549
0.7	.2580	.2611	.2642	.2673	.2704	.2734	.2764	.2794	.2823	.2852
0.8	.2881	.2910	.2939	.2967	.2995	.3023	.3051	.3078	.3106	.3133
0.9	.3159	.3186	.3212	.3238	.3264	.3289	.3315	.3340	.3365	.3389
1.0	.3413	.3438	.3461	.3485	.3508	.3531	.3554	.3577	.3599	.3621
1.1	.3643	.3665	.3686	.3708	.3729	.3749	.3770	.3790	.3810	.3830
1.2	.3849	.3869	.3888	.3907	.3925	.3944	.3962	.3980	.3997	.4015
1.3	.4032	.4049	.4066	.4082	.4099	.4115	.4131	.4147	.4162	.4177
1.4	.4192	.4207	.4222	.4236	.4251	.4265	.4279	.4292	.4306	.4319
1.5	.4332	.4345	.4357	.4370	.4382	.4394	.4406	.4418	.4429	.4441
1.6	.4452	.4463	.4474	.4484	.4495	.4505	.4515	.4525	.4535	.4545
1.7	.4554	.4564	.4573	.4582	.4591	.4599	.4608	.4616	.4625	.4633
1.8	.4641	.4649	.4656	.4664	.4671	.4678	.4686	.4693	.4699	.4706
1.9	.4713	.4719	.4726	.4732	.4738	.4744	.4750	.4756	.4761	.4767
2.0	.4772	.4778	.4783	.4788	.4793	.4798	.4803	.4808	.4812	.4817
2.1	.4821	.4826	.4830	.4834	.4838	.4842	.4846	.4850	.4854	.4857
2.2	.4861	.4864	.4868	.4871	.4875	.4878	.4881	.4884	.4887	.4890
2.3	.4893	.4896	.4898	.4901	.4904	.4906	.4909	.4911	.4913	.4916
2.4	.4918	.4920	.4922	.4925	.4927	.4929	.4931	.4932	.4934	.4936
2.5	.4938	.4940	.4941	.4943	.4945	.4946	.4948	.4949	.4951	.4952

(continued)

THE STANDARD NORMAL DISTRIBUTION *(Continued)*

z	.00	.01	.02	.03	.04	.05	.06	.07	.08	.09
2.6	.4953	.4955	.4956	.4957	.4959	.4960	.4961	.4962	.4963	.4964
2.7	.4965	.4966	.4967	.4968	.4969	.4970	.4971	.4972	.4973	.4974
2.8	.4974	.4975	.4976	.4977	.4977	.4978	.4979	.4979	.4980	.4981
2.9	.4981	.4982	.4982	.4983	.4984	.4984	.4985	.4985	.4986	.4986
3.0	.4987	.4987	.4987	.4988	.4988	.4989	.4989	.4989	.4990	.4990

EXAMPLES:

(1) The area between $Z = 0$ and $Z = 2.14$ is .4838.
(2) The area between $Z = 0$ and $Z = -1.62$ is .4474.
(3) The area between $Z = -1.31$ and $Z = 2.05$ is $.4049 + .4798 = .8847$.
(4) The area between $Z = 1.76$ and $Z = 2.54$ is $.4945 - .4608 = .0337$.

THE t DISTRIBUTION

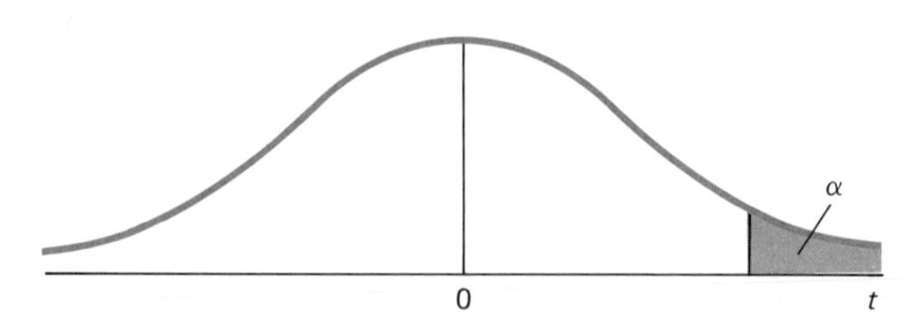

Degrees of Freedom	Critical values for the upper α area				
	.10	.05	.025	.01	.005
1	3.078	6.314	12.706	31.821	63.657
2	1.886	2.920	4.303	6.965	9.925
3	1.638	2.353	3.182	4.541	5.841
4	1.533	2.132	2.776	3.747	4.604
5	1.476	2.015	2.571	3.365	4.032
6	1.440	1.943	2.447	3.143	3.707
7	1.415	1.895	2.365	2.998	3.499
8	1.397	1.860	2.306	2.896	3.355
9	1.383	1.833	2.262	2.821	3.250
10	1.372	1.812	2.228	2.764	3.169
11	1.363	1.796	2.201	2.718	3.106
12	1.356	1.782	2.179	2.681	3.055
13	1.350	1.771	2.160	2.650	3.012
14	1.345	1.761	2.145	2.624	2.977

(continued)

THE t DISTRIBUTION *(Continued)*

Degrees of Freedom	*Critical values for the upper α area*				
	.10	.05	.025	.01	.005
15	1.341	1.753	2.131	2.602	2.947
16	1.337	1.746	2.120	2.583	2.921
17	1.333	1.740	2.110	2.567	2.898
18	1.330	1.734	2.101	2.552	2.878
19	1.328	1.729	2.093	2.539	2.861
20	1.325	1.725	2.086	2.528	2.845
21	1.323	1.721	2.080	2.518	2.831
22	1.321	1.717	2.074	2.508	2.819
23	1.319	1.714	2.069	2.500	2.807
24	1.318	1.711	2.064	2.492	2.797
25	1.316	1.708	2.060	2.485	2.787
26	1.315	1.706	2.056	2.479	2.779
27	1.314	1.703	2.052	2.473	2.771
28	1.313	1.701	2.048	2.467	2.763
29	1.311	1.699	2.045	2.462	2.756
30	1.310	1.697	2.042	2.457	2.750
40	1.303	1.684	2.021	2.423	2.704
60	1.296	1.671	2.000	2.390	2.660
120	1.289	1.658	1.980	2.358	2.617
∞	1.282	1.645	1.960	2.326	2.576

EXAMPLES:

(1) $P[t_{(9)} > 1.383] = .10$
(2) $P[t_{(15)} > 2.131] = .025$
(3) $P[t_{(23)} < -2.069] = .025$
(4) $P[-2.763 < t_{(28)} < 2.763] = .99$

THE χ^2 DISTRIBUTION

Degrees of Freedom	Critical values for the upper α area									
	0.995	0.990	0.975	0.950	0.900	0.100	0.050	0.025	0.010	0.005
1	0.0000393	0.000157	0.000982	0.00393	0.0157908	2.70554	3.84146	5.02389	6.63490	7.87944
2	0.0100251	0.0201007	0.0506356	0.102587	0.210721	4.60517	5.99146	7.37776	9.21034	10.5966
3	0.0717218	0.114832	0.215795	0.351846	0.584374	6.25139	7.81473	9.34840	11.3449	12.8382
4	0.206989	0.297109	0.484419	0.710723	1.063623	7.77944	9.48773	11.1433	13.2767	14.8603
5	0.411742	0.554298	0.831212	1.145476	1.61031	9.23636	11.0705	12.8325	15.0863	16.7496
6	0.675727	0.872090	1.23734	1.63538	2.20413	10.6446	12.5916	14.4494	16.8119	18.5476
7	0.989256	1.239043	1.68987	2.16735	2.83311	12.0170	14.0671	16.0128	18.4753	20.2777
8	1.34441	1.64650	2.17973	2.73264	3.48954	13.3616	15.5073	17.5345	20.0902	21.9550
9	1.73493	2.08790	2.70039	3.32511	4.16816	14.6837	16.9190	19.0228	21.6660	23.5894
10	2.15586	2.55821	3.24697	3.94030	4.86518	15.9872	18.3070	20.4832	23.2093	25.1882
11	2.60322	3.05348	3.81575	4.57481	5.57778	17.2750	19.6751	21.9200	24.7250	26.7568
12	3.07382	3.57057	4.40379	5.22603	6.30380	18.5493	21.0261	23.3367	26.2170	28.2995
13	3.56503	4.10692	5.00875	5.89186	7.04150	19.8119	22.3620	24.7356	27.6882	29.8195
14	4.07467	4.66043	5.62873	6.57063	7.78953	21.0641	23.6848	26.1189	29.1412	31.3194
15	4.60092	5.22935	6.26214	7.26094	8.54676	22.3071	24.9958	27.4884	30.5779	32.8013
16	5.14221	5.81221	6.90766	7.96165	9.31224	23.5418	26.2962	28.8454	31.9999	34.2672
17	5.69722	6.40776	7.56419	8.67176	10.0852	24.7690	27.5871	30.1910	33.4087	35.7185
18	6.26480	7.01491	8.23075	9.39046	10.8649	25.9894	28.8693	31.5264	34.8053	37.1565
19	6.84397	7.63273	8.90652	10.1170	11.6509	27.2036	30.1435	32.8523	36.1909	38.5823

THE χ^2 DISTRIBUTION *(Continued)*

Degrees of Freedom	*Critical values for the upper α area*									
	0.995	0.990	0.975	0.950	0.900	0.100	0.050	0.025	0.010	0.005
20	7.43384	8.26040	9.59078	10.8508	12.4426	28.4120	31.4104	34.1696	37.5662	39.9968
21	8.03365	8.89720	10.28293	11.5913	13.2396	29.6151	32.6706	35.4789	38.9322	41.4011
22	8.64272	9.54249	10.9823	12.3380	14.0415	30.8133	33.9244	36.7807	40.2894	42.7957
23	9.26043	10.19567	11.6886	13.0905	14.8480	32.0069	35.1725	38.0756	41.6384	44.1813
24	9.88623	10.8564	12.4012	13.8484	15.6587	33.1962	36.4150	39.3641	42.9798	45.5585
25	10.5197	11.5240	13.1197	14.6114	16.4734	34.3816	37.6525	40.6465	44.3141	46.9279
26	11.1602	12.1981	13.8439	15.3792	17.2919	35.5632	38.8851	41.9232	45.6417	48.2899
27	11.8076	12.8785	14.5734	16.1514	18.1139	36.7412	40.1133	43.1945	46.9629	49.6449
28	12.4613	13.5647	15.3079	16.9279	18.9392	37.9159	41.3371	44.4608	48.2782	50.9934
29	13.1211	14.2565	16.0471	17.7084	19.7677	39.0875	42.5570	45.7223	49.5879	52.3356
30	13.7867	14.9535	16.7908	18.4927	20.5992	40.2560	43.7730	46.9792	50.8922	53.6720
40	20.7065	22.1643	24.4330	26.5093	29.0505	51.8051	55.7585	59.3417	63.6907	66.7660
50	27.9907	29.7067	32.3574	34.7643	37.6886	63.1671	67.5048	71.4202	76.1539	79.4900
60	35.5345	37.4849	40.4817	43.1880	46.4589	74.3970	79.0819	83.2977	88.3794	91.9517
70	43.2752	45.4417	48.7576	51.7393	55.3289	85.5270	90.5312	95.0232	100.425	104.215
80	51.1719	53.5401	57.1532	60.3915	64.2778	96.5782	101.879	106.629	112.329	116.321
90	59.1963	61.7541	65.6466	69.1260	73.2911	107.565	113.145	118.136	124.116	128.299
100	67.3276	70.0649	74.2219	77.9295	82.3581	118.498	124.342	129.561	135.807	140.169

EXAMPLES:

(1) $P(\chi^2_{(9)} > 4.17) = .90$

(2) $P(\chi^2_{(17)} > 6.41) = .99$

(3) $P(\chi^2_{(23)} > 35.17) = .05$

(4) $P(\chi^2_{(28)} < 15.31) = .025$

THE F DISTRIBUTION

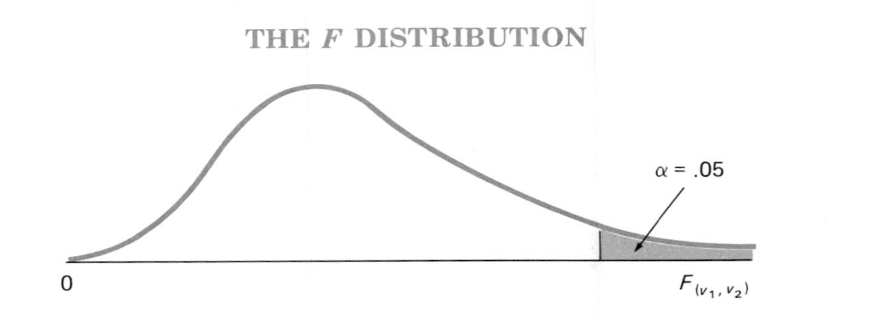

Critical values for the upper .05 area
v_1 = Degrees of Freedom—Numerator

v_2 = Degrees of Freedom—Denominator	1	2	3	4	5	6	7	8	9	10	12	15	20	24	30	40	60	120	∞
1	161.4	199.5	215.7	224.6	230.2	234.0	236.8	238.9	240.5	241.9	243.9	245.9	248.0	249.1	250.1	251.1	252.2	253.3	254.3
2	18.51	19.00	19.16	19.25	19.30	19.33	19.35	19.37	19.38	19.40	19.41	19.43	19.45	19.45	19.46	19.47	19.48	19.49	19.50
3	10.13	9.55	9.28	9.12	9.01	8.94	8.89	8.85	8.81	8.79	8.74	8.70	8.66	8.64	8.62	8.59	8.57	8.55	8.53
4	7.71	6.94	6.59	6.39	6.26	6.16	6.09	6.04	6.00	5.96	5.91	5.86	5.80	5.77	5.75	5.72	5.69	5.66	5.63
5	6.61	5.79	5.41	5.19	5.05	4.95	4.88	4.82	4.77	4.74	4.68	4.62	4.56	4.53	4.50	4.46	4.43	4.40	4.36
6	5.99	5.14	4.76	4.53	4.39	4.28	4.21	4.15	4.10	4.06	4.00	3.94	3.87	3.84	3.81	3.77	3.74	3.70	3.67
7	5.59	4.74	4.35	4.12	3.97	3.87	3.79	3.73	3.68	3.64	3.57	3.51	3.44	3.41	3.38	3.34	3.30	3.27	3.23
8	5.32	4.46	4.07	3.84	3.69	3.58	3.50	3.44	3.39	3.35	3.28	3.22	3.15	3.12	3.08	3.04	3.01	2.97	2.93
9	5.12	4.26	3.86	3.63	3.48	3.37	3.29	3.23	3.18	3.14	3.07	3.01	2.94	2.90	2.86	2.83	2.79	2.75	2.71
10	4.96	4.10	3.71	3.48	3.33	3.22	3.14	3.07	3.02	2.98	2.91	2.85	2.77	2.74	2.70	2.66	2.62	2.58	2.54
11	4.84	3.98	3.59	3.36	3.20	3.09	3.01	2.95	2.90	2.85	2.79	2.72	2.65	2.61	2.57	2.53	2.49	2.45	2.40
12	4.75	3.89	3.49	3.26	3.11	3.00	2.91	2.85	2.80	2.75	2.69	2.62	2.54	2.51	2.47	2.43	2.38	2.34	2.30
13	4.67	3.81	3.41	3.18	3.03	2.92	2.83	2.77	2.71	2.67	2.60	2.53	2.46	2.42	2.38	2.34	2.30	2.25	2.21
14	4.60	3.74	3.34	3.11	2.96	2.85	2.76	2.70	2.65	2.60	2.53	2.46	2.39	2.35	2.31	2.27	2.22	2.18	2.13

THE F DISTRIBUTION *(Continued)*

v_2 = Degrees of Freedom— Denominator	Critical values for the upper .05 area v_1 = Degrees of Freedom—Numerator																		
	1	2	3	4	5	6	7	8	9	10	12	15	20	24	30	40	60	120	∞
15	4.54	3.68	3.29	3.06	2.90	2.79	2.71	2.64	2.59	2.54	2.48	2.40	2.33	2.29	2.25	2.20	2.16	2.11	2.07
16	4.49	3.63	3.24	3.01	2.85	2.74	2.66	2.59	2.54	2.49	2.42	2.35	2.28	2.24	2.19	2.15	2.11	2.06	2.01
17	4.45	3.59	3.20	2.96	2.81	2.70	2.61	2.55	2.49	2.45	2.38	2.31	2.23	2.19	2.15	2.10	2.06	2.01	1.96
18	4.41	3.55	3.16	2.93	2.77	2.66	2.58	2.51	2.46	2.41	2.34	2.27	2.19	2.15	2.11	2.06	2.02	1.97	1.92
19	4.38	3.52	3.13	2.90	2.74	2.63	2.54	2.48	2.42	2.38	2.31	2.23	2.16	2.11	2.07	2.03	1.98	1.93	1.88
20	4.35	3.49	3.10	2.87	2.71	2.60	2.51	2.45	2.39	2.35	2.28	2.20	2.12	2.08	2.04	1.99	1.95	1.90	1.84
21	4.32	3.47	3.07	2.84	2.68	2.57	2.49	2.42	2.37	2.32	2.25	2.18	2.10	2.05	2.01	1.96	1.92	1.87	1.81
22	4.30	3.44	3.05	2.82	2.66	2.55	2.46	2.40	2.34	2.30	2.23	2.15	2.07	2.03	1.98	1.94	1.89	1.84	1.78
23	4.28	3.42	3.03	2.80	2.64	2.53	2.44	2.37	2.32	2.27	2.20	2.13	2.05	2.01	1.96	1.91	1.86	1.81	1.76
24	4.26	3.40	3.01	2.78	2.62	2.51	2.42	2.36	2.30	2.25	2.18	2.11	2.03	1.98	1.94	1.89	1.84	1.79	1.73
25	4.24	3.39	2.99	2.76	2.60	2.49	2.40	2.34	2.28	2.24	2.16	2.09	2.01	1.96	1.92	1.87	1.82	1.77	1.71
26	4.23	3.37	2.98	2.74	2.59	2.47	2.39	2.32	2.27	2.22	2.15	2.07	1.99	1.95	1.90	1.85	1.80	1.75	1.69
27	4.21	3.35	2.96	2.73	2.57	2.46	2.37	2.31	2.25	2.20	2.13	2.06	1.97	1.93	1.88	1.84	1.79	1.73	1.67
28	4.20	3.34	2.95	2.71	2.56	2.45	2.36	2.29	2.24	2.19	2.12	2.04	1.96	1.91	1.87	1.82	1.77	1.71	1.65
29	4.18	3.33	2.93	2.70	2.55	2.43	2.35	2.28	2.22	2.18	2.10	2.03	1.94	1.90	1.85	1.81	1.75	1.70	1.64
30	4.17	3.32	2.92	2.69	2.53	2.42	2.33	2.27	2.21	2.16	2.09	2.01	1.93	1.89	1.84	1.79	1.74	1.68	1.62
40	4.08	3.23	2.84	2.61	2.45	2.34	2.25	2.18	2.12	2.08	2.00	1.92	1.84	1.79	1.74	1.69	1.64	1.58	1.51
60	4.00	3.15	2.76	2.53	2.37	2.25	2.17	2.10	2.04	1.99	1.92	1.84	1.75	1.70	1.65	1.59	1.53	1.47	1.39
120	3.92	3.07	2.68	2.45	2.29	2.17	2.09	2.02	1.96	1.91	1.83	1.75	1.66	1.61	1.55	1.50	1.43	1.35	1.25
∞	3.84	3.00	2.60	2.37	2.21	2.10	2.01	1.94	1.88	1.83	1.75	1.67	1.57	1.52	1.46	1.39	1.32	1.22	1.00

THE F DISTRIBUTION

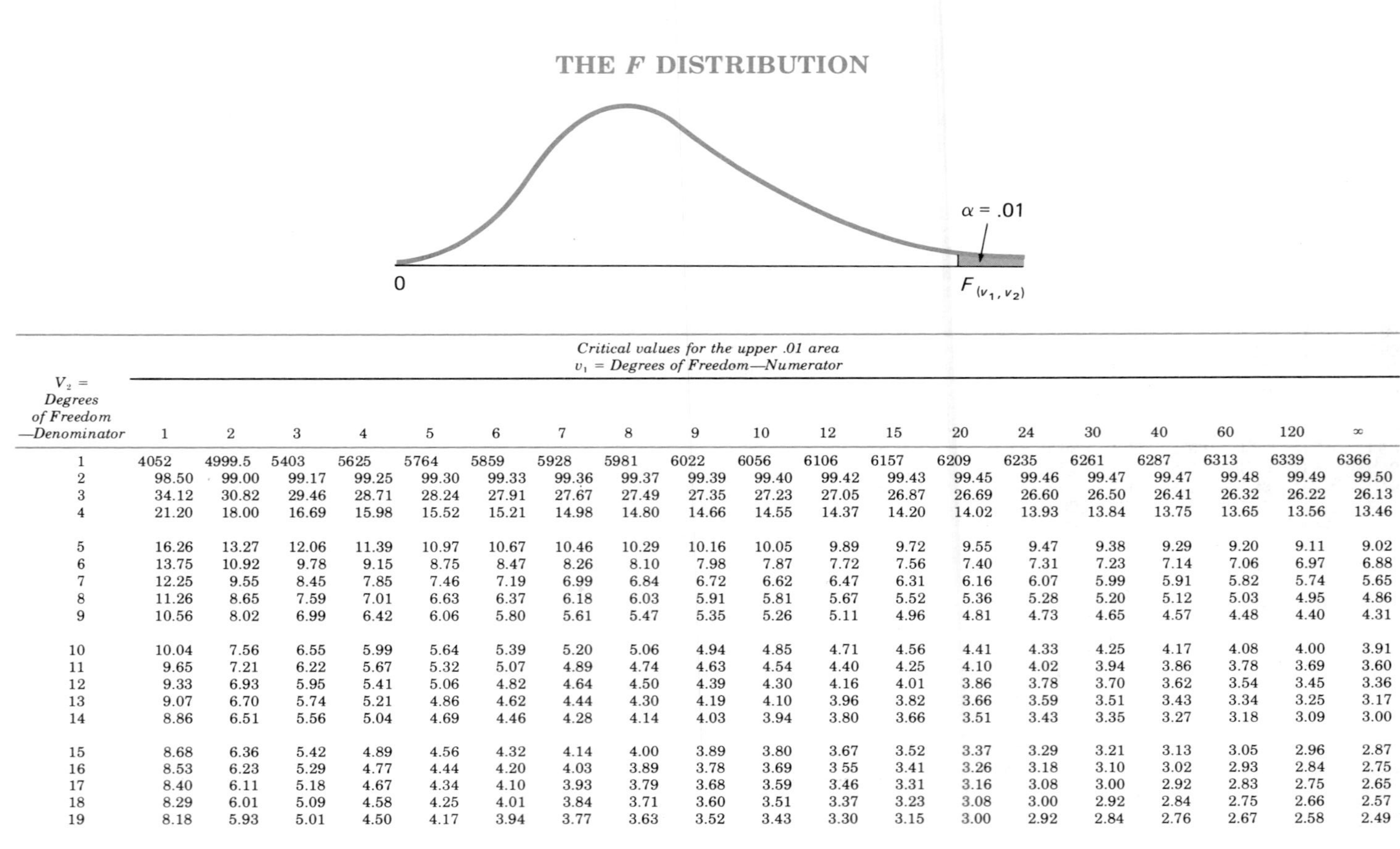

Critical values for the upper .01 area
v_1 = Degrees of Freedom—Numerator

V_2 = Degrees of Freedom —Denominator	1	2	3	4	5	6	7	8	9	10	12	15	20	24	30	40	60	120	∞
1	4052	4999.5	5403	5625	5764	5859	5928	5981	6022	6056	6106	6157	6209	6235	6261	6287	6313	6339	6366
2	98.50	99.00	99.17	99.25	99.30	99.33	99.36	99.37	99.39	99.40	99.42	99.43	99.45	99.46	99.47	99.47	99.48	99.49	99.50
3	34.12	30.82	29.46	28.71	28.24	27.91	27.67	27.49	27.35	27.23	27.05	26.87	26.69	26.60	26.50	26.41	26.32	26.22	26.13
4	21.20	18.00	16.69	15.98	15.52	15.21	14.98	14.80	14.66	14.55	14.37	14.20	14.02	13.93	13.84	13.75	13.65	13.56	13.46
5	16.26	13.27	12.06	11.39	10.97	10.67	10.46	10.29	10.16	10.05	9.89	9.72	9.55	9.47	9.38	9.29	9.20	9.11	9.02
6	13.75	10.92	9.78	9.15	8.75	8.47	8.26	8.10	7.98	7.87	7.72	7.56	7.40	7.31	7.23	7.14	7.06	6.97	6.88
7	12.25	9.55	8.45	7.85	7.46	7.19	6.99	6.84	6.72	6.62	6.47	6.31	6.16	6.07	5.99	5.91	5.82	5.74	5.65
8	11.26	8.65	7.59	7.01	6.63	6.37	6.18	6.03	5.91	5.81	5.67	5.52	5.36	5.28	5.20	5.12	5.03	4.95	4.86
9	10.56	8.02	6.99	6.42	6.06	5.80	5.61	5.47	5.35	5.26	5.11	4.96	4.81	4.73	4.65	4.57	4.48	4.40	4.31
10	10.04	7.56	6.55	5.99	5.64	5.39	5.20	5.06	4.94	4.85	4.71	4.56	4.41	4.33	4.25	4.17	4.08	4.00	3.91
11	9.65	7.21	6.22	5.67	5.32	5.07	4.89	4.74	4.63	4.54	4.40	4.25	4.10	4.02	3.94	3.86	3.78	3.69	3.60
12	9.33	6.93	5.95	5.41	5.06	4.82	4.64	4.50	4.39	4.30	4.16	4.01	3.86	3.78	3.70	3.62	3.54	3.45	3.36
13	9.07	6.70	5.74	5.21	4.86	4.62	4.44	4.30	4.19	4.10	3.96	3.82	3.66	3.59	3.51	3.43	3.34	3.25	3.17
14	8.86	6.51	5.56	5.04	4.69	4.46	4.28	4.14	4.03	3.94	3.80	3.66	3.51	3.43	3.35	3.27	3.18	3.09	3.00
15	8.68	6.36	5.42	4.89	4.56	4.32	4.14	4.00	3.89	3.80	3.67	3.52	3.37	3.29	3.21	3.13	3.05	2.96	2.87
16	8.53	6.23	5.29	4.77	4.44	4.20	4.03	3.89	3.78	3.69	3.55	3.41	3.26	3.18	3.10	3.02	2.93	2.84	2.75
17	8.40	6.11	5.18	4.67	4.34	4.10	3.93	3.79	3.68	3.59	3.46	3.31	3.16	3.08	3.00	2.92	2.83	2.75	2.65
18	8.29	6.01	5.09	4.58	4.25	4.01	3.84	3.71	3.60	3.51	3.37	3.23	3.08	3.00	2.92	2.84	2.75	2.66	2.57
19	8.18	5.93	5.01	4.50	4.17	3.94	3.77	3.63	3.52	3.43	3.30	3.15	3.00	2.92	2.84	2.76	2.67	2.58	2.49

THE F DISTRIBUTION *(Continued)*

V_2 = Degrees of Freedom —Denominator	*Critical values for the upper .01 area* *v_1 = Degrees of Freedom—Numerator*																		
	1	2	3	4	5	6	7	8	9	10	12	15	20	24	30	40	60	120	∞
20	8.10	5.85	4.94	4.43	4.10	3.87	3.70	3.56	3.46	3.37	3.23	3.09	2.94	2.86	2.78	2.69	2.61	2.52	2.42
21	8.02	5.78	4.87	4.37	4.04	3.81	3.64	3.51	3.40	3.31	3.17	3.03	2.88	2.80	2.72	2.64	2.55	2.46	2.36
22	7.95	5.72	4.82	4.31	3.99	3.76	3.59	3.45	3.35	3.26	3.12	2.98	2.83	2.75	2.67	2.58	2.50	2.40	2.31
23	7.88	5.66	4.76	4.26	3.94	3.71	3.54	3.41	3.30	3.21	3.07	2.93	2.78	2.70	2.62	2.54	2.45	2.35	2.26
24	7.82	5.61	4.72	4.22	3.90	3.67	3.50	3.36	3.26	3.17	3.03	2.89	2.74	2.66	2.58	2.49	2.40	2.31	2.21
25	7.77	5.57	4.68	4.18	3.85	3.63	3.46	3.32	3.22	3.13	2.99	2.85	2.70	2.62	2.54	2.45	2.36	2.27	2.17
26	7.72	5.53	4.64	4.14	3.82	3.59	3.42	3.29	3.18	3.09	2.96	2.81	2.66	2.58	2.50	2.42	2.33	2.23	2.13
27	7.68	5.49	4.60	4.11	3.78	3.56	3.39	3.26	3.15	3.06	2.93	2.78	2.63	2.55	2.47	2.38	2.29	2.20	2.10
28	7.64	5.45	4.57	4.07	3.75	3.53	3.36	3.23	3.12	3.03	2.90	2.75	2.60	2.52	2.44	2.35	2.26	2.17	2.06
29	7.60	5.42	4.54	4.04	3.73	3.50	3.33	3.20	3.09	3.00	2.87	2.73	2.57	2.49	2.41	2.33	2.23	2.14	2.03
30	7.56	5.39	4.51	4.02	3.70	3.47	3.30	3.17	3.07	2.98	2.84	2.70	2.55	2.47	2.39	2.30	2.21	2.11	2.01
40	7.31	5.18	4.31	3.83	3.51	3.29	3.12	2.99	2.89	2.80	2.66	2.52	2.37	2.29	2.20	2.11	2.02	1.92	1.80
60	7.08	4.98	4.13	3.65	3.34	3.12	2.95	2.82	2.72	2.63	2.50	2.35	2.20	2.12	2.03	1.94	1.84	1.73	1.60
120	6.85	4.79	3.95	3.48	3.17	2.96	2.79	2.66	2.56	2.47	2.34	2.19	2.03	1.95	1.86	1.76	1.66	1.53	1.38
∞	6.63	4.61	3.78	3.32	3.02	2.80	2.64	2.51	2.41	2.32	2.18	2.04	1.88	1.79	1.70	1.59	1.47	1.32	1.00

EXAMPLES:

(1) $P[F_{(6,15)} > 2.79] = .05$

(2) $P[F_{(6,15)} > 4.32] = .01$

(3) $P[F_{(15,6)} > 3.94] = .05$

(4) $P[F_{(6,15)} < (1/3.94) \text{ or } .25] = .05$

TABLE OF RANDOM NUMBERS

67245	57739	71894	05092	98422	66427	44532	99528	98140	28542
16668	92606	61965	80165	49762	38869	56878	21188	60837	15300
81072	42106	11961	45102	24938	47764	78635	93276	37506	12058
30978	25139	26356	79764	32142	41757	21431	02019	26488	59223
29627	83125	17542	04131	65456	40501	97604	58716	92269	66697
81962	75304	22151	09897	38030	79085	28701	41588	22546	12761
26296	88598	73403	96617	43268	01470	98074	19969	22792	85476
11146	25544	84381	98928	42862	01967	04583	28670	88746	48857
57117	90192	25254	78992	27324	75203	83820	84260	56712	06536
97513	00339	78752	08299	59886	34316	60136	44376	33010	87203
34249	49500	33957	94626	80843	79329	56928	67173	05498	94094
77756	61009	60548	15162	66132	65045	93348	10605	02498	48439
40571	36272	93886	93664	68719	80015	28345	51392	09187	28382
90087	24569	14500	45689	32876	56768	71861	90872	85153	02809
68470	72812	59247	92965	36492	01564	82282	66677	78747	91349
92314	92521	96195	23104	47846	03038	70660	38955	07479	58041
20675	77855	25127	41707	53922	60349	11610	32152	64094	26517
83013	86452	36206	77551	44833	75023	83774	13586	34596	49473
35944	83776	57641	11694	76808	18707	02818	25940	22639	89168
41641	49817	35066	84171	64106	61938	39751	71367	14302	45560
59131	24022	88481	84407	07186	76409	77997	99118	76609	85909
50483	20272	97072	12145	14267	11918	48839	88105	94849	08017
92044	49651	39029	58146	98605	39318	05544	06006	99686	86441
49084	25574	63204	73486	13897	03045	33080	67900	46838	68163
32447	67437	83344	38746	89235	51922	07933	17686	21388	93225
35656	83624	16225	10824	30288	76696	95626	17603	27278	20472
33939	90576	60557	17891	85294	18528	33618	23047	21159	41620
53132	71864	37661	78843	34824	95848	18205	68886	89177	32559
48656	42723	41890	41573	66283	87294	96486	53435	76962	31992
60131	37548	25942	20221	21199	52813	85833	97845	39473	52592
87753	86939	91368	37994	01473	52708	51653	53636	03576	35186
31519	49224	42553	29513	14715	06673	92863	48713	87600	03697
67784	32191	40336	15042	22340	27932	44842	29116	84322	81967
38817	71055	76042	45593	13220	72254	30991	61345	04309	67486
59398	59634	13215	57218	26355	48081	77237	18034	76210	61453
55215	44403	59066	79667	83179	79595	98577	03862	55429	13817
04460	95197	25214	51106	20173	17018	08238	14692	99356	68749
14252	62973	60027	08104	56222	82763	36385	20833	29628	10087
40240	83556	74334	06092	58657	95385	22749	03571	96578	99525
98743	01514	03616	56372	78053	87064	70998	97591	16926	65779

(continued)

TABLE OF RANDOM NUMBERS *(Continued)*

85240	83785	31102	57306	36277	01340	81035	42910	09632	17791
45574	46659	27270	53948	93560	58240	32977	03306	70135	34785
42465	36649	33992	31040	79312	59165	81152	36392	48492	19199
73627	35535	79488	79938	07219	57037	02070	22286	75668	54172
34886	86421	01357	67274	27030	71650	65300	23664	01896	69378
52998	63612	19651	16074	04575	70509	95420	51569	87284	41693
60937	24831	20441	98220	39065	95945	82663	49286	50481	19663
97357	55869	29861	50831	03127	19918	21486	42788	38729	63074
28874	21369	20578	06009	21097	94368	51062	91612	20575	29354
23613	09659	83689	72036	41942	57834	29378	03434	43779	69085

Source: The RAND Corporation, *A Million Random Digits with 100,000 Normal Deviates* (Glencoe, Ill.: The Free Press, 1955), p. 337.

TABLE OF SQUARES AND SQUARE ROOTS

n	n^2	$\sqrt{n}$	$\sqrt{10\,n}$
1	1	1.000	3.162
2	4	1.414	4.472
3	9	1.732	5.477
4	16	2.000	6.325
5	25	2.236	7.071
6	36	2.449	7.746
7	49	2.646	8.367
8	64	2.828	8.944
9	81	3.000	9.487
10	100	3.162	10.000
11	121	3.317	10.488
12	144	3.464	10.954
13	169	3.606	11.402
14	196	3.742	11.832
15	225	3.873	12.247
16	256	4.000	12.649
17	289	4.123	13.038
18	324	4.243	13.416
19	361	4.359	13.784
20	400	4.472	14.142
21	441	4.583	14.491
22	484	4.690	14.832
23	529	4.796	15.166
24	576	4.899	15.492

(continued)

TABLE OF SQUARES AND SQUARE ROOTS *(Continued)*

n	n^2	$\sqrt{n}$	$\sqrt{10\,n}$
25	625	5.000	15.811
26	676	5.099	16.125
27	729	5.196	16.432
28	784	5.292	16.733
29	841	5.385	17.029
30	900	5.477	17.321
31	961	5.568	17.607
32	1,024	5.657	17.889
33	1,089	5.745	18.166
34	1,156	5.831	18.439
35	1,225	5.916	18.708
36	1,296	6.000	18.974
37	1,369	6.083	19.235
38	1,444	6.164	19.494
39	1,521	6.245	19.748
40	1,600	6.325	20.000
41	1,681	6.403	20.248
42	1,764	6.481	20.494
43	1,849	6.557	20.736
44	1,936	6.633	20.976
45	2,025	6.708	21.213
46	2,116	6.782	21.448
47	2,209	6.856	21.679
48	2,304	6.928	21.909
49	2,401	7.000	22.136
50	2,500	7.071	22.361
51	2,601	7.141	22.583
52	2,704	7.211	22.804
53	2,809	7.280	23.022
54	2,916	7.348	23.238
55	3,025	7.416	23.452
56	3,136	7.483	23.664
57	3,249	7.550	23.875
58	3,364	7.616	24.083
59	3,481	7.681	24.290
60	3,600	7.746	24.495
61	3,721	7.810	24.698
62	3,844	7.874	24.900

(continued)

TABLE OF SQUARES AND SQUARE ROOTS *(Continued)*

n	n^2	$\sqrt{n}$	$\sqrt{10\,n}$
63	3,969	7.937	25.100
64	4,096	8.000	25.298
65	4,225	8.062	25.495
66	4,356	8.124	25.690
67	4,489	8.185	25.884
68	4,624	8.246	26.077
69	4,761	8.307	26.268
70	4,900	8.367	26.458
71	5,041	8.426	26.646
72	5,184	8.485	26.833
73	5,329	8.544	27.019
74	5,476	8.602	27.203
75	5,625	8.660	27.386
76	5,776	8.718	27.568
77	5,929	8.775	27.749
78	6,084	8.832	27.928
79	6,241	8.888	28.107
80	6,400	8.944	28.284
81	6,561	9.000	28.460
82	6,724	9.055	28.636
83	6,889	9.110	28.810
84	7,056	9.165	28.983
85	7,225	9.220	29.155
86	7,396	9.274	29.326
87	7,569	9.327	29.496
88	7,744	9.381	29.665
89	7,921	9.434	29.833
90	8,100	9.487	30.000
91	8,281	9.539	30.166
92	8,464	9.592	30.332
93	8,649	9.644	30.496
94	8,836	9.695	30.659
95	9,025	9.747	30.822
96	9,216	9.798	30.984
97	9,409	9.849	31.145
98	9,604	9.899	31.305
99	9,801	9.950	31.464
100	10,000	10.000	31.623

ANSWERS TO SELECTED EXERCISES

CHAPTER 1

1–2. hypothesis testing
1–4. probability
1–6. descriptive statistics
1–8. estimation
1–10. probability
1–12. estimation
1–14. descriptive statistics

CHAPTER 2

1. nominal
3. (a) 7; (c) −4; (e) 28; (g) 35; (i) −24
6. $\sum_{i=2}^{3} (X_i - Y_i + 6)$
7. (a) $\Sigma XY = 6$; $(\Sigma X)(\Sigma Y) = (4)(10) = 40$; $6 \neq 40$
   (c) $\Sigma(X/Y) = 3.3$; $\Sigma X/\Sigma Y = 4/10 = .4$; $3.3 \neq .4$
9. mean = 107.5′; median = 86′; mode = 94′
11. $$\text{Average salary for quarterbacks and kickers combined} = \frac{\$10,312,006}{147} = \$70,149.70$$
13. (a) $s^2 = 31,858/3 = 10,619.33$
    (b) $s^2 = \dfrac{(4)(2,928,662) - (3,404)^2}{(4)(3)} = 10,619.33$
14. $\mu = 4.76$; range = 2.9; mean absolute deviation = 1.112; $\sigma^2 = 6.892/5 = 1.378$; $\sigma = 1.174$
17. Coefficient of variation for $X = \dfrac{6.05}{80.75}(100) = 7.49\%$

    Coefficient of variation for $Y = \dfrac{4.31}{30.08}(100) = 14.33\%$

    since 14.33% > 7.49%, Y is more erratic.
19. mean = 3.24; median = 2.8
20. (a) $\mu_x = 125.33$; $\sigma_x^2 = 1,059.89$
    (b) $\mu_y = 37.33$; $\sigma_y^2 = 1,059.89$
    (c) $Y = X - 88 = a + bX = -88 + (1)X$; $\mu_y = a + b\mu_x$. Therefore, $37.33 = -88 + (1)\mu_x$ or $\mu_x = 125.33$. Inasmuch as $\sigma_y^2 = b^2\sigma_x^2$, $\sigma_x^2 = 1,059.89/(1)^2 = 1,059.89$.
22. (a) 30 to 62 is the same as $\mu - 2\sigma$ to $\mu + 2\sigma$. Since $k = 2$, $[1 - (1/k^2)](100) = 75\%$. Therefore, at least .75(1,000) = 750 employees fall in the interval.
    (b) 34 to 58 is the same as $\mu - 1.5\sigma$ to $\mu + 1.5\sigma$. Since $k = 1.5$, $[1 - (1/k^2)](100) = 55.6\%$. Therefore, at least .556(1,000) = 556 employees fall in the interval.
24. Y = price is 1.3 times greater than X = cost. Hence, $\mu_y = a + b\mu_x = 0 + 1.3(\$38) = \$49.40$ and $\sigma_y^2 = b^2\sigma_x^2 = (1.3)^2(\$6)^2 = 60.84$.

CHAPTER 3

1–1. a. $g^* = (2{,}486 - 784)/5 = 340.4$. Using $g = 350$, one possibility is

X	f
784–1,133	12
1,134–1,483	4
1,484–1,833	3
1,834–2,183	4
2,184–2,533	2
	25

1–3. $g^* = (42 - 3)/5 = 7.8$. Using $g = 8$, one possibility is

X	f
3–10	9
11–18	5
19–26	4
27–34	3
35–42	3
	24

1–5. $g^* = (50 - 5)/5 = 9$. Using $g = 10$, one possibility is

X	f
5–14	5
15–24	8
25–34	8
35–44	6
45–54	1
	28

2–1. (a) $\overline{X} = 3.48$; median $= 3.2$. The sample variance for grouped data is $s^2 \approx \dfrac{\Sigma(m - \overline{X})^2 f}{n - 1}$. Hence, $s^2 = 3.12$ and $s = 1.77$.
(c) .47

2–4. (a) $\overline{X} = 4.5$, median $= 3.7$, $s^2 = 14.03$, and $s = 3.75$.
(c) .64

3–1. $\overline{X} = 6.25$; $s = 4.31$

3–3. $\overline{X} = 3.64$; $s = .94$

4–2. median $= 9.375$. The modal class is 8–10.

CHAPTER 4

2. (b) .2; (c) .4; (d) .6; (e) .6
4. (a) 4/52 or .077; (c) .5; (e) 2/52 or .038
6. .4
7. (a) no; (b) .5; (c) .167
9. 1
11. 120
13. (a) .58; (c) .13; (e) 13/49 or .27; (g) 36/58 or .62
14. (a) .6; (c) .1; (e) .25
16. (a) .3; (b) .5; (c) .5
18. (b) .7; (d) .75
19. 126
21. (a) 1/64 or .016; (b) 1/8 or .125; (c) 1/52 or .019
23. .875

CHAPTER 5

1. $E(X) = 1.7$; $V(X) = 1.61$
3. $E(U) = 5 + 2E(X) = 8.4$; $V(U) = 2^2V(X) = 6.44$
6. $V(X) = \Sigma X^2P(X) - [\Sigma XP(X)]^2$
8. (a) 1.6; (b) .24; (c) 2.6; (d) .24; (e) .04
9. $E(U) = 2E(X) + 3E(Y) = 27.4$; $V(U) = 2^2V(X) + 2(2)(3)\text{Cov}(X,Y) + 3^2V(Y) = 3.04$
11. (a)

X	$P(X\|Y = 1)$
0	.282
1	.077
2	.128
3	.513
	1.000

13. (a)

		X			
		1	2	3	*Marginal Probability of Y*
Y	1	.2	.1	.2	.5
	2	.2	.1	.2	.5
Marginal Probability of X		.4	.2	.4	

14. (b) $X_2Y_3 + X_3Y_3 + X_2Y_4 + X_3Y_4 = 77$
16. $E(Y) = \$170$; $\sigma_y = \$100.50$
19. (a)

r *Number of Purple Balls*	$P(r)$
0	.064
1	.288
2	.432
3	.216
	1.000

(b) $E(r) = 1.8$; $V(r) = .72$

20–2. (a) .0548; (b) .2007

CHAPTER 6

1. (b) area = (5)(1/5) = 1.0; (d) .4
2. (a) area = (1/2)(6)(6/18) = 1.0; (c) .8264; (e) .6944
4. (a) no; (c) yes
5. (b) .1995
7. (b) .8021; (d) .9979
8. (b) .9987; (d) .7905
10. 10.52
12. 8.8
13. (a) 2.88; (c) .4876; (e) .0188
16. (b) .6319
17. (b) .3085

19–1. .1401

CHAPTER 7

4. 75,287,520
6. 30
7. (a)

$\bar{X}$	$P(\bar{X})$
3.25	1/15
3.50	1/15
3.75	1/15
4.00	1/15
4.25	1/15
4.50	2/15
4.75	2/15
5.00	1/15
5.25	2/15
5.50	1/15
5.75	1/15
6.00	1/15
	15/15 = 1.0

(b) $\mu_{\bar{x}} = 4.67$; $\sigma_{\bar{x}} = .79$

8. 3/6
9. (a)

Sample Median	P(Sample Median)
3	4/20 = .2
4	6/20 = .3
5	6/20 = .3
6	4/20 = .2
	20/20 = 1.0

(b) E(sample median) = 4.5; V(sample median) = 1.05
(c) .5
(d) E(sample median) = population median

11. (b) .4840; (d) .3707
12. (b) .4013; (d) .0228
13. (b) .3372; (d) 0
14. (a) $\mu_{\bar{x}} = 34$; $\sigma_{\bar{x}} = 5/7$ or .71
15. (a) .0125

CHAPTER 8

1. (a) biased; (c) biased; (e) unbiased
4. E(sample median) = 6.5; μ = 6.25. Since $6.5 \neq 6.25$, the sample median is a biased estimator.
5. E(sample median) = 6.0; μ = 6.0. Since 6.0 = 6.0, the sample median is an unbiased estimator.
8. (a) .1586; (c) .9948; (e) .9964
9. (a) .3108; (c) .9750; (e) .9976
11. 3,440
13. .84
14. (b) 12.72%
16. 1,067
18. (a) .9544
20. (a) 29.95 minutes $< \mu <$ 34.05 minutes
    (b) 29.55 minutes $< \mu <$ 34.45 minutes
    (c) 28.77 minutes $< \mu <$ 35.23 minutes
22. (a) 43.80 pounds $< \mu <$ 44.20 pounds
    (b) 43.74 pounds $< \mu <$ 44.26 pounds
23. (a) 14.35 hours $< \mu <$ 15.65 hours

CHAPTER 9

1. (b) .0376; (d) .0086
2. (b) −1.75
3. (c) 1.34
5. .0301
10. $Z^* = 1.64$; computed $Z = 8.97$. Since $8.97 > 1.64$, reject H_0: $\mu \leq \$30,000$.

12. $Z^* = -1.64$; computed $Z = -2.63$. Since $-2.63 < -1.64$, reject H_0: $\mu \geq 600$ words.
13. (b) .8038; (d) .7422
14. (a) .9021; (c) .0605
16. $Z^* = -2.33$; computed $Z = -10.00$. Since $-10.00 < -2.33$, reject H_0: $\mu \geq 120$ minutes.
18. $Z^* = 1.64$; computed $Z = 11.11$. Since $11.11 > 1.64$, reject H_0: $\mu \leq 10$ gallons.

CHAPTER 10

1. (a) 1.350; (c) 1.711
2. (b) .98
3. (b) .005
4. (a) $\$2{,}100.75 < \mu < \$2{,}199.25$
   (c) $\$2{,}066.24 < \mu < \$2{,}233.76$
6. $t^* = 1.796$; $-t^* = -1.796$; computed $t = -6.00$. Since -6.00 isn't in the interval -1.796 to 1.796, reject H_0: $\mu = 1.6$ ounces.
8. (a) $\$270{,}710.87 < \mu < \$285{,}289.13$
9. (b) 263.9 minutes $< \mu <$ 288.1 minutes

CHAPTER 11

1. (a) .2515; (c) .6293; (e) .1646
2. (b) .6591; (d) .2146
4. 104
5. (a) .3108; (c) .9282; (e) .7924
7. (a) $.69 < \pi < .91$; (c) $.63 < \pi < .97$
9. $Z^* = 2.33$; computed $Z = 3.29$. Since $3.29 > 2.33$, reject H_0: $\pi \leq .50$.
10. $Z^* = 1.64$; computed $Z = 3.08$. Since $3.08 > 1.64$, reject H_0: $\pi \leq .30$.
12. 748
14. (a) $.33 < \pi < .41$

CHAPTER 12

2. $$\begin{aligned} E(p^*) &= E\left(\frac{n_1p_1 + n_2p_2}{n_1 + n_2}\right) \\ &= \frac{1}{n_1 + n_2} E(n_1p_1 + n_2p_2) \\ &= \frac{1}{n_1 + n_2} [E(n_1p_1) + E(n_2p_2)] \\ &= \frac{1}{n_1 + n_2} [n_1E(p_1) + n_2E(p_2)] \\ &= \frac{1}{n_1 + n_2} (n_1\pi + n_2\pi) \\ &= \frac{1}{n_1 + n_2} (n_1 + n_2)\pi = (1)\pi = \pi \end{aligned}$$
7. $Z^* = 1.64$; computed $Z = 0.70$. Since $0.70 < 1.64$, accept H_0: $\pi_1 - \pi_2 \leq 0$.
8. $Z^* = -1.64$; computed $Z = -19.65$. Since $-19.65 < -1.64$, reject H_0: $\mu_1 - \mu_2 \geq 0$.

10. $Z^* = 1.96$; $-Z^* = -1.96$; computed $Z = -1.07$. Since -1.07 is in the interval -1.96 to 1.96, accept H_0: $\pi_1 - \pi_2 = 0$.
14. $Z^* = 1.96$; $-Z^* = -1.96$; computed $Z = -3.36$. Since -3.36 isn't in the interval -1.96 to 1.96, reject H_0: $\mu_1 - \mu_2 = 0$.

CHAPTER 13

1. (a) a_2; (b) a_3; (c) a_1
3. a_2
5. (a) $EOL_1 = \$22.20$; $EOL_2 = \$14.80$
   (b) \$187.40; (c) \$14.80
8. .047
9. .75
11.

		Number of Cakes Baked				
		$a_1 = 1$	$a_2 = 2$	$a_3 = 3$	$a_4 = 4$	$a_5 = 5$
Number of Cakes Demanded	$S_0 = 0$	−\$5	−\$10	−\$15	−\$20	−\$25
	$S_1 = 1$	5	0	−5	−10	−15
	$S_2 = 2$	5	10	5	0	−5
	$S_3 = 3$	5	10	15	10	5
	$S_4 = 4$	5	10	15	20	15
	$S_5 = 5$	5	10	15	20	25

12. (a) 1; (b) 5; (c) 3
14. (a) Introduce the game. (b) \$54,000
16. $P(\pi = .02|r = 18) = 0$; $P(\pi = .04|r = 18) = .758$; $P(\pi = .06|r = 18) = .236$; $P(\pi = .08|r = 18) = .005$

CHAPTER 14

1. (a) 26.30; (c) 18.48
2. (b) .025
3. (b) .80
6.

$$P\left(14.57 < \frac{27s^2}{\sigma^2} < 43.19\right) = .95$$

$$P\left(\frac{1}{14.57} > \frac{\sigma^2}{27s^2} > \frac{1}{43.19}\right) = .95$$

$$P\left(\frac{27s^2}{14.57} > \sigma^2 > \frac{27s^2}{43.19}\right) = .95$$

$$P\left(\frac{27s^2}{43.19} < \sigma^2 < \frac{27s^2}{14.57}\right) = .95$$

Hence, the 95 percent confidence interval estimator of σ^2 is $.625s^2 < \sigma^2 < 1.853s^2$.

8. $\chi^{2*} = 11.07$; computed $\chi^2 = 5.8752$. Since $5.8752 < 11.07$, accept H_0: The probability distribution of r is binomial with $\pi = .5$.
10. $\chi^{2*} = 4.60$; $\chi^{2**} = 32.80$; computed $\chi^2 = 22.82$. Since 22.82 is in the interval 4.60 to 32.80, accept H_0: $\sigma^2 = .09$.
14. $\chi^{2*} = 11.07$; computed $\chi^2 = 3.7630$. Since $3.7630 < 11.07$, accept H_0: The University Q distribution is identical to the national distribution.
16. $\chi^{2*} = 5.99$; computed $\chi^2 = 8.5648$. Since $8.5648 > 5.99$, reject H_0: Opinion and type of company are independent.

CHAPTER 15

1. (a) 2.32; (c) 2.15; (e) 3.07
2. (c) 0.244
3. $F^* = .36$; $F^{**} = 2.25$; computed $F = 1.16$. Since 1.16 is in the interval .36 to 2.25, accept H_0: $\sigma_1^2 = \sigma_2^2$.
6. $F^* = 3.55$; computed $F = 65.13$. Since $65.13 > 3.55$, reject H_0: $\mu_1 = \mu_2 = \mu_3$.
7.

Source of Variation	*Degrees of Freedom*	*Sum of Squares*	*Mean Squares*	*F*
Between samples	2	6.122	3.061	$\frac{3.061}{.047} = 65.13$
Within samples	18	.840	.047	
Total	20	6.962		

10. $F^* = 3.89$; computed $F = 18.08$. Since $18.08 > 3.89$, reject H_0: $\mu_1 = \mu_2 = \mu_3$.
11. (b) $-2.50 < \mu_1 - \mu_3 < -1.14$
15. $F^* = .35$; $F^{**} = 2.74$; computed $F = .48$. Since .48 is in the interval .35 to 2.74, accept H_0: $\sigma_1^2 = \sigma_2^2$.

CHAPTER 16

1. $\Sigma Y = an + b\Sigma X$

$$\frac{\Sigma Y}{n} = a + b\frac{\Sigma X}{n}$$

$$\overline{Y} = a + b\overline{X}$$

$$a = \overline{Y} - b\overline{X}$$

4. $Y^* = 11.366 - 1.633X$
6. $SS_T = 28.87$; $SS_E = 8.87$; $SS_R = 20.00$; and $28.87 = 8.87 + 20.00$
8. (a) $Y^* = -11.095 + 7.799X$; (b) 261.87; (c) .974

11. $Y^* = 2.571 + 6.000X$
13. (a) $Y^* = -8.838 + 0.798X$; (b) 23.082
15. (a) $Y^* = 11.531 + 0.399X$; (b) 51.431

CHAPTER 17

1. (a) $Y^* = 1.406 + 1.218X$
   (b) $b = (-4/78)(3) + (0/78)(10) + (-2/78)(7) + (-4/78)(4) + (1/78)(9) + (4/78)(12) + (5/78)(16) = 95/78 = 1.218$
3. (a) 4.668; (b) 2.802
5. (b) 2.57536; (d) 3.27617
7. (b) −0.72241; (d) 48.33981
9. (a) $Y^* = 93.485 - 4.493X$
11. (a) 39.57 minutes; (c) 40.74 minutes $< \mu_{y|x=11} <$ 47.38 minutes
13. (a) $Y^* = 1.632 + 2.072X$
14. (b) 3.911

CHAPTER 18

2. −0.22
4. −0.03

5.
$$\frac{\Sigma(bX - b\overline{X})(cY - c\overline{Y})}{\sqrt{\Sigma(bX - b\overline{X})^2}\sqrt{\Sigma(cY - c\overline{Y})^2}}$$
$$= \frac{\Sigma(bc)(X - \overline{X})(Y - \overline{Y})}{\sqrt{\Sigma b^2(X - \overline{X})^2}\sqrt{\Sigma c^2(Y - \overline{Y})^2}}$$
$$= \frac{(bc)\Sigma(X - \overline{X})(Y - \overline{Y})}{\left[b\sqrt{\Sigma(X - \overline{X})^2}\right]\left[c\sqrt{\Sigma(Y - \overline{Y})^2}\right]}$$
$$= \frac{\Sigma(X - \overline{X})(Y - \overline{Y})}{\sqrt{\Sigma(X - \overline{X})^2}\sqrt{\Sigma(Y - \overline{Y})^2}}$$

8. .99
12. (a) $Y^* = -1.76073 + 0.44227X_1 + 0.85472X_2$
    (c) 2.16006; (e) 2.18208
13. $F^* = 9.55$; computed $F = 29.91626$. Since $29.91626 > 9.55$, reject H_0: $B_1 = B_2 = 0$.
15. .95

CHAPTER 19

1. (a) $T = 8.684 - 0.006X$ (1971 is $X = 0$; one year = one unit of X; T is in millions of cars.)
   (b) 8.624
3. (a) $T = 1.008 + 0.148X$ (1971 is $X = 0$; one year = one unit of X; T is in \$ trillions.)
   (c) 2.488
5.

Year	*Real Earnings*
1975	\$118.36
1976	122.77
1977	126.12
1978	127.57

7. \$2,145,922.70
10. (a)

Year	*Price Index (1972 = 100)*
1972	100.0
1973	128.6
1974	130.6
1975	118.0
1976	117.7

11. (a) 163.9; (c) 154.5
15. (a)

Month	*Seasonal Index*
August	99.8
September	97.2
October	93.6
November	95.8
December	95.2
January	109.3
February	110.3
March	105.0
April	93.7
May	88.6
June	108.0
July	104.4
	≈ 1,200.0

(c) 5.37 million

CHAPTER 20

1. $Z^* = 1.96$; $-Z^* = -1.96$; computed $Z = 1.79$. Since 1.79 is in the interval -1.96 to 1.96, accept H_0: $\mu = \$7{,}000$.
2. 43,758
5.

R_+	$P(R_+)$
0	1/16
1	1/16
2	1/16
3	2/16
4	2/16
5	2/16
6	2/16
7	2/16
8	1/16
9	1/16
10	1/16
	16/16 = 1.0

7. $Z^* = 1.96$; $-Z^* = -1.96$; computed $Z = -4.95$. Since -4.95 isn't in the interval -1.96 to 1.96, reject H_0: The data are randomly selected.
11. (b) $Z^* = 1.96$; $-Z^* = -1.96$; computed $Z = -1.15$. Since -1.15 is in the interval -1.96 to 1.96, accept H_0: $\mu = 54$.
13. (a) -0.09
14. 7,726,160

Index

Symbol	Interpretation	Chapter or Equation Reference
$P(A\|B)$	Conditional probability of event A given event B (i.e., the probability that an outcome belongs to event A given that event B has occurred)	4–9, 4–13
$P(S_i\|E)$	Posterior probability (i.e., the probability of the ith state of nature given effect E)	13–5
$P(h_i)$	Probability of a particular outcome	4–1, 4–2
p^*	Pooled sample proportion	12–18
π	Pi (i.e., the probability of a "success" outcome in a Bernoulli trial)	Five
	Constant in the normal curve probability density function, approximately equal to 3.141	Six
	Population proportion	Eleven
R	Number of rows in a contingency table	Fourteen
R_1	Sum of the ranks for the observations in Sample 1	Twenty
R_+	Sum of the positive ranks derived for the Wilcoxon test	Twenty
R^2	Sample coefficient of multiple determination	18–12
r	Number of successes in n independent Bernoulli trials	Five
	Sample correlation coefficient	18–3
r_S	Spearman rank correlation coefficient	20–10
r/n	Sample proportion	11–1
r^2	Sample coefficient of determination	16–13, 16–14
ρ	Rho (i.e., the population coefficient of correlation)	18–2
S	Event "sample space"	Four
	Seasonal index	Nineteen
S_i	ith state of nature in a payoff table	Thirteen
SS_B	Between-sample sum of squares	Fifteen
SS_E	Error sum of squares	Sixteen
SS_R	Regression sum of squares	Sixteen
SS_T	Total sum of squares	Fifteen
SS_W	Within-sample sum of squares	Fifteen
s	Sample standard deviation	2–15
s_a	Estimator of σ_a	Seventeen
s_b	Estimator of σ_b	17–8
s_d	Estimator of σ_d	12–13
$s_{y^*\|x_g}$	Estimator of $\sigma_{y^*\|x_g}$	Seventeen
$s_{y_f^*\|x_g}$	Estimator of $\sigma_{y_f^*\|x_g}$	Seventeen
$s_{\bar{x}_1-\bar{x}_2}$	Estimator of $\sigma_{\bar{x}_1-\bar{x}_2}$	12–4
s^2	Sample variance	2–13
s_p^2	Pooled sample variance	12–6, 15–8
$s_{\bar{x}}^2$	Estimator of $\sigma_{\bar{x}}^2$	15–6
$s_{y\|x}^2$	Estimator of $\sigma_{y\|x}^2$	16–9, 16–10
$s_{y\|x_1,x_2}^2$	Estimator of the variance of the X_1, X_2 subpopulation of Ys	18–11
$s_{y^*\|x_g}^2$	Estimator of $V(Y^*\|X_g)$	Seventeen
Σ	Sigma (i.e., the summation sign)	Two
$\Sigma\Sigma$	Double summation sign	Five